Renewable Bioresources

Renewable Bioresources
Scope and Modification for Non-food Applications

Editors

CHRISTIAN V. STEVENS

Department of Organic Chemistry, Ghent University, Belgium

with

ROLAND VERHÉ

Department of Organic Chemistry, Ghent University, Belgium

John Wiley & Sons, Ltd

Other Wiley Editorial Offices

John Wiley & Sons Inc., 111 River Street, Hoboken, NJ 07030, USA

Jossey-Bass, 989 Market Street, San Francisco, CA 94103-1741, USA

Wiley-VCH Verlag GmbH, Boschstr. 12, D-69469 Weinheim, Germany

John Wiley & Sons Australia Ltd, 33 Park Road, Milton, Queensland 4064, Australia

John Wiley & Sons (Asia) Pte Ltd, 2 Clementi Loop #02-01, Jin Xing Distripark, Singapore 129809

John Wiley & Sons Canada Ltd, 22 Worcester Road, Etobicoke, Ontario, Canada M9W 1L1

Wiley also publishes its books in a variety of electronic formats. Some content that appears in print may not be
available in electronic books.

Library of Congress Cataloging-in-Publication Data

Renewable bioresources : scope and modification for non-food applications / editors
Christian V. Stevens, Roland Verhé.
    p.   cm.
Includes bibliographical references and index.
ISBN 0-470-85446-4 (cloth : alk. paper) — ISBN 0-470-85447-2 (pbk. : alk. paper)
1. Environmental chemistry—Industrial applications.   2. Renewable natural resources.
3. Chemical industry—Environmental aspects.   4. Biopolymers.   I. Stevens, Christian V.
II. Verhé, Roland.
TP155.2.E58R46 2004
333.95′3—dc22

                                                                              2004005082

British Library Cataloging in Publication Data

A catalogue record for this book is available from the British Library

ISBN 0-470-85446-4 (HB)
ISBN 0-470-85447-2 (PB)

Typeset in 10/12pt Times by Integra Software Services, Pvt. Ltd, Pondicherry, India

This book is printed on acid-free paper responsibly manufactured from sustainable forestry
in which at least two trees are planted for each one used for paper production.

Contents

List of Contributors

Mehrdad Arshadi Unit of Biomass Technology and Chemistry, Swedish University of Agricultural Sciences, P.O. Box 4097, SE-904 03, Umeå, Sweden

Christian O. Brämer Institut für Mikrobiologie, University of Münster, Corrensstrasse 3, 48149 Münster, Germany

Pascale De Caro Institut National Polytechnique de Toulouse (INPT), Laboratory of Agro-industrial Chemistry, 118 route de Narbonne, 31077 Toulouse Cedex 04, France

Jan Demyttenaere European Flavour and Fragrance Association, Square Marie-Louise, 49, 1000 Brussels, Belgium

Sevim Erhan Food and Industrial Oil Research, National Center for Agricultural Utilization Research, Midwest Area, Peoria, United States Department of Agriculture, 1815 N. University Street, Peoria, IL 61604, United States of America

Valérie Eychenne Cognis France, Usine d'Estarac, 31360 Boussens, France

Eero Forss University of Joensuu, P.O. Box 111, 80101 Joensuu, Finland

David Gritten University of Joensuu, P.O. Box 111, 80101 Joensuu, Finland

Jeffrey Hardy Green Chemistry Group, Department of Chemistry, University of York, York, YO10 5DD, United Kingdom

Bernd Honermeier Institute of Crop Science and Plant Breeding, Justus-Liebig-University Giessen, Germany

Anton Huber Institut für Chemie, Kolloide, Polymere, University of Graz, Heinrichstrasse 28, A-8010 Graz, Austria

Piotr Janas Department of Food Technology and Storage, Agricultural University, Skromna 8, 20-950 Lublin, Poland

Wieslaw Kopec   Department of Animal Products Technology, Agricultural University of Wroclaw, Norwida Str. 25, 50-375 Wroclaw, Poland

Monica Kordowska-Wiater   Department of Food Technology and Storage, Agricultural University, Skromna 8, 20-950 Lublin, Poland

Paul Kosma   Institut für Bodenkultur Wien, Christian-Doppler-Laboratory "Pulp Reactivity", Institute of Chemistry, Muthgasse 18, A-1190 Vienna, Austria

Sandrine Mateo   ENSIACET (Ecole Nationale Supérieure des Ingénieurs en Arts Chimiques et Technologiques), UMR 1010-INRA/INP-ENSIACET, Laboratoire de Chimie Agro-industrielle, 118 route de Narbonne, 31077 Toulouse Cedex 04, France

Martin Mittelbach   Institut für Chemie, University of Graz, Heinrichstrasse, 28, A-8010 Graz, Austria

Zéphirin Mouloungui   ENSIACET (Ecole Nationale Supérieure des Ingénieurs en Arts Chimiques et Technologiques), UMR 1010-INRA/INP-ENSIACET, Laboratoire de Chimie Agro-industrielle, 118 route de Narbonne, 31077 Toulouse Cedex 04, France

Elias T. Nerantzis   Department of Oenology and Spirits Technology, Faculty of Food Science and Nutrition, Technical Educational Institute Athens, Ag. Spyridona Street, 12210 Aegaleo, Athens, Greece

Jan Oszmianski   Department of Fruit and Vegetable Technology, Faculty of Food Sciences, Agricultural University of Wroclaw, ul. Norwida 25, 50-375 Wroclaw, Poland

Paavo Pelkonen   University of Joensuu, P.O. Box 111, 80101 Joensuu, Finland

Joseph M. Perez   Chemical Engineering Department, The Pennsylvania State University, 105 fenske laboratory, University Park, PA 16802-4400, United States of America

Jozef Poppe Department of Crop Protection, Faculty of Agricultural and Applied Biological Sciences, Ghent University, Coupure Links 653, 9000 Gent, Belgium

Werner Praznik   Institute of Chemistry, Institut für Bodenkultur Wien, Muthgasse 18, A-1190 Vienna, Austria

Dominik Röser   University of Joensuu, P.O. Box 111, 80101 Joensuu, Finland

Waldemar Rymowicz   Department of Biotechnology and Food Microbiology, Agricultural University of Wroclaw, Norwida 25, 50-375 Wroclaw, Poland

Tanja Schaefer   Institute of Crop Science and Plant Breeding, Justus-Liebig-University Giessen, Germany

Simone Siebenborn   Institute of Crop Science and Plant Breeding, Justus-Liebig-University Giessen, Germany

Alexander Steinbüchel   Institut für Mikrobiologie, University of Münster, Corrensstrasse 3, 48149 Münster, Germany

Christian V. Stevens   Department of Organic Chemistry, Faculty of Agricultural and Applied Biological Sciences, Ghent University, Coupure links 653, 9000 Gent, Belgium

Liisa Tahvanainen   Silva Network, P.O. Box 111, 80101 Joensuu, Finland

Zdzislaw Targonski   Department of Food Technology and Storage, Agricultural University, Skromna 8, 20-950 Lublin, Poland

Carlos Vaca Garcia   Institut National Polytechnique de Toulouse, Laboratory of Agro-Industrial Chemistry, 118 route de Narbonne, 31077 Toulouse, Cedex 4, France

Roland Verhé   Department of Organic Chemistry, Faculty of Agricultural and Applied Biological Sciences, Ghent University, Coupure Links 653, 9000 Gent, Belgium

List of Abbreviations

AAS	Atomic absorption spectroscopy
ACTH	Adrenal-cortex stimulating hormone
AFM	Atomic force microscopy
AMU	Atomic Mass Units
AOAC	Association of Analytical Communities
BCA	Biological control agents
BOD	Biological oxygen demand
BPH	Benign prostatic hyperplasma
BSE	Bovine spongiform encephalopathy
BTU	British Thermal Unit
CAFE	Corporate Average Fuel Economy
CAP	Common Agricultural Policy
CH	Carbohydrate
CS	Complex system
DSC	Differential scanning calorimetry
E	Energy
ECF	Elemental chlorine free
EEE	Electrical and electronic equipment
EPA	(US) Environment Protection Agency
EPS	Exopolysaccharide
EORRA	Edible Oil Regulatory Reform Act
ETBE	Ethyl tertiary butyl ether
FA	Fatty acids
FAME	Fatty acid methyl esters
FFA	Free fatty acids
FID	Flame ionisation detector
Fru	Fructose
GA	Gibberilic acid
GH	Growth promoting hormone
GLA	γ-Linolenic acid
Glc	Glucose
GLC	Gas–liquid chromatography

HDPE	High density polyethylene
HPAEC	High pressure anion exchange chromatography
HPLC	High performance liquid chromatography
ICP-MS	Inductively coupled plasma-mass spectroscopy
IL	Ionic liquid
IPPC	Integrated Pollution and Prevention Control
LCA	Life cycle analysis
lcb	Long-chain branched
LCI	Life Cycle Inventory
LCIA	Life cycle impact analysis
LPLA	Low molecular weight polylactic acid
MBP	Microbial biomass protein
MS	Mass spectrometry
MTBE	Methyl tertiary butyl ether
MW	Molecular weight
nb	non-branched
NMMO	N-methylmorpholine-N-oxide
PAD	Pulsed amperometric detection
PCS	Photon correlation spectroscopy
PFS	Process Flow Sheet
PHA	Polyhydroxyalkanoates
PI	Process intensification
PLA	Polylactic acid
PS	Polystyrene
PS	Polysaccharide
RCRA	Resource Conservation and Recovery Act
RES	Renewable energy sources
RME	Rape methyl ester
RRMs	Renewable raw materials
scb	Short-chain branched
SCB	Single cell biomass
SCF	Supercritical fluid
SCP	Single cell protein
SDR	Spinning disc reactor
SEC	Size-exclusion chromatography
SEM	Scanning electron microscopy
SME	Small-Medium-sized Enterprise
SmF	Submerged fermentation
SRF	Short rotation forestry
SRM	Specified risk material
SSF	Solid state fermentation
SUV	Sports utility vehicle
TAPPI	Technical Association of the Pulp and Paper Industry
TBL	Triple Bottom Line
TCBD	Toxic chlorinated bensodioxines
TCF	Totally chlorine free

TLC	Thin layer chromatography
TSH	Thyroid stimulating hormone
USDA	United States Department of Agriculture
WEEE	Waste Electrical and Electronic Equipment
% AE	% Atom Economy

Foreword

The 21st century offers enormous challenges but also exciting opportunities at social, economic and environmental levels. As the world population grows we seek to bring an increasing proportion of that population up to an acceptable quality of life. At the same time people and governments are becoming increasingly concerned at the levels of resource depletion and global pollution that we have inherited from the 20th century as we brought a quite small proportion of the world's population up to reasonable levels of healthcare, food, housing and standard of living. The challenge for scientists, economists and environmentalists is to work together to achieve sustainable growth. This cannot be achieved using the industry of the last century.

The economy of the 20th century was largely based on the industrial revolution of the 1800s and the exploitation of fossil resources. The availability of cheap and abundant petroleum resources was the platform for economic growth by providing readily available energy for industry and a select proportion of the planet. The energy-driven petroleum industry sought ways to add value to its abundant feedstock and was able to exploit the organic chemistry discovered in the 19th century to produce an ever-increasing range of new products for society. Industrial growth was accelerated by the Second World War during which manufacturing industry stepped up a gear. After the war, the over capacity of industry was satisfied by a marketing led growth in consumerism with wealthy westerners being encouraged to buy more and to buy more often. Arguably, this was the beginning of the 'throwaway' society whereby consumers were encouraged to replace items of clothing, household goods and 'lifestyle goods' long before the end of their useful life. This is now witnessed in many modern product lines; for example mobile phones, which are now marketed as much on appearance and add-on gimmicks as on actually performing the essential function of easier communication. The consequences of 50 years of uncontrolled consumerism driven by ever more sophisticated marketing is both an accelerated consumption of mostly non-renewable resources and a disturbing growth in often complex and hazardous waste, the long-term effects of which we do not understand. There are complex social and socio-economic issues underlying these issues but while we can anticipate or at least hope for fundamental shifts in human expectations and attitudes these will not occur in a short time frame. Indeed the problems can be expected to grow in the foreseeable future as billions of people in the developing world demand an increasing quality of life including the luxuries that the privileged few have taught the rest of the world to regard as being essential in today's society. We are therefore faced with the enormous challenge of meeting the needs and, at least to some extent, the demands of an increasing number of people with an industrial base that is not sustainable and a level of pollution which is unacceptable.

We should always learn from history. In the second half of the 19th century in the USA the price of oil – a relatively scarce commodity at that time – and the price of corn were equal on a weight-by-weight basis. Energy was largely supplied by burning coal and the chemical products that were available (e.g. dyes) were derived from coal and plants. The foundations of modern organic chemistry were established in that period including the great "named" reactions such as Friedel–Crafts.

When petroleum became widely available in the first half of the 20th century, a substantial part of the First World's energy needs were switched to this new industry and the new organic chemistry was used to add value to a proportion of the petroleum which became the basis of the petrochemical industry and many of its healthcare, food, household and other products that we enjoy today. However, we now know that petroleum is a limited, non-renewable resource. It is remarkable to realize that almost 100 tonnes of prehistoric material is required for every gallon of gasoline we burn in our vehicles. Since we now burn almost 10^{11} kg of carbon every year to maintain our current lifestyles, this means that we annually consume the equivalent of 400 years worth of all the plant matter grown on earth. These figures pay striking witness to the unsustainable and highly inefficient ways we have been consuming the planet's finite resources. What can replace petroleum for energy and other products and is sustainable and non-environmentally threatening? The renewable bioresources! The best calculations show that it is not unreasonable to expect to see the proportion of energy supplied by biomass increase. We can do this and feed the world's population as much through an increase in the efficiency of the agricultural industry as in the adoption of new feedstocks and the associated new technology by the energy industries. In parallel with the growth in the chemical industry in the 20th century we can also anticipate a new biomass-based chemical industry growing in the 21st century as the increasingly feedstock-starved and legislatively challenged petroleum-based industry declines. Another interesting parallel is the availability of biotechnology as a new science enabling us to exploit biomass for making useful molecules in the same way as organic chemistry was available when we had petroleum to exploit in the last century. There should be significant environmental advantages for increasing the application of carbon-neutral and renewable bioresources. This will be encouraged by legislation and taxation (e.g. "carbon taxes") and by social appreciation of an associated reduction in the build-up of greenhouse gases and the manufacture of more sustainable and biodegradable products. Economic factors will remain a problem for the foreseeable future – petroleum-based products are very cheap! However, penalties on the use of environmentally harmful processes and products, and increasing fossil fuel costs will slowly reduce existing price gaps. It is interesting to note that on a weight-for-weight basis the price of oil and corn are once again the same in the US – after more than 100 years of (often much) cheaper oil.

In this book we seek to address the fundamental issues underlying the properly managed transition to a bioeconomy. The development of greener technologies to convert new renewable resources into valuable products in a sustainable manner, often referred to as "Green Chemistry", are described in Chapter 1. Environmental issues must be addressed alongside economic and social issues – we must see the triple bottom line as being essential to future development. Socio-economical aspects and policy of renewable resources are addressed in Chapter 2. While raw material costs are becoming more favourable, it is vital that we learn to efficiently utilize plant resources. This will enable us to both feed the planet and use plants as raw materials for the manufacturing and energy industries. One gallon of useful fuel product from 100 tonnes of biomass is not an acceptable level of efficiency! Integral valorization of agricultural products including the use of agricultural waste is discussed in Chapter 3, while the primary production of the new raw materials largely based on plant-derived chemicals, but including a brief consideration of raw materials of animal origin, is described in Chapter 4. As I have tried to explain, substantial and broadly successful biomass industries for manufacturing valuable products such as chemicals will depend on the parallel growth in the bio-energy industries. It is appropriate therefore, that bio-energy is covered in considerable detail in Chapter 5. This chapter includes a brief overview of the wide range of new energy sources currently under consideration. The effective identification and quantification of renewable crop materials is also fundamental to this whole area and analytical approaches for raw materials from crops are discussed in Chapter 6. The next four chapters deal with industrial products based on renewable resources. Chapter 7 considers products from carbohydrates, wood and fibres, based on the use of polysaccharides, oligosaccharides, disaccharides and monosaccharides. In Chapter 8 we consider a more specific but very large application area – non-carbohydrate biopolymers. A number of interesting materials are considered with polyamides being discussed in most detail. The production of appropriate monomers is also discussed. The development of industrial products from lipids and proteins is considered in Chapter 9. Sections in this chapter include ones on oleochemistry, oils and fats, and proteins. In Chapter 10, higher value uses for renewable resources are considered. These include areas with long histories but small manufacturing bases such as perfumes and dyes, as well as newer areas with real potential for the future. The book ends with a chapter on "Renewable Resources: Back to the Future" which overviews the major issues of sustainability, policy, integral valorization, production, and energy and other products.

We are at the beginning of a revolution which, if managed correctly, can lead us into a new age of health, wealth and prosperity for the world's population achieved through the sustainable and environmentally acceptable exploitation of the planet's resources. However, we must do it correctly and we must learn from the mistakes of a century of growth that used technology based on incomplete and short-sighted economic and social evaluations to feed a society driven more by greed than by need.

James Clark
York, England

Preface

The use of renewable resources is becoming more and more important in our society. It is intensively connected to natural bio-resources, agricultural production and new developments in the global agricultural policy. In addition, environmental concerns especially about climate changes and about the use of fossil fuels and raw materials have increased the impact of renewable resources.

More and more companies become aware of the importance of this new tendency and realise that investment in new technology based on renewable resources will be important in order to develop their business in a sustainable way. Therefore, the interaction between chemistry, biology, biochemistry, agricultural sciences, environmental technology and economy is very intense and will have to be optimal for a successful, economically feasible application. To be an expert in all these different fields and its complicated interconnection is impossible. Therefore, close collaboration and interdisciplinary work is essential for the advancement of this field. This first handbook on renewable resources that aims at giving a broad overview of this emerging field is also written by an extensive group of scientists, all connected to one of the aspects of renewable resources.

The mission to give a good overview of all aspects of renewable resources without missing out on certain points is possibly one of the reasons that such a book was not written before.

In an effort to organise a joint European Masters programme on renewable resources, several universities became partner in this trial to bring together the information for this overview. We want to thank all the contributors for their efforts in trying to overcome the administrative problems to organise the European Masters programme and for their patience to put together this manuscript.

This overview cannot be specific on all topics since it would result in a complete series of books. Therefore, the subjects discussed in this book are an invitation to people interested in this area to go deeper into the subjects and look for more specific information.

The interconnection of the whole subject also makes it difficult to arrange the topics in different chapters. Certain subjects will therefore be discussed in different chapters from a different point of view.

Anyway, we hope that students, doctoral students and people interested in this broad area get some basic information and enjoy the message that working in this area still needs a lot of interdisciplinary efforts.

Christian V. Stevens
Roland Verhé

1

Green Chemistry and Sustainability

Jeffrey Hardy

1.1 Introduction

Sustainable development is a concept that has become increasingly important over the last 20 years and many companies have become aware of the necessity to run their business from another perspective. This chapter tries to give some fundamental aspects of this change of thinking especially seen from the point of view of the chemical industry, which is often blamed for leaving a large environmental footprint. In this chapter, a complete view is given on process and material development for non-food applications and it highlights that renewable raw materials (RRMs) can be an important tool, but not the only instrument required in developing a more sustainable way of creating useful materials for society. The chapter covers:

- the concept of the Green Chemistry;
- a selection of methods which are available to increase the sustainability of processes;
- methods how the sustainability of processes can be measured; and
- some conclusions.

1.2 The Chemical Industry in Context

It is difficult to imagine how life in the 21st century would be were it not for the advances made by the chemical industry throughout the 19th and 20th centuries. Virtually every aspect of modern day life is affected by the chemical industry in some way. The production of pharmaceuticals such as antibiotics and painkillers has increased life expectancy and improved our quality of life. Synthetic fibres and advanced composite

Renewable Bioresources: Scope and Modification for Non-food Applications. Edited by C.V. Stevens and R. Verhé
© 2004 John Wiley & Sons, Ltd   ISBNs: 0-470-85446-4 (HB); 0-470-85447-2 (PB)

materials have found applications in clothing, performance sports equipment, transport, safety equipment (e.g. cycle helmets) and also in medical applications including blood bags and artificial joints. Other miscellaneous uses of chemicals arise in refrigerants, fuels, liquid crystal displays, plastics for electronic goods, detergents, fertilisers, pesticides...the list is almost endless. The chemical industry is a big business; the annual turnover of the European chemical industry is more than €400 billion, greater than that of USA or Japan.

So why, when the chemical industry is responsible in many ways for the standard of life we are used to and is also so vital for the economy, is the industry perceived by the public as doing more harm than good? This can in part be attributed to the lack of public connection between the chemical industry and the end products; however, a major reason is that the industry is perceived as being polluting and causing significant environmental damage. The physical appearance of chemical plants as well as the historic events such as the release of foul-smelling chemicals, chemical fires, foaming rivers and disasters such as Bhopal have all added to this negative impression. This negative view is particularly virulent in the younger generation (16–24 years of age) and has resulted in applications for chemistry and chemical engineering courses in universities across Europe falling to critically low levels over recent years.

It is also true that at present the chemical industry is inexorably linked with the petroleum industry as its primary source of raw materials (petrochemicals). Citizens in the European Union (EU) are being made increasingly aware of the need for sustainable development. Current thinking regarding sustainability and sustainable development came out at a United Nations Commission on Environment and Development in 1987 (The World Commission on Environment and Development, 1987), which defined sustainable development as:

> Sustainable development is a social development which fulfils the needs of present generations without endangering the possibilities or fulfilment of the needs of future generations.

This presents a challenge for the chemical industry in the 21st century. Can the industry continue to produce the chemicals and products it does today in an economically viable manner without an unacceptable environmental impact? Two of the most important questions that have to be addressed by the chemical industry are:

- Where will the feedstocks for chemical production come from when fossil fuels are exhausted?
- How much waste/pollution can the environment safely deal with as part of its natural cycle?

In this chapter we will explore some of the concepts of how and why the chemical industry will have to change in the future. Also presented will be some of the developing technologies and philosophies relating to the concept of Green Chemistry.

1.2.1 Waste and the Chemical Industry

We all naturally produce waste as part of our living; an average adult produces 1.3 kg of faecal and urinary matter per day. Although this seems like a substantial amount, compared to the average waste in a typical developed country, sewerage sludge accounts for less than 1% of the total waste. Mining and agriculture account for almost 50% of waste production (an important figure when we consider greater use of biomass for chemical production); industry and commerce account for a further 18% (some 76×10^6 t per annum in the UK).

As well as the inefficient use of materials, waste is now recognised by the general populace as having intrinsic risks to both the welfare and the environment. Many countries now have active programmes aimed at reducing the waste streams to land, air and water. Typical initiatives target increased recycling and the introduction of waste minimisation practices. Although it is possible to treat most waste streams after they have been formed, it is considered better to tackle them at source either by reduction or avoidance altogether.

The chemical industry, in a given country, is responsible for up to 70% of the total emissions of volatile organic chemicals (VOCs); these emissions are accrued through industrial processes and solvent usage in surface coatings, etc. Volatile organic compounds are organic compounds with vapour pressure of at least 0.01 kPa at 20 °C or having a corresponding volatility under the particular conditions of use. As well as causing a number of health problems by their presence alone, VOCs increase formation of ozone in the lowest layers of atmosphere. Increased ozone concentrations relate to respiratory diseases and affect vegetation. The chemical industry is also responsible for the formation of a huge amount of inorganic salts, resulting from the widespread use of mineral acids and alkalis.

Paints – Case Study

Traditionally, paints have relied on the presence of VOCs within their formulation in order to facilitate rapid drying. The VOCs were totally lost to the environment as part of the drying process and accounted for a substantial percentage of total VOC emissions. As well as the respiratory health problems associated with being present in a room full of paint fumes, the formation of ozone in the lower layers of the atmosphere was considered to be a significant concern. More recently, water-based paints have been developed which substantially reduce the amount of VOCs in paint. In fact, such formulations can also offer practical advantages such as requiring less coats for a required finish. This final point is particularly pertinent when we discuss sustainable development and especially Green Chemistry. If we substitute a product for a more "sustainable" or "environmentally friendly" alternative, we must be certain we are not introducing a product that has an inferior technical performance. For instance, if the water-based paints required several more coats than the traditional formulation, then it could easily be viewed as less "environmentally friendly" because much more raw material was required to affect the same results. This is an equally key point when we look towards increased utilisation of RRMs, although products derived from such materials can be considered renewable, if their technical performance is poor they cannot be considered an appropriate substitute for materials derived from fossil fuels.

How do the various branches of the chemical industry compare to each other with respect to waste generation? Roger Sheldon (1994) undertook a study of the waste produced by various sectors of the chemical industry. He coined the term "E" factor as the ratio of kg

Table 1.1 *kg waste produced per kg of product – The 'E' factor (Sheldon, 1994, 1997)*

Industry sector	Annual production (Te)	E factor	Approximate total annual waste (Te)
Oil refining	10^6–10^8	ca. 0.1	10^6
Bulk chemicals	10^4–10^6	<1 5	10^5
Fine chemicals	10^2–10^4	5 >50	10^4
Pharmaceuticals	10–10^3	25 >100	10^3

of by-product to kg of product; the results for the chemical industry sector are summarised in Table 1.1 (Sheldon, 1997).

This data can be viewed in several ways. First, the oil refining industry could be viewed favourably because it produces so little waste compared to its product. However, because of the huge scale of the industry, the actual amount of waste produced annually is substantial. Secondly, the pharmaceutical industry is the opposite of the oil refining industry, because of the complex (and often multi-step) synthesis of pharmaceuticals, and requirement for extremely pure final products, the industry generates many times more waste than they make product. However, because the pharmaceutical industry manufactures on a relatively small scale, the total amount of waste generated annually is actually small compared to other industries.

The waste implications of even a simple reaction can be substantial. The synthesis of benzophenone, a classical Friedel–Crafts catalysed acylation reaction, has been broken down into starting materials, products and waste as shown in Table 1.2 (Furniess *et al.*, 1989).

Upon examination of the data in the table several things become apparent. First, the reaction is quite wasteful, for every gram of product produced almost 18 g of waste are generated; this figure does not represent all the waste, as the acidic and alkaline water produced will require neutralisation before release to a water-effluent stream. Water contaminated with organic species is very difficult to purify and may have to be treated as special waste (and in extreme circumstances may have to be incinerated). Secondly, aluminium chloride is used in stoichiometric quantities, even though it is generally

Table 1.2 *The reagents, products and waste implications of the formation of benzophenone*

Starting materials	Products	Waste
105 g dry benzene	30 g benzophenone (66% yield)	99 g benzene
35 g redistilled benzoyl chloride		11.9 g benzoyl chloride
37 g anhydrous $AlCl_3$		21.5 g $Al(OH)_3$
		300 g acidified water
		50 g alkaline water
		50 g organic contaminated water
		2 g magnesium sulphate
Total 177 g	30 g	534.4 g

considered a catalyst because it co-ordinates strongly with the product. During the quench step of the reaction the $AlCl_3$ is hydrolysed to $Al(OH)_3$ and cannot be reused. Finally, in the reaction benzene is used as both a solvent and a reactant. Benzene is an undesirable chemical to be used in reactions because of its carcinogenic properties. In this reaction scheme unreacted benzene has been classed as waste simply because in most cases the solvent would not be recovered on a laboratory scale.

As we can see even a simple one-step reaction can have large waste issues associated with it. If we were to consider this reaction as part of a much more complex synthesis, which is quite likely as benzophenone is an important intermediate, we can envisage a very wasteful process. It should be noted that this data is based on a laboratory experiment and not on an optimised industrial process, but as an example it shows clearly how important waste issues can be. As will be seen later in this chapter several improvements could be made to this reaction, such as the use of a heterogeneous catalyst, which could dramatically reduce the waste.

1.2.2 The Cost of Waste

Waste costs companies money through several mechanisms:

- loss of raw materials;
- wasted energy;
- low reactor utilisation;
- end-of-pipe clean-up technology;
- fines for pollution;
- cost of waste disposal.

The cost of waste also varies depending upon the nature of the waste. For instance, aqueous or organic waste contaminated with highly toxic species cannot be released into the environment without substantial (and expensive) treatment and in many cases may simply be incinerated. Whilst not acceptable in terms of environmental impact, for high-value products large quantities of waste per mass unit of product (in some cases accounting for almost half the cost of the product) are tolerable because of high profit margins. This is clearly a situation that needs to be addressed.

In realistic terms, a wasteful process can reduce the economic competitiveness of a chemical process. However, current industrial thinking is not focussed purely on economic aspects. The Triple Bottom Line (TBL) concept is used as an indicator of business performance (Elkington, 1999). The TBL concept takes into account the economic, environmental and societal performance of a company. The environmental performance is of great importance because tightening legislation is financially punishing companies for poor environmental performance. A polluting process can cost a substantial amount of money in fines and waste disposal costs, and can make it uneconomical. The social issues are of equal importance; the effect of bad media coverage on a company that has a poor environmental record cannot be underestimated. In the past there have been many campaigns led by NGOs such as Friends of the Earth and Greenpeace against companies and this is now extending to the use of hazardous chemicals.

1.2.3   Waste Minimisation

Legislation plays an important role in driving companies towards implementing new strategies such as waste minimisation techniques. The Waste Electrical and Electronic Equipment (WEEE) directive from the EU Commission is an example of a legislative driver. Under the legislation it will become the responsibility of the producing company to take back and recycle WEEE from its customers. Also, certain hazardous chemicals, such as mercury, cadmium, lead and brominated flame-retardants, used in the manufacture of Electrical and Electronic Equipment (EEE) will have to be substituted where technically possible. It is perceived that the additional cost of taking back and recycling the WEEE will encourage companies dealing in EEE to adopt a design for recycling strategy (COM(2000) 347 final, 2000).

Legislation for the chemical industry has severely restricted the use of certain noxious and harmful chemicals such as benzene and banned the use of others outright. This legislation encourages the uptake of cleaner technologies through environmental taxes and regulations such as Integrated Pollution and Prevention Control (IPPC). However, legislation is really designed to deter production of waste, it does not itself describe a waste minimisation policy for companies to adopt and therefore it is very much up to individual corporations to initiate a waste minimisation policy. A successful waste minimisation initiative can lead to substantial reductions in waste streams and energy for a company and consequently a more profitable process.

Waste minimisation is a fairly simple concept; examine a process and identify where waste is being generated, then take action to reduce or eliminate this waste. This can be carried out for a theoretical process or for an actual process. Perhaps the most sensible and thorough method of examining a chemical process to identify areas that could be improved upon is to represent it as a flow sheet. By representing all the interconnecting parts of a process combined with the complete material and energy flows (balanced across process), areas of inefficiency and areas where high levels of waste are generated can be identified and action can be taken. Process Flow Sheets (PFSs) can also be used to compare two similar processes, if all the data for both processes are accurate and complete.

Many of the waste issues in chemical processes arise simply because the different parties at the various stages of process development do not communicate well. Therefore, for instance, an engineering problem that could have been identified at the laboratory scale (such as a small exotherm), deemed unworthy of note by the chemist, might have been noticed and solved at source instead of being solved as a large-scale problem much further down the line. A more efficient system would be to allocate a multi-disciplinary team to the project at the start. The team might typically comprise of chemists, chemical engineers, production personnel, SHE (Safety, Health and Environmental Issues) advisors and possibly a representative from the business. By involving all parties at regular meetings, problems that would otherwise manifest much later in the development process can be addressed very early on and could potentially be avoided. A number of the Green Chemical tools and techniques discussed in this chapter are valuable instruments for the reduction of waste in chemical manufacture.

1.3   Green Chemistry: An Introduction

Green Chemistry is both a philosophy and a methodology for progressing towards sustainability within the chemical industry. The US Environment Protection Agency (EPA) coined the phrase "Green Chemistry" in the early 1990s, describing it as:

> To promote innovative chemical technologies that reduce or eliminate the use or generation of hazardous substances in the design, manufacture and use of chemical products.

During the development of the concept of Green Chemistry, Paul Anastas and his co-workers developed the Twelve Principles of Green Chemistry. These principles are common-sense concepts for a paradigm shift in the way chemistry is carried out and still hold complete relevance, 13 years after their conception. It is strongly recommended that any interested party study the Twelve Principles of Green Chemistry (Anastas and Warner, 1998).

There are several ways to visualise Green Chemistry; perhaps one of the most provocative is to view it as striving towards the perfect reaction. The perfect synthesis would be one that goes completely and selectively to the desired product with a 100% atom economical yield at room temperature and pressure, requiring no solvent and utilising non-toxic and completely sustainable reagents. For most reaction systems the idea of a perfect reaction is almost pure fantasy, however, as a philosophy, striving towards a perfect synthesis can lead only to step improvements in any given process.

In this chapter we shall focus on Green Chemistry by viewing it as a series of reductions. This is demonstrated in Figure 1.1.

As can be seen in the reduction model above, reducing the use of non-renewable feedstocks is one of the key concepts. There are several obvious reasons for moving away from a fossil fuel-based chemical feedstock, but of key importance are:

- Petroleum-based feedstocks are finite and consequently our major chemical feedstock will in the future run out. Therefore, we need to carry out fundamental research into alternative feedstocks now, as opposed to in the future when it may be too late.
- Petroleum feedstocks contain carbon that has been trapped for millions of years. Consequently, petrochemicals on their ultimate destruction at the end of life will lead to carbon emissions directly to the biosphere, which are extra to the natural carbon cycle. Strong evidence suggests that this leads directly to global warming.

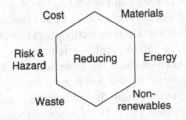

Figure 1.1   *Green Chemistry conceptualised as a series of reductions*

However, Green Chemistry should be viewed as a holistic approach to sustainable development within the chemical industry and therefore the subject should never be approached from a single angle. Regardless of the pressing need, merely utilising a renewable feedstock does not automatically constitute a "Greener" process. If the synthesis of the desired end product still remains energy intensive, wasteful and/or dangerous, then the process cannot be considered Green even if it has a feedstock based on a renewable resource.

It could be said that all the reductions listed above are based on common sense. However, in numerous chemical processes carried out today many of the principles are not taken into account. Later in this chapter there will be several examples of current Green Chemical Technology which all demonstrate how key reductions can be made. Perhaps, first we should concentrate on the need for the reduction of materials and waste in the chemical industry.

1.3.1   The Concept of Atom Economy (Reducing Materials and Waste)

The success of traditional synthetic chemical reactions has generally been measured by obtaining a % yield value – a reaction with a high % yield is considered a good reaction. This is an unsatisfactory treatment of a reaction system from a Green Chemistry standpoint. Atom economy is the Green Chemistry equivalent of % yield. It is essentially a measure of how many atoms of reactants end up in the final product and how many end up as by-products (Trost, 1998). The calculations for % yield and % Atom Economy (%AE) are shown below in (1.1) and (1.2) respectively.

Traditional method $\% \text{ Yield} = \dfrac{\text{Actual yield}}{\text{Theoretical yield}} \times 100$ (1.1)

Atom Economy $\% \text{ AE} = \dfrac{\text{MW of atoms utilised}}{\text{MW of all reactants used}} \times 100$ (1.2)

It is possible for a reaction to have a yield approaching 100% and yet be extremely wasteful. An example of this would be the traditional Wittig reaction using triphenylphosphine as reagent, which is not incorporated in the final product. Although the reaction has a high selectivity to the desired alkene and a high % yield, the phosphine oxide is waste at the end of the reaction.

A typical Wittig reaction is shown in Figure 1.2, which gives an excellent example of the point made previously (Furniess *et al.*, 1989). In this reaction we can view it simply as the transformation of cyclohexanone into methylenecyclohexane. In order to carry out this reaction we have produced 280 molecular weight units of waste in order to facilitate a transformation which ultimately reduces the molecular weight of the starting ketone by 2 AMU. If we calculate the %AE for this reaction (assuming 100% yield, stoichiometric use of reagents and that the base is catalytic and hence not consumed – a generous set of assumptions), we find the process to be 21.1% atom efficient. Therefore, almost 80% of the molecular mass used in the reaction has been wasted. This is not to suggest that the Wittig reaction should not be used, as it is an incredibly useful organic transformation.

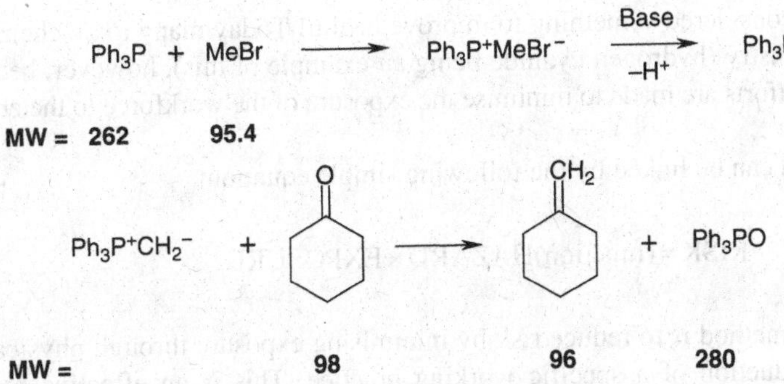

$$Ph_3P \ + \ MeBr \ \longrightarrow \ Ph_3P^+MeBr^- \ \xrightarrow[-H^+]{Base} \ Ph_3P^+CH_2^-$$

MW = 262   95.4

$$Ph_3P^+CH_2^- \ + \ \underset{}{\overset{O}{\bigcirc}} \ \longrightarrow \ \underset{}{\overset{CH_2}{\bigcirc}} \ + \ Ph_3PO$$

MW =   98   96   280

Figure 1.2 *The formation of methylenecyclohexane via the Wittig reaction*

In fact, for many cases it is a substantial problem replacing wasteful reagents with a more atom economical synthetic route.

Atom economy gives by no means a complete picture of the inherent greenness of a reaction as it does not take into account factors such as solvent and energy usage. There are a number of other chemical metrics (including E-factors which have already been discussed) that must be taken into account for a more complete picture of the "Greenness" of a reaction. Curzons *et al.* have completed a good practical review of a number of chemical metrics which have been developed and which is recommended for further reading (Constable, Curzons and Cunningham, 2002).

It is not a simple task to categorise reactions in terms of how atom economical they are. However, as a sensible generalisation, reactions involving rearrangement (e.g. the Claisen rearrangement) or addition (e.g. the Diels–Alder reaction) are generally atom economical. Reactions involving substitution (e.g. S_N1 and S_N2) or elimination (e.g. the Hofmann elimination reaction) can generally be considered atom uneconomical.

In addition to inefficient reactions, many chemical syntheses involve the use of blocking groups or protecting/deprotecting steps. This involves protecting (or blocking) a sensitive group within a group so that chemistry can be carried out on another group (or regioselective area of the molecule). At the end of the reaction the protecting/blocking group has to be removed. If the chemistry could be carried out using stereoselective/ regioselective reagents, or more preferentially catalysts, then the protecting/deprotecting steps could be avoided, making the reaction more atom economical.

1.3.2 Reducing Risk and Hazard

The chemical industry does not deliberately attempt to produce toxic chemicals to harm the environment or the general populace (with the exception of chemical and biological weapons). Problems arise because it is difficult to predict the toxicity of a chemical simply by examining its structure, it is only when the chemical is actually synthesised that its toxicity can be assessed. Unfortunately, the first signs of toxicity generally manifest themselves through ill health or indeed the tragic death of workers. We should bear in mind that bathing in a solution of radioactive salts or indeed benzene, in the not too distant past, has been considered something to improve health! Today many toxic chemicals are still used in industry (hydrogen cyanide being an example of this), however, because the risk is known, efforts are made to minimise the exposure of the workforce to the chemical.

Risk and hazard can be linked by the following simple equation:

$$RISK = (function)HAZARD \times EXPOSURE$$

The traditional method is to reduce risk by minimising exposure through physical means or by the introduction of a specific working practice. This is an effective method of controlling toxic chemicals, but no system is infallible and risk of exposure is always a threat. Green Chemistry approaches the problem from the opposite direction and addresses the issue of reducing the hazard ("what you don't have can't harm you"). Recent legislation is beginning to mirror this approach. The Control of Substances Hazardous to Health (COSHH) regulations require that chemical processes should be assessed as to whether hazardous materials could be avoided. The use of personal protective equipment for workers to prevent exposure to a hazardous material should be considered only as a final resort.

The global production of chemicals has increased from 1×10^6 t in 1930 to 400×10^6 t today. In the EU there are 30000 different substances produced in amounts greater than 1 t. Recently, a White Paper, Strategy for a Future Chemicals Policy, has been drafted (COM (2001) 88 final, 2001) which states that current EU chemicals policy provides insufficient protection. The paper recommends that a new system called REACH (Registration, Evaluation and Authorisation of Chemicals) be put in place. The requirement of this system would be that all chemicals produced in volumes over 1 t be registered in a central database. Of these, the substances that lead to high exposure or have dangerous properties should be tested within 5 years to measure their impact upon human health. Particularly hazardous chemicals will have to be authorised for usage in chemical processes. In time it is recommended that all 30000 chemicals produced in amounts over 1 t should be assessed. The cost of the assessment is not trivial, and for all 30000 substances it has been estimated at €2.1 billion.

The replacement of hazardous materials in chemical reactions is an area where a great deal of Green Chemical research is currently being carried out. For example, chromium(VI)-compounds are used in a range of industrially important oxidation reactions. Such reactions result in the production of large volumes of typically acidic waste containing toxic chromium species. It is always a risk that the chromium species could be accidentally released into the local water supply, leading to a distinct toxicological hazard. Research is being carried out into developing chromium(VI)compounds immobilised on inert support materials such as silica. After the oxidation reaction, the supported chromium species can be removed from the reaction by filtration, thus removing chromium from the waste stream altogether (the catalyst may also be reusable adding another benefit).

A different approach is to examine the system from a different direction; is it always necessary to carry out oxidation chemistry at all? Oxidation chemistry is generally aimed at adding value and functionality to rather inert hydrocarbons, which is a difficult and wasteful process. Many of the chemical feedstocks available from renewable resources are highly oxygenated (e.g. glucose). In future, it may be the case that oxygenated products are formed preferentially from oxygenated starting materials.

1.3.3 Design for Degradation

Chemical products should be designed so that at the end of their function they break down into innocuous degradation products and do not persist in the environment. – Twelve Principles of Green Chemistry.

Anastas and Warner, 1998

The application of Green Chemistry concepts such as atom economy, reduced environmental burden and reduction of waste may lead to reductions in the environmental footprint and the cost of a given chemical manufacturing process. However, many of the products of chemical manufacture end up in the environment as waste, e.g. surfactants, total loss lubricants, agrochemicals and consumer plastics. Chemicals that do not degrade through typical biodegradation pathways will persist in the environment for many years. This can lead to a range of problems; plastics are taking up too much room in landfills, surfactants can cause foaming in rivers and persistent organic chemicals (POCs) may have long-term toxic effects and are often found to bioaccumulate in plants, animals and human beings. With these issues in mind it is increasingly obvious that chemicals and products of the chemical industry must be designed with degradation in mind.

What follows are short case studies of detailing the examples cited above, explaining the nature of the current problem and the alternatives that Green Chemistry and RRMs can offer.

Surfactants

In the late 1950s and early 1960s it was noticed that problems with excessive foaming were developing in sewerage works and in turbulent parts of rivers and streams. The problem was traced to the lack of biodegradation of alkylbenzene sulphonates used in detergents. More specifically, the problem was found to arise due to the branched nature of the alkyl chain. It was discovered that replacing the branched chain with a linear chain resulted in a sulphonate with much-improved biodegradability and consequently reduced foaming problem (structures are shown in Figure 1.3).

Surfactants for detergents can also be made using renewable resources. This has been identified as a potential growth area for renewable resources in the EU. The most common type of surfactants from renewable resources are alkylpolyglucosides. These can essentially be viewed in their purest form as surfactants formed by the reaction of glucose and a fatty acid. Examples of the surfactants are shown in Figure 1.4. The surfactants have uses in cosmetics, textile finishing and industrial cleaning applications. The latter two arise from the surfactants stability in alkaline solution (Soderman and Johansson 2000).

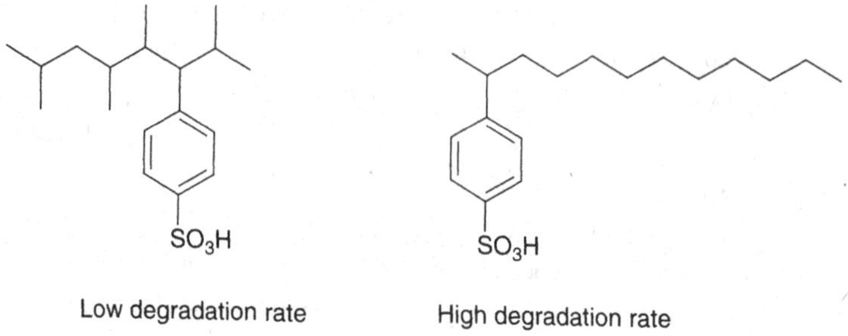

Low degradation rate   High degradation rate

Figure 1.3   *Relationship of degradation rate to structure of alkylbenzene sulphonates*

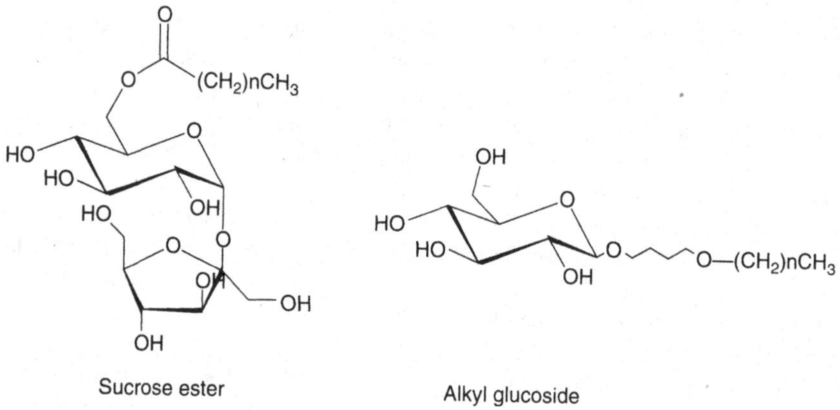

Sucrose ester   Alkyl glucoside

Figure 1.4   *Examples of alkylpolyglucosides*

Total Loss Lubricants

Total loss lubricants are those that are ultimately completely lost to the environment during their routine usage; examples would be chainsaw, bike chain and agricultural machinery lubricants and hydraulic fluids. Traditionally, such lubricants have been based upon mineral oil-based formulations. When lost to the environment, such lubricants were found to persist because microbial action cannot break down their structure. Compounds that persist in the environment can find their way into groundwater supplies and into plants (and ultimately animals) through soil contamination, which can cause a number of health issues.

Vegetable oils can be used to replace mineral oil-based lubricants in total loss applications. The total loss lubricants based on RRMs are completely biodegradable within a short timeframe and do not have any adverse toxicological effects on the environment. This had led to a number of key organisations such as the UK environment agency and the Forestry Commission, replacing mineral oil-based total loss lubricants with those based on vegetable oils. In most cases the performance of vegetable oils is acceptable compared to the analogues and they are reasonably priced. Several companies are now looking into the possibility of extending the use of vegetable oils to other lubrication purposes, such as engine oils and hydraulic fluids. Currently, however, there is no standard test to measure the effectiveness of vegetable-based oils such as engine oils, which has limited their application (IENICA, 2000).

Fertilisers

The need for fertiliser is obvious, however, the problems associated with the overuse of fertilisers are not so clear. When too much fertiliser is applied to a field (or when heavy rainfall washes it away), the crop does not benefit from all of the nutrients and a proportion of them will find their way into the groundwater, and ultimately into rivers, streams and lakes. A high concentration of nutrients in water leads to a rapid growth of algae which forms a layer over the surface of the water. This reduces the ability of oxygen to enter the water, leading to gradual asphyxiation of species relying on dissolved oxygen content (such as fish). It also leads to the promotion of anaerobic bacteria and the generation of sulphurous compounds, leading to a stench. This process is known as eutrophication.

To address this issue, a range of controlled-release fertilisers have been developed (Shaviv, 1999). The idea is quite simple: a biodegradable polymer, such as starch, is reacted with a source of nutrient (such as urea) and placed on the field to be fertilised. During the biodegradation cycle of the polymeric material, nutrients are slowly released as a function of time into the field. It is possible to chemically alter the nature of the polymer in order to increase or decrease the rate of biodegradation and consequently the rate at which the nutrients enter the soil. By utilising a controlled-release fertiliser, it is possible to supply the nutrients to the field at the rate at which the plants will uptake them. This reduces the possibility of the nutrients entering a water source and causing problems such as eutrophication.

Plastics

The vast majority of consumer plastics produced are non-degradable (polyethylene, polypropylene, polystyrene, etc.). In the UK in particular, there is a poor record on recycling (on average less than 10% of plastics are recycled), meaning that the majority of these plastics end up in a landfill, which is causing a problem because landfill space is finite. There are three key ways to tackle this issue:

1. Use less plastics.
2. Recycle more plastics.
3. Make biodegradable plastics (either through a petrochemical or renewable resources route).

Use less plastics

It is the simplest of the ideas and has been adopted by a number of countries. Simple measures such as charging a tax (€15 cents) for plastic carrier bags at supermarkets has reduced their usage in countries such as the Republic of Ireland by up to 95%. Other alternatives are products like paper and linen bags. Biodegradable polyethylene bags can also be made and some supermarket chains are now exclusively using these.

Recycling of more plastics

There are three ways by which plastics can be recycled:

1. Incineration to recover energy.
2. Mechanical recycling to lower-grade products.
3. Chemical recycling to monomers.

Hydrocarbon-based polymers have an energy content similar to that of heating oil (twice that of paper-based waste) and hence plastic waste is a potentially valuable form of fuel. Incineration is especially suitable for small items, which would otherwise be impractical to collect and sort for recycling. However, the public view incineration as an environmentally unfriendly practice. They associate it with the production of ash containing heavy metals and the possibility of dioxin formation from chlorine containing waste (such as PVC). Dioxins are known to accumulate in fatty tissue and can be highly toxic to humans.

With mechanical and chemical recycling there is an issue relating to waste separation; it is generally necessary to have a feedstock based only on one type of polymer for recycling. A mixed polymer feedstock could have detrimental effects on the recycled products' physical properties. Manual techniques are largely too expensive and have been replaced with other techniques. Common techniques are based on the difference between density of the polymers, such as the Float–Sink method. More recent techniques have been developed, which use spectroscopy to differentiate between polymers based on their chemical properties.

Mechanical recycling of pure polymer feedstock generally involves a process of cleaning and drying the polymer feedstock (generally flakes) to remove contaminants. The polymer is then processed by an existing technology such as injection moulding, blow moulding or extrusion into the desired product. Polyethylene terephthalate (PET) is typically recycled to granules by an extrusion process to make fibres for clothing or carpets. HDPE (High Density Polyethylene) is generally recycled into low-value applications such as bin-liners and drainage pipes. PVC, a high production volume polymer, can be recycled from sources such as mineral water bottles and can be processed by co-extrusion into products such as window frames. Mechanically recycled polymer is almost always of lower quality than a polymer made from virgin monomers and consequently of lower value.

Chemical recycling to monomers is advantageous compared to mechanical methods because the monomer can be used to make new polymers which are of higher value than mechanically recycled polymers. However, the processing can be cost-intensive and is consequently not widely used on the commercial scale for most polymers. For polymers containing C−O bonds the process generally entails some form of hydrolysis or alcoholysis.

It should always be borne in mind that processes involving recycling might have energy requirements greater than that required during the entire life cycle involved in making the initial polymer. Factors such as this should be taken into account when determining the true environmental impact of a recycling (or indeed any) process.

The use of biodegradable polymers

Nature provides us with a number of biodegradable polymers such as starch, cellulose and chitin. Polymers such as starch have been used as fillers in non-degradable polymers like polyethylene (PE). In such composites, when starch is degraded the remaining PE cage collapses and also degrades. Other biodegradable polymers can be made via a petrochemical route such as poly(ε-caprolactone) and polyesters. However, in this section we shall concentrate on polymers made from feedstocks based on renewable resources (Stevens, 2002) and in particular polylactic acid (PLA) (polyhydroxy alkanoates, PHAs, are also an example of such polymers and are covered in detail in Chapter 8).

Polylactic acid is a polymer that has recently been put into mass production (140 000 t/yr) by Cargill-Dow in Blair, Nebraska in America (marketed as Natureworks™). PLA is an interesting polymer as the monomer (lactic acid) can be made through a petrochemical route or through a fermentation route. Both routes have advantages and disadvantages, which are summarised in Table 1.3.

For a comparison of which of these routes has the lowest environmental impact, a full Life Cycle Analysis (LCA) would have to be carried out. However, for the sake of this discussion, it is assumed that there is a source of lactic acid of sufficient purity to be used as feedstock.

The polymerisation of lactic acid to make low molecular weight PLA (LPLA) has been known since 1932, however, the expense of the lactic acid monomer and the lack of applications for LPLA have limited its manufacture. New processes to form cheap lactic acid from both renewable and petrochemical feedstocks have resulted in renewed interest in PLA. Part of the interest in PLA arises because it is a completely biodegradable polymer.

Recent developments in both the separation of the lactic acid from the fermentation process and the polymerisation process of lactic acid itself have lead to high-quality, reasonably priced polymer. Cargill-Dow utilise a solvent-free and a novel distillation process to produce a range of polymers. The process involves the controlled depolymerisation

Table 1.3 Comparison of the petrochemical and renewable feedstock routes to the manufacture of lactic acid

Petrochemical route		Renewable route	
Advantages	Disadvantages	Advantages	Disadvantages
Highly atom efficient	Uses HCN as reagent – handling problems	Renewable feedstock	High dilution (problems with product separation)
Good yield	Product requires purification	Inexpensive feedstock (corn starch)	Slow reaction (4–6 days)
		Low-temperature reaction (biocatalysis)	Calcium sulphate produced as waste
		High selectivity	Product requires purification

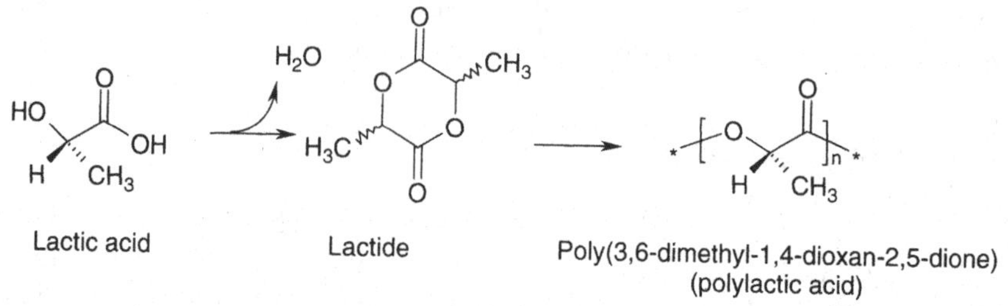

Figure 1.5 *Conversion of lactic acid to high molecular weight polylactic acid via a lactide intermediate*

of LPLA to produce the cyclic dimer, commonly referred to as lactide (this is represented in Figure 1.5). This lactide is maintained in liquid form and purified by distillation. Catalytic ring opening of the lactide intermediate results in the production of PLAs with controlled molecular weights.

Lactic acid exists as D- and L-optical isomers; the lactide has the potential to have L-, D- or *meso*-stereochemistry. The properties of the final polymer depend not only on the MW but also significantly on the optical ratios of the lactides used. In the Cargill-Dow method, first the lactide is catalytically produced in specific optical ratios, followed by highly controlled distillation processes to isolate the correct optical isomer and to recycle unwanted lactide for racemisation. Therefore, polymers with controlled optical ratio, hence controlled properties (Figure 1.6), can be produced. Using this technology, Cargill-Dow is producing PLA fibres for clothing, films for food packaging and agrochemical use and bottles.

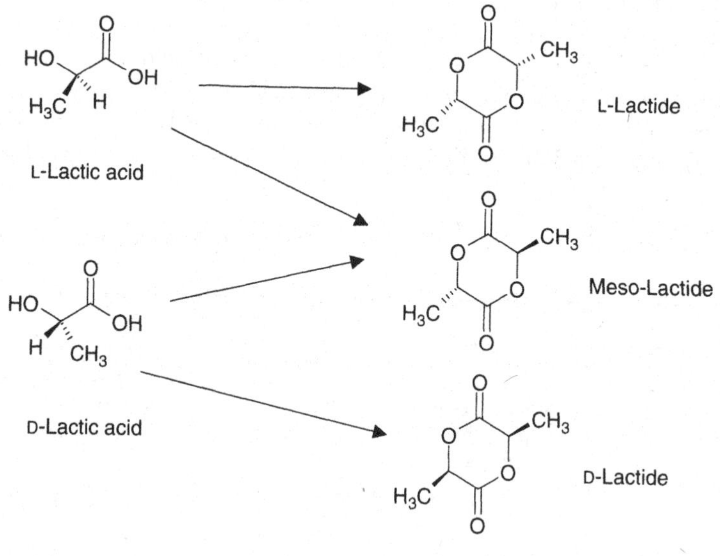

Figure 1.6 *Production of lactide intermediates with controlled optical ratios*

1.4   The Clean Technology Pool

Over recent years many technologies that can be considered Green have been developed. These are the technologies that can now be considered the technology pool for Green Chemistry. It is impossible to go into any of the subjects in depth and for further information it is recommended that the key references cited be studied.

1.4.1   Catalysis (Reducing the Use of Stoichiometric Reagents)

Catalysts are used to facilitate a transformation without being consumed as part of the reaction or without being incorporated into the final product. The way in which a catalyst improves a reaction system can take several forms. The catalyst can lower the energy of a given reaction pathway, meaning that lower temperatures can be used. This is especially relevant for reactions of industrial scale where reducing the reaction temperature of a large volume reaction process by 10 °C would represent an enormous saving in energy (and production of CO_2). Catalysts can also offer advantages with respect to selectivity. This could be viewed as a steering effect, where the catalyst steers the reactants towards a desired product. Selectivity enhancement could be represented by regioselectivity, mono-additions compared to multiple additions, stereoselectivity and indeed enantioselectivity (important in the pharmaceutical industry).

Catalysts are superior to stoichiometric reagents because in a stoichiometric reaction for every mole of product formed at least one mole of reagent is used. A catalyst may catalyse the formation of many molecules of the product before it becomes deactivated. There are several broad families of catalysts, some of which will be discussed in the following section.

Homogeneous Catalysis

Homogeneous catalysts are difficult to be removed/recycled from the reaction mixture after the reaction, but still find application in sectors of the chemical industry including polymers and pharmaceuticals. Homogeneous catalysts, by definition, are in the same phase as the reactants and consequently are not subject to diffusion limitation (meaning the reaction rate is often faster than heterogeneous systems). Typical acids and bases such as sulphuric acid and sodium hydroxide can be considered homogeneous catalysts. They can be removed from organic products by aqueous work-up, followed by neutralisation. However, this process cannot be described as Green Chemistry due to the stoichiometric formation of salt during neutralisation. Areas of the industry where homogeneous catalysts are used include alkene hydrogenation (Wilkinson-type catalysts), hydroformylation (rhodium and cobalt catalysts), polyethylene formation (metallocene-type catalysts) and Friedel–Crafts reactions (Gates, 1992).

Heterogeneous Catalysis

Heterogeneous catalysts typically (but not always) consist of an inorganic (sometimes organic, e.g. polystyrene) polymeric support, which may have acidic, basic or other synthetically useful functionality (which may be naturally present or may have been specifically incorporated). Bound to the inorganic support by some mechanism, there may additionally be a metal ion (typically Pt, Pd, Fe, Ni, Co, etc.). The nature of the support material is vitally important for heterogeneous catalysts. Typically, high surface area materials with controlled pore dimensions are used. High surface area is important because reactants need to diffuse to and from the surface of the catalyst (as they are in a different phase). The catalytic cycle involves the absorption, reaction and desorption of substrates on a heterogeneous catalyst. The rate of heterogeneous catalysis is often described as diffusion limited, which means the rate is controlled by the speed at which the substrate and the product can absorb and desorb from the surface of the catalyst.

There are many kinds of support material in heterogeneous catalysis. The choice of support material can be as important as the catalytic species. Typical support materials include zeolites, clays, zirconia, alumina, silica and polymers such as polystyrene. Zeolites are interesting microporous support materials as they have 3-D aluminosilicate structures (often referred to as cage structures). The size of the holes through the structure is often of molecular dimension, in fact zeolites can be used to separate molecules based on molecular dimensions. A good example of this is in the separation of *para*-xylene from a mixed xylene stream. The linear *p*-xylene diffuses through the holes in the zeolite, HZSM5, 10 000 times faster than the *ortho*- and *meta*-xylene and can be collected as an almost pure feedstock (Figure 1.7). For this reason many zeolites are known as molecular sieves. In chemical synthesis zeolites are particularly useful for vapour phase reactions involving small molecules. Zeolites are also used heavily in petrochemical cracking processes and as calcium- and magnesium-sequestering agents in detergents (Macquarrie, 2000).

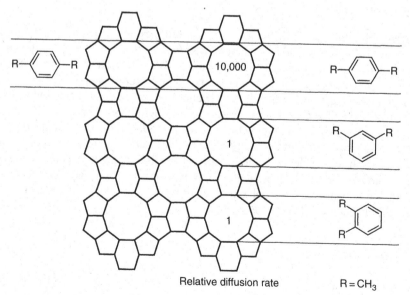

Figure 1.7 *Relative rates of diffusion of different xylenes through HZSM5*

Table 1.4 A summary of some species, which have been supported upon silica

Acidic	Basic	Other
$AlCl_3$	NH_2, NHR, NR_2	Ni^{2+}
BF_3	Imine	Cr^{3+}
$FeBr_3$	Alkoxide	Fe^{3+}
SbF_3	Phenolate	Co^{3+}
HSO_3	Binolate	Photoinitiator

There is broad application in chemical synthesis for all classes of heterogeneous catalysts. The catalysts are used industrially in gaseous and liquid phase reactions. Heterogeneous catalysts are often less toxic and easier to handle than the homogeneous analogues; they can also offer selectivity improvements in reactions. The biggest advantage of heterogeneous catalysis is the ease of separation of the catalyst from a solution of reactants. At the end of the reaction the catalyst can simply be filtered from the reaction system or removed in some other way (e.g. through immobilisation as part of the actual reactor such as a catalytic membrane plate, etc.). In many cases this avoids an aqueous work-up, which is often used in homogeneous systems to separate (or destroy) the catalyst from the products. After the catalyst is separated, it is often possible to reuse it in further reactions, although it may require reactivation (Sheldon and van Bekkum, 2001).

Silica is a common support material for heterogeneous catalysts because it is very tunable with respect to surface area and pore dimensions. Much research has been carried out on a multitude of catalytic species supported on silica. A number of species which have been successfully supported on silica in Table 1.4 are summarised.

Biocatalysis

Biocatalysis can be considered as the catalysis of reactions by microorganisms or enzymes. Typically, enzymes carry out specific transformations selectively (and mostly enantioselectively); the active site of an enzyme typically contains a transition metal such as Mg^{2+}, Fe^{2+}, Fe^{3+}, Ca^{2+} and Zn^{2+} (Davis and Borer, 2001). The range of reactions of enzymes is shown below:

- Oxireductases include enzymes such as dehydrogenases, oxidases and peroxidases, which catalyse transformations such as oxidation of alcohols to carbonyls and dehydrogenation of functionalised alkanes to alkenes.
- Hydrolases such as the digestive enzymes, amylase and lactase catalyse hydrolysis of glycosides, esters, anhydrides and amides.
- Transferases include transmethylases and transaminases and transfer a group (e.g. acyl) from one molecule to another.
- Isomerases catalyse reactions such as *cis–trans* isomerisation or more complex transformations such as D-glucose to D-fructose.
- Lyases catalyse group removal such as decarboxylation.
- Ligases catalyse bond-forming reactions, typified by condensation reactions.

The diversity of reactions carried out by enzymes is phenomenal, however, as Green Chemistry tools they do suffer certain drawbacks. They are only stable in mild conditions, and changes in pH and temperature effectively kill most microbes or enzymes. Micro-organisms, whilst cheap, are often unselective to a desired product because multiple products are formed as part of their metabolic cycle. Enzymes are selective to single products, but are extremely expensive and often difficult to recycle (although immobilised enzymes are available). Biocatalysis often needs to be carried out in dilute aqueous solution, which means separation of the products can be energy-intensive.

Others

There are many other kinds of catalysis that can be considered as examples of Green Chemistry. Examples include phase-transfer catalysis (Starks, Liotta and Halpern, 1994), water-tolerant Lewis acids (Kobayashi *et al.*, 1999) and photocatalysis (Maldotti, Molinari and Amadelli 2002).

1.4.2   Alternative Solvents (Reducing Auxiliaries)

Many organic reactions are carried out in an organic solvent (hydrocarbon, haloalkane, alcohol, etc.). Solvents are very useful because they bring immiscible reagents together in a common phase, which speeds up the reaction, they can also help to remove heat from exothermic reactions. Many common solvents are VOCs, which have all the negative environmental and health impacts as discussed earlier. In the chemical industry this means that companies are seriously considering alternative solvents. In this section an overview of alternative Green solvents is given.

1.4.3   Solventless Reactions

Many industrial organic processes are in fact carried out without an auxiliary solvent. In most cases one of the reactants will be used as the solvent; even if the reactants are only partially soluble, it may be enough for the reaction to occur. In many cases a solvent is used in a process simply because it has been directly scaled up from the laboratory reaction where solvents are normally used. In a chemical laboratory the chemist should not ask "which solvent is best for this reaction?" but rather "do I need a solvent for this reaction?".

Solvents are still commonplace in the manufacture of fine chemicals, as many of the reactants are high melting point solids. Recent work by Raston and Scott (2000) has shown that it is possible to make some solids react, almost quantitatively simply by grinding them together (in a pestle and mortar), with a catalyst. For example, complex pyridines can be synthesised by sequential aldol and Michael reactions. The yield is almost quantitative, which compared to the reaction in the solvent (50% yield) is a great improvement. The exceptional purity of the final product in the solid phase reaction also avoids the need for isolation of the aldol product by crystallisation compared to the process where solvent is used. The range of reactions that can be carried out under solventless conditions is growing, and now examples of Friedel–Crafts, transesterification, rearrangement and condensation reactions have been reported.

Water

Water is considered a poor solvent for many reactions because of its high polarity (rendering many organic compounds insoluble) or due to the fact that reactants may be hydrolytically unstable. However, for certain processes using water as a solvent, water can actually offer an advantage over traditional organic solvents. Breslow found that the Diels–Alder reaction between cyclopentadiene and butenone was 700 times faster in water than in organic solvents, despite the insolubility of one of the substrates (Breslow and Maitra, 1984). It was suggested that the hydrophobic nature of the reactants was forcing the droplets to cohere, thus increasing collisions and reaction rate. Water can also be advantageous when the required products are flavours or fragrances, where trace amounts of organic solvents would be unacceptable from an odour perspective. Water is odourless and benign, so small amounts present in the product will not be a quality issue.

Ionic Liquids

Ionic liquids (ILs) can essentially be considered to be similar to common ionic materials such as sodium chloride, except that due to poor packing of the respective ions they are in the liquid state at room temperature. To achieve this poor packing, it is usual for one of the ions to be organic in nature; there are almost limitless combinations of ions that could make ILs, an example is shown in Figure 1.8. The synthesis of ILs can be simple in some cases, whereas in others it can be a complicated multi-step synthesis. When considering the "Greenness" of a process using ILs, the synthesis of the IL should be taken into account. It should also be noted that at this stage toxicological data for ILs are lacking.

There are several general points that can be made about the properties of ILs:

- ILs have no vapour pressure and hence they are not lost through evaporation.
- ILs can be used as solvents and they can also act as catalysts.
- Tunability – by varying the cation/anion ratio, type and alkyl chain length properties such as acidity/basicity, melting temperature and viscosity can be varied to meet particular demands.
- Many ILs are stable at temperatures over 300 °C, providing the opportunity to carry out high-temperature reactions at low pressure.
- ILs that are not miscible with organic solvents or water may be used to aid product separation or may be used in liquid–liquid extraction processes.
- For a given cation the density and viscosity of an IL are dependent on the anion in general. Density increases in the order $BF_4^- < PF_6^- < (CF_3SO_2)_2N$ and viscosity increases in the order $(CF_3SO_2)_2N < BF_4^- < PF_6^- < NO_3^-$.

ILs are very good solvents when using homogeneous catalysts as the catalyst generally stays in the IL, thus avoiding tricky separation methods. Typically, at the end of the reaction the products can be removed by decanting or distillation, however, in some cases it is

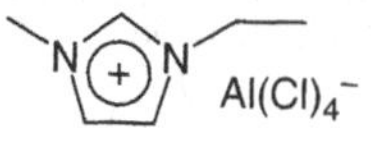

1-Ethyl-3-methylimidazolium chloride-aluminium(III)chloride ([emim]AlCl$_4^-$)

Figure 1.8   *An example of an ionic liquid*

difficult to remove the product from the IL. Many catalytic systems have been examined including hydrogenation (Rh catalysed), Heck and Suzuki (Pd catalysed) reactions (Welton, 1999).

Supercritical Fluids

A supercritical fluid (SCF) can be defined as a compound that is above its critical pressure (P_c), and above its critical temperature (T_c). Above T_c and P_c the material is in a single condensed state with properties between those of a gas and a liquid. Typically, the key advantages of carrying out a process under supercritical conditions include:

- Improved heat and mass transfer due to high diffusion rates and low viscosities.
- The possibility of fine-tuning solvent properties by varying temperature and pressure.
- A potentially large operating window in supercritical region.
- Easy solvent removal and recycling.

There are many substances that can be made supercritical, although some may require exceptionally high temperatures and pressures. The T_c (31.1 °C) and P_c (73.8 bar) of CO$_2$ are relatively easy to achieve in simple apparatus and the properties of the supercritical fluid have resulted in an enormous amount of research in this area. Supercritical CO$_2$ (scCO$_2$) has two major uses; as an extraction solvent and as an in-process solvent. First, we shall deal with scCO$_2$ as an extraction solvent.

The decaffeination of coffee is one of the biggest industrial uses of scCO$_2$; previously coffee was decaffeinated using the solvent dichloromethane, which was clearly unacceptable on health grounds. The success of decaffeination has led to a number of projects looking into the selective removal of flavours, fragrances and essential oils from plants. The dry-cleaning industry, to very limited extent, has also started adopting scCO$_2$ processes as a replacement for perchloroethylene. The dry-cleaning results can be good if surfactants are used to dissolve certain stains during the process (Taylor, 1996).

As a reaction solvent, scCO$_2$ is especially useful for several types of reaction. For polymerisation reactions scCO$_2$ has shown excellent and controllable solvation properties. Controlled molecular weight polymers can be produced by changes in temperature/pressure and also by the addition of surfactants. As well as polymerisation, a number of reactions can be carried out using scCO$_2$. Perhaps the most widely studied, due to the high solubility of H$_2$ in scCO$_2$, are hydrogenation reactions where high conversions and high selectivity (including enantioselectivity) can be achieved. It should be noted, however, that CO$_2$ itself can be hydrogenated to make formic acid. Other reactions that have been carried out using scCO$_2$ include the Heck and Suzuki reactions (although for solubility in scCO$_2$ the Pd catalyst requires perfluorinated ligands that are difficult and expensive to synthesise), hydroformylation, Friedel–Crafts, esterification, chlorination, Diels–Alder and oxidation (Jessop and Leitner, 1999).

1.4.4   Process Intensification and Innovative Engineering

Process intensification (PI) is commonly defined as "Technologies and strategies that enable the physical sizes of conventional process engineering unit operations to be significantly reduced." This is achieved through:

- Improving mass transfer rates to match that of the reaction.
- Improving heat transfer rates to match the exothermicity of a reaction.
- Having an appropriate residence time for the reaction.

Originally devised as a cost reduction concept, as a result of the development of novel smaller reactors and ancillary equipment, PI is now recognised as a way of providing safety improvements, greater throughput, and improved product quality through better control. All these features are important in the development of more sustainable processes. As a general rule major equipment such as reactors and distillation columns account for only 20% of the price of a manufacturing plant, the remainder being pipework, instrumentation, labour and engineering charges, etc. The concept behind PI was that even though novel pieces of key equipment may be a little more expensive, the overall plant cost would be reduced as a result of simplification and size reduction.

Spinning Disc Reactors (SDRs)

SDRs have been proposed as an efficient alternative for fast reactions. As its name suggests the SDR consists of a disc rotating at a speed of 5000 rpm or more. The disc may be smooth or may contain ridges to aid mixing; it may simply be a stainless steel plate acting as a source of heat or have a catalytic surface on which the reaction is carried out. The reactant liquids are pumped onto the centre of the disc, the resulting flow patterns causing intense mixing as the liquids move towards the edge of the disc, where the products are collected. Because the liquid forms a thin film on the disc surface heat transfer is rapid (heating or cooling), this together with the intense mixing overcomes any heat and mass transfer limitations, allowing the reaction to run under kinetic control (Brechtelsbauer, Lewis and Oxley, 2001).

Micro-Reactors

The phrase "lab-on-a-chip" was coined after the development of tiny reactors that resembled printed circuit boards; this was a micro-reactor. In its simplest form a micro-reactor device could be a capillary column (similar to that used in a GLC) through which reactant(s) are pumped (by syringe) and where reaction occurs at catalytic sites previously introduced to the capillary wall. The development of micro-reactors has been driven by the need for controllable, information-rich, high-throughput, environmentally friendly methods of producing products with a high degree of selectivity. The greatest thrust of micro-reactor research has been in the field of analytical chemistry, and in particular towards the development of total analytical systems. The biomedical field leads the way in this arena especially relating to the analysis of DNA and proteomics, and has resulted in the release of the first commercial analytical micro-reactor device. For reaction chemistry the pharmaceutical industry is particularly interested in the development of micro-reactors that enable the rapid synthesis and testing of numerous drug candidates on a micro-scale. The key concept is that for promising drug candidates the product could quickly be made in larger quantities for testing by simply running numerous micro-reactors in parallel. This has been referred to as the scale-out concept (as opposed to the scale-up concept). For a thorough review the following reference is recommended (Fletcher *et al.*, 2000).

We have now seen and discussed a number of tools, techniques and concepts of Green Chemistry. The question now is, how do we measure whether we are actually making a significant difference in the environmental impact of the process. If we change a step early on in a chemical process, how do we know we have not simply shifted a problem further downstream? In order to assess the environmental impact of a process from cradle to grave, a tool of choice is Life Cycle Assessment (LCA). In this next section, the concept of LCA will be discussed and examples relevant to RRMs will be briefly highlighted.

1.5   Life Cycle Assessment

1.5.1   Introduction to Life Cycle Assessment

Product A is made from a feedstock based on fossil fuels. Product B is made from a feedstock based on biomass. Given that A and B are similar products used for the same purpose, how can we determine which of them has the smallest environmental footprint based on the origin of their feedstock?

Life Cycle Assessment (LCA) is a process that attempts to measure the environmental impact of a product or function on its entire life cycle (Gradel, 1998). Ultimately, a life cycle should be a cradle-to-grave assessment, that is from the extraction of raw materials for manufacture through to the products' fate at the end of its useful life. The life cycle can also include refurbishment, remanufacture or recycling (into useful material or into energy). An example of a general life-cycle scheme is shown in Figure 1.9.

As mentioned earlier (section 1.2.3), the concept of mass and energy balance in PFSs is currently used by chemical engineers, however, this process starts at the plant entrance and ends at the plant exit. LCA takes this concept much further and requires all the inputs and outputs (energy, materials and waste) to be calculated for each unit operation during the full life cycle of a given product. The origin and method of extraction of the raw materials is taken into account for the energy, materials and waste concerns. Transportation issues are also considered, which can have an impact on the final environmental impact; for instance for a process requiring precious metals, the metals may be imported from a distant country such as South Africa. That also means that the outcome of an LCA might

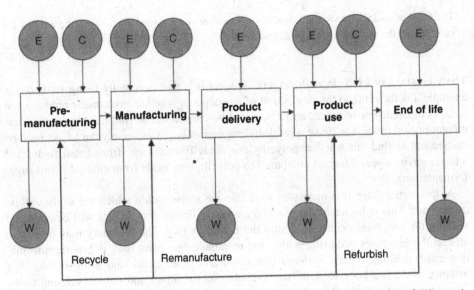

Figure 1.9 *Representation of a product life cycle (E is energy, C is chemicals (raw materials) and W is waste)*

be different in terms of the location where the process is evaluated (e.g. Western World or developing country). The energy and material demands of a product during its operational life is also taken into account; for a product such as a washing machine, around 80% of its energy usage may be accounted for at this stage. Finally, the fate of the product is addressed, that is whether the product biodegrades in landfill, is burned for energy or in some way recycled for materials.

To carry out a full LCA requires a large amount of time and patience and access to high-quality data, but it is the only method backed by ISO for assessing the environmental impact of a product. One of the most important features of a full LCA is that it should be completely transparent. This means if someone is studying an LCA carried out by a company they are not connected with, the rationale behind every single step in the LCA should be obvious and all the findings and claims should be fully justified and should not be merely personal judgements.

The process of carrying out a full LCA can conveniently be broken down into four interconnecting steps:

1. Goal and scope definition.
2. Life Cycle Inventory analysis (LCI).
3. Life Cycle Impact Analysis (LCIA).
4. Life Cycle Interpretation.

1.5.2 Goal and Scope Definition

This stage can essentially be regarded as the planning stage; it can also be regarded as the most important stage of the LCA.

The **Goal** of an LCA study shall unambiguously state the intended application, the reasons for carrying out the study and the intended audience.

ISO 14041

The *goal* stage of LCA depends largely on what it is that is actually being studied, for example it is the route to the production of a product or whether two similar products are to be compared. For instance, an LCA could be carried out to find the lowest environmental impact process to make a polyethylene carrier bag. Conversely, an LCA could be carried out to find out whether polyethylene made from ethane (from fossil fuels) had a lower environmental impact compared to polyethylene made from ethanol (from sugar fermentation).

At this initial stage it is necessary to define the *scope* within which the study will be carried out. This is basically the boundary where a line is drawn; data will be collected only on the processes occurring within the boundary line. The boundary may be drawn around the processes occurring within the manufacturing plant only if generic information exists relating to raw material extraction and product use and ultimate fate. For instance, there is a large bank of data available on the inputs and outputs relating to the extraction of fossil fuels. If the data are not available for a process outside the scope of the LCA, then this could be a problem if this is deemed a necessary part of the LCA.

For comparative studies it is essential that the same *functional unit* is chosen. For instance, if the study is related to a storage medium for milk, then the functional unit would be an appropriate receptacle for storing $500\,cm^3$ of milk. This could be a glass, a cardboard or a plastic bottle. For a chemical product it could be the production of $1\,t$ of the desired final product.

The strategy for data collection is also decided upon during this initial planning stage. It may not be possible to collect all the necessary data, however, any assumptions that are made must be recorded and justified. Goal and scope definition is an iterative process and needs to be part of a constant review and refinement cycle throughout the LCA.

1.5.3   Life Cycle Inventory (LCI) Analysis

A Life Cycle Inventory Analysis is concerned with the data collection and calculation procedures necessary to complete the inventory.

ISO 14041

This stage consists of accounting for all of the energy and material inputs and outputs across the entire scope of the chosen process. Typically, this involves breaking the process into a series of unit operations and collecting data on each operation. A typical unit operation could be a fermentation reactor, a distillation column, etc. Ideally, data should be as precise as possible and consequently directly from the producers. Typical data sources are direct measurements, interviews with experts, literature and database searches, theoretical calculations and best guess estimates (guesstimates). As LCA becomes a more widely used tool, more databases containing generic data are becoming widely available.

It is important that the data that is used can stand up to scrutiny. Assumptions made and guesstimates should have transparent explanations; any uncertainty in the data should be reported. The data can be tested by sensitivity analysis. If an assumption is made that something does not have an impact on a process, studying the effect of a change of the parameter can test this. If the impact is negligible then the assumption holds, and if the effect is large the assumption is invalid.

1.5.4 Life Cycle Impact Assessment (LCIA)

The LCIA phase aims to examine the product system from an environmental perspective using category indicators, derived from the LCI results. The LCIA phase also provides information for the interpretation phase.

ISO 14042

During this phase the data from the LCI phase are collated into impact categories, for instance CO_2 and N_2O emissions would be categorised in the global warming potential category:

- Abiotic Depletion (This takes account of depletion of all non-renewable resources)
- Acidification Potential
- Aquatic Toxicity
- Eutrophication Potential
- Global Warming Potential
- Human Toxicity Potential
- Ozone Depletion Potential
- Photochemical Oxidants Creation Potential.

A point of note is that the impact categories do not take into account the nature of a particular emission, but just the total amount of it. What this means is that two processes may have a similar global warming potential according to the LCA assessment, as over the course of the study almost exactly the same amount of CO_2 was released into the atmosphere. However, in one process the CO_2 was released gradually over 1 year; and in the other process the CO_2 may have been released intensively over a short period of time. This would have a different effect on global warming for each system. Issues like this can be addressed in the interpretation phase.

1.5.5 Life Cycle Interpretation

Life Cycle Interpretation is a systematic technique to identify, qualify, check, and evaluate information from the results of the Life Cycle Inventory (LCI) analysis and/or LCIA of a product system, and present them in order to meet the requirements of the application as described in the goal and scope of the study.

ISO 14043

It is in this phase of the LCA that the results obtained are scrutinised in order to try and determine where the major environmental impacts occur within a process, and how, if possible, improvements to the environmental impact of the process could be made.

1.5.6   Case Study – Environmental Impact of the Manufacture of PHA

It is often assumed that chemicals and materials derived from RRMs have a lower environmental impact when compared to those derived from fossil fuels. In a LCA recent study by Gerngross (Gerngross, 1999), a polymer synthesised from a RRM (PHA) was compared to a polymer synthesised from a fossil fuel feedstock (polystyrene, PS). In both cases the scope of the study was from the extraction of raw materials through to the synthesis of the polymer resin.

The study showed that the production of PHA from a corn feedstock by a fermentation process had surprisingly high-energy requirements. In fact, it was found that the fermentation process required 22% more steam, 19-fold more electricity and 7-fold more water than that required for the production of PS. In terms of fossil fuel consumption 1 kg of PHA required 2.39 kg of fossil fuel, whereas 1 kg of PS required 2.26 kg. Although this does not seem a great difference, we should consider that for PS while 1.26 kg of the fossil fuel requirement is used as a feedstock for the final product, only 1 kg is used for energy; for PHA all the fossil fuel requirements are for energy.

This does not appear to offer a promising future for PHA, but we should consider that the technology is relatively new compared to that for PS, and therefore can be optimised further. One further suggestion is that some of the energy requirements could be addressed by burning corn-waste (such as straw) as an energy source. Theoretically, burning such biomass has been shown to reduce the environmental impact of the PHA process, however, burning biomass as an energy source would be equally beneficial for the production of PS. As an example, this study does show the need to study the overall environmental impact of any new product or process involving RRMs; although it may look favourable for one aspect, the overall environmental impact compared to that of a contemporary analogue may not be an improvement. This does not mean that research in RRMs should not be carried out, simply that we should as scientists be aware of the "big picture" and not simply shift an environmental issue to somewhere else in the process.

1.6   Conclusions

This chapter tried to provide a broad overview of the necessity of the chemical industry in modern society and tried to put it into perspective against the negative public perception and the negative points of the industry. The concept of Green Chemistry has been introduced as a series of reductions, of which reducing the use of non-renewable materials is an important issue. However, when we consider substituting a fossil fuel-based product with one derived from RRMs, we should be extremely careful not to introduce a technically inferior product, as this is unacceptable. The substitution of a non-renewable feedstock by a renewable feedstock does not guarantee that the end product will be one with a reduced environmental impact. To explore environmental impact of a product or process in depth, it is necessary to apply a method such as an LCA study. By being aware of the need for designing and evaluating new products and processes from a holistic perspective, it is much more likely that superior and sustainable chemical products will be produced.

References

Anastas, P.T. and Warner, J.C. (1998). *Green Chemistry Theory and Practice*, Oxford University Press, Oxford.

Brechtelsbauer, C., Lewis, N. and Oxley, P. (2001). *Org. Process Res. Dev.*, **5**, 65–68.

Breslow, R. and Maitra, U. (1984). *Tetrahedron Lett.*, **25**, 1239–1240.

COM(2000) 347 final (2000). *Proposal for a Directive of the European Parliament and of the Council on Waste Electrical and Electronic Equipment and Proposal for a Directive of the European Parliament and of the Council on the Restriction of the use of certain Hazardous Substances in Electrical and Electronic Equipment*, Brussels, 13.06.2000.

COM(2001) 88 final, 2001, *White Paper, Strategy for a future Chemicals policy.*

Constable, D.J.C., Curzons, A.D. and Cunningham, V.L. (2002). *Green Chem.*, **4**, 521–527.

Davis, B.G. and Borer, V. (2001). *Nat. Prod. Rep.*, **18**, 618–640.

Elkington, J. (1999). *Australian CPA*, **69**, 18–21.

Fletcher, P.D.I., Haswell, S.J., Pombo-Villar, E., Warrington, B.H., Watts, P., Wong, S.Y.F. and Zhang, X.L. (2000). *Tetrahedron*, **58**, 4735–4757.

Furniess, B.S., Hannaford, A.J., Rogers, V., Smith, P.W.G. and Tatchell, A.R. (1989). *Vogels Textbook of Practical Organic Chemistry*, Fifth Edition, Chapter 4, p. 338 (Wittig reaction) and p. 775 (benzophenone). Longman Scientific & Technical.

Gates, B.C. (1992). *Catalytic Chemistry*, Chapter 2, pp. 15–142. John Wiley & Sons, New York.

Gerngross, T.U. (1999). *Nature Biotechnology*, **17**, 541–544.

Gradel, T.E. (1998). *Streamlined Life-Cycle Assessment*, Prentice-Hall, New Jersey.

IENICA (Ref 1495) (August 2000). *Summary Report Oil crops.doc.*

Jessop, P.G. and Leitner, W. (1999). *Chemical Synthesis Using Supercritical Fluids*, Wiley-VCH, Weinheim.

Kobayashi, S., Mori, Y., Nagayama, S. and Manabe, K. (1999). *Green Chem.*, **1**, 175–177.

Macquarrie, D.J. (2000). *Phil. Trans. R. Soc. Lond. A*, **358**, 419–430.

Maldotti, A., Molinari, A. and Amadelli, R. (2002). *Chem. Rev.*, **102**, 3811–3836.

Raston, C.L. and Scott, J.L. (2000). *Green Chem.*, **2**, 49–52.

Shaviv, A. (1999). *Preparation Methods and Release Mechanisms of Controlled Release Fertilisers: Agronomic Efficiency and Environmental Significances (Proceedings of the International Fertilizer Society)*, The International Fertiliser Society.

Sheldon, R.A. (1994). *CHEMTECH*, **24**, 38–47.

Sheldon, R.A. (1997). *Chemistry and Industry*, 6th January, 12–15.

Sheldon, R.A. and van Bekkum, H. (eds) (2001). *Fine Chemicals through Heterogeneous Catalysis*, Wiley-VCH, Germany.

Soderman, O. and Johansson, I. (2000). *Current Opinion in Colloid & Interface Science*, **4**, 391–401.

Starks, C.M., Liotta, C.L. and Halpern, M. (1994). *Phase Transfer Catalysis: Fundamentals, Applications and Industrial Perspectives*, Chapman & Hall, New York.

Stevens, E.S. (2002). *Green Plastics*, Princeton University Press, Princeton.

Taylor, L.T. (1996). *Supercritical Fluid Extraction*, John Wiley & Sons, Incorporated, New York.

The World Commission on Environment and Development (1987). *Our Common Future*, Oxford University Press, New York.

Trost, B. (1998). Received the Presidential Green Chemistry Challenge Award – website www.epa.gov/greenchemistry.

Welton, T. (1999). *Chem. Rev.*, **99**, p. 2071.

2

Socio-Economical Aspects and Policy of Renewable Resources

Liisa Tahvanainen and Christian V. Stevens
In co-authorship with *Paavo Pelkonen, David Gritten, Eero Forss, Sevim Erhan and Joseph M. Perez*

2.1 Introduction

Building a more sustainable society by using more renewable resources instead of non-renewables is not only connected with technical aspects and chemical conversions, but is also very much connected with socio-economical aspects and with policy. In order to induce changes, politicians need to be convinced and a policy sustaining the concept must be developed and applied. Since the concept has a global impact, global policies would be favourable, although the development of global policies are extremely difficult since they need to be accepted and negotiated by all parties. Despite the global nature of the concept, most renewable resources are local and are effecting numerous local impacts particularly on rural areas. This emphasizes not only the significance of rural and regional policy, but also the importance of the acceptability of the concept to local inhabitants, producers and consumers.

In this chapter the following aspects are discussed:

- The socio-economical aspects such as rural development, employment, security of supply and ethical questions.
- The policy towards renewables in the European Union (EU) and in the US.
- General conclusions on the socio-economical aspects and policy.

Renewable Bioresources: Scope and Modification for Non-food Applications. Edited by C.V. Stevens and R. Verhé
© 2004 John Wiley & Sons, Ltd   ISBNs: 0-470-85446-4 (HB); 0-470-85447-2 (PB)

2.2 General Concepts

The concept of renewable resources has become an important factor in agriculture, forestry and related trade and industry. Biomass is not only used traditionally as food, animal feed and paper, but also as an energy carrier or as a raw material in various industries. Parts and components of crops can be used in an increasing number of new applications. In addition, an increasing demand for renewable biomass may contribute to climate change mitigation and also to rural development in the form of additional income and environmental benefit. During recent years the focus has shifted from pure agricultural policy, for example the reduction of agricultural surpluses or the creation of new job opportunities, to more and more environmental aspects such as the reduction of greenhouse gas emissions. The following are some examples of the driving forces behind the increasing interest in renewable resources: decrease in global non-renewable resources; growth in the world population; concern about the environment with resulting environmental taxes; market-strategic reasons (consumer awareness); effects on rural livelihood; and also the advantages of biological raw materials over synthetic ones.

Biodegradable materials are opening up a comparatively large market potential for raw materials from agriculture and forestry, yielding a high added value for agricultural production. There has been increasing interest in the use of renewable resources in the chemical industry and also for energy production. The chemical industry requires basic materials as well as specialty chemicals on a renewable basis to meet the consumer demand of ecologically sound products. An ecologically sound construction industry is promoting recycling and refining of biomaterials in various ways. In energy production, the so-called 'green energy' label has been used in marketing for a couple of years. However, in most cases these are still small niche markets, which invariably means that they are more expensive. The RES-e (promotion of electricity produced from renewable energy sources in the internal electricity market) directive of the EU (2001/77/EC) is opening new market opportunities for power production based on renewable resources. The requirements of the demand (consumers, industry, agriculture and environmental protection) are increasingly important to the aim of creating economically profitable market possibilities for renewable commodities.

The use of biomass and organic waste focuses on a number of advantages such as reduction of agricultural surpluses, new jobs and income for agriculture and forestry, contribution to greenhouse gas mitigation and decreased dependence on imported crude oil. However, in order to achieve these possible benefits major political, strategic and also marketing challenges should be faced. Trossero (2000) has defined these challenges in bioenergy production – but these can be applied to any renewable biomass production in general – as follows: (1) to enhance the role of agriculture and forestry in rural economic development and to provide a substitute for fossil fuel; (2) to increase political awareness, social appreciation and to promote cultural changes about bioenergy roles in energy, environment and agricultural sectors; (3) to build national capacities, research and demonstration projects, and to remove constraints for the development of wider biofuel markets at commercial prices; and (4) to prepare strategic studies that illustrate the potential of agriculture as a cost-effective and market-oriented energy producer and specifically on the issues of land availability, economics, technology development and transfer, assessment, planning and participation.

The beneficial effects of renewables on society and the advantages for industry are well defined but the possible disadvantages have not been clearly defined so far. The vision of utilizing considerable amounts of renewable resources in the future might also incur negative effects, for example on forests or on the agricultural environment, or lead to ethical and ecological problems by creating a demand for an increase in productivity (e.g. genetic modification), to competition for wood and increased utilization of forests or to competition for agricultural land use. This emphasizes the real importance of political responsibility. To reach a sustainable, economically profitable, and regionally and globally acceptable way to utilize the potential of renewable resources, close cooperation between agriculture, research, industry, policy-makers and administration is of huge importance.

2.3 Socio-Economical Impacts

2.3.1 Rural Development

The income potential involved in the use of renewable biomass and the connected beneficial environmental effects lie especially in rural areas. It is seen that the intensified and new uses of biomass could help to stabilize rural areas and open up new or additional income opportunities for agriculture and forestry. Agriculture can start to supply renewable agricultural commodities, for example starting materials for chemicals and construction or even for the tourism industry.

The use of waste generated by agriculture, forestry and the processing industry can help to enhance the added value in basic production of rural communities. Furthermore, sustainable production and the production of ecologically sound commodities can create a more positive image of the countryside that may, for example, attract more tourists. This is an important additional income opportunity for many remote rural areas.

Locality is one of the major preconditions of profitable production for rural development. A good example of a locally available product is biofuel that can provide income not only for farmers (the biomass producer), but also for local businesses (small-scale energy production). This substitution of imported fossil fuel can be beneficial for the local economy.

For example, the FAO's bioenergy programme bases its operations on the following concepts: (1) bioenergy can stimulate diversification of agricultural and forestry activities, for example, through the establishment of energy plantations with trees and crops; (2) biofuels can provide locally the necessary energy to improve agriculture and forestry productivity; and (3) bioenergy can attract investment to rural areas where most of the biofuels are produced (Trossero, 2000).

However, particularly for rural areas, the profitability of using and producing biomass for different new purposes has to be carefully analyzed in order to reach the most economical and competitive land use alternative for private landowners. To secure the rural development, the feasibility of all new alternatives for agriculture and forestry should also be proven in private economy, regardless of the benefits for the national economy, renewability or other positive environmental effects. The possible conflict between environmental benefits for society and income for rural entrepreneurs, landowners, etc. may require political issues to be resolved.

Production of agriculture and forestry has to be based to a great extent on entrepreneurship of private individuals, e.g. farmers in Small–Medium sized Enterprises (SMEs). Networks of small enterprises and cooperatives have proven to be strong competitors even in a competitive market economy. The networking of mechanical wood industries in northern Italy is one of the most impressive examples of opportunities in areas outside the large urban areas. Specialization together with well-targeted education will provide good preconditions for new methods of non-wood production in rural areas.

High unemployment, especially in the fringe regions in the European Union (EU), is unacceptable. Radical steps have to be taken in order to offer working opportunities to unemployed people in many rural areas in the EU. The aim has to be the creation of new jobs instead of protection of existing ones. People have to be encouraged to develop new production opportunities through life-long learning. Continuous improvement of non-competitive production concepts is going to lead to increasing subsidies and to a clash between the rural and urban cultures. New, ecologically, economically and socially sustainable production is a key factor in order to combine rural and urban policies in the European societies.

2.3.2   Employment

Renewable resources, especially biomass, are seen as a major possibility and challenge to secure rural employment and to constrain excessive migration from rural to urban areas. This is an important aspect for industrialized countries as well as for developing countries. Therefore, most studies and estimates are being done for the bioenergy sector. Study of other sectors would be even more difficult.

For the local economy, products and commodities that can be produced and refined locally are especially valuable. In addition to the agricultural and forestry sectors, such commodities also bring job opportunities for small-scale industry, reduce cash flow outside the area and bring indirect employment and welfare improvements for rural areas.

The progression towards bioenergy is a good example of such a process. Renewable energy resources and technologies are increasingly valued because they can stimulate rural development and agricultural diversification, as well as deliver local environmental improvements. Utilization of local biomass energy sources to fulfil local energy needs creates small- and medium-sized enterprises as well as local markets, where additional incomes are supporting the community. Production, procurement and transportation of raw materials (in addition to small-scale entrepreneurship and industry producing heating energy, electricity or, for example, pellets) provide an important source of new jobs, even when compared with other renewable energies. Indirect benefits arising from the utilization of biomass energy may also moderate excessive migration from rural to urban areas (Wisniewski, 1998; Grassi, 2000).

Poland is one example of a country with a very high unemployment rate in rural areas, reaching 50% in some areas. Additionally, rural areas show a negative economic balance (Wisniewski, 1998). Therefore, identifying and exploiting renewable energy sources is emphasized to avoid the need to pay additionally for transporting fuel and electricity over long distances.

In countries with a promising biomass potential and high unemployment, bioenergy could provide a significant contribution to alleviate this problem, especially as job creation in this field would require little investment. It is calculated that the costs of investment in bioenergy per year per person are clearly less than the average cost of unemployment per year per person in Germany (Grassi, 2000). On the European level it has been estimated that, for example, in primary energy production the targeted doubling of the share of renewables (from 6 to 12%) could create an estimated gross figure of 500 000–900 000 new jobs, providing annual savings (in 2010) of €3 billion in fuel costs and a total of €21 billion for the period 1997–2010, reducing imported fuels by 17.4% and CO_2 emissions by 402×10^6 t/yr by 2010 (The European Commission, 1997).

It is estimated that in Germany, for example, the domestic supply, processing and energy use of biomass and organic wastes could create an additional 25 000–35 000 full-time jobs in the long term (which means an additional employment effect of 3–5%). In addition, there will be an indirect employment effect (Leibe, 1998).

An example of a region with estimated effects of increased bioenergy use is the North Karelian area (170 000 inhabitants, unemployment rate 15%) in eastern Finland. According to the regional bioenergy strategy, procurement of energy wood and small-scale energy entrepreneurship would directly lead to the employment of 200 persons in the year 2010 (the initial number of jobs was 60), partly as a secondary occupation in agriculture and forestry. In addition, 130 indirect jobs are expected to be created (1160 m^3 of wood provides one full-time job – production and transportation of forest chips) (Tahvanainen, Asikainen and Puhakka, 1998).

On the basis of long-lasting and large-scale forest biofuel production in Sweden the employment impact on various elements of production can be estimated. Two hundred man-years are required for collecting and harvesting logging residues for the production of 1 TWh, but only 40 are required for energy production from by-products of the wood processing industry, such as sawdust or bark. Production of briquettes requires 140 man-years and that of pellets requires 220 man-years for 1 TWh. In the local small-scale use of biofuels, around 200 man-years are needed and district heating entails about 80 (Energi från skogen, 1999).

It is not possible to reach exact conclusions as to the possible cumulative job creation which would derive from investment the various forms of renewable resources. The economical effects of production depend on the opportunities to further process the material in the area, location and size of markets, and the nature (substitutive, additional, totally new) and extent of production. When the markets are not local, especially if the raw material has to be transported, the optimization of the logistic chain will be crucial for profitability. However, it is quite clear that the move towards renewable resources, especially biomasses, will lead to significant new employment opportunities in rural areas. This is especially important as it coincides with the decline in traditional agricultural production as an income for farmers.

An important additional economic benefit not included above is the potential growth of the European industry in international markets. In the renewable energy sector, a €17 billion annual export business is projected for 2010, creating potentially as many as 350 000 additional jobs (The European Commission, 1997).

In spite of remarkable positive effects on global and local environments and on regional economy, unfortunately there is also another side of this transition when considering it on a global scale. According to the Intergovernmental Panel on Climate Change (IPCC), other areas will incur significant costs due to the heavily increased use of renewable resources in energy production: (1) reduced economic activity in coal and oil producing nations as a result of reduced sales; (2) significant job losses in the extraction, transport and processing of fossil fuels; (3) possible slowdown in economic growth during the transition from fossil fuels to renewables, particularly in countries with a very heavy reliance on non-renewables.

These impacts will, in many cases, be felt outside the energy sector and should be taken into account in a global context when designing the renewable resources strategies. In many cases, the countries producing fossil fuels and raw materials are not those leading research and using new substitutive materials. Even though negative impacts (e.g. the offset of jobs) could be covered by the growth of the renewables sector in the long run, the time span could be crucially and problematically long for some areas if not considered early enough.

2.3.3 Security of Supply and Use

One of the major political reasons (in addition to the environmental reasons) for promoting the use of renewables is that in most cases renewable energy sources are indigenous and can therefore contribute to the reduction of dependency on energy imports and to increasing the security of supply. In particular, the dependency on imports that are occasionally unstable can be reduced. The development of renewables can be a key feature in stable regional development with the aim of achieving greater social and economic cohesion within the community. In local, regional and national economies, versatile energy production offers especially economical stability and security. In addition to economic security, local energy production can secure the steady supply of energy in many remote areas.

The development of bio-based energy, chemicals and materials is one of the most important activities we can undertake to expand the global economy whilst also protecting our shared environment. The profitability of the use of renewable resources includes all the direct and indirect private, regional, national, environmental and also global economic benefits and advantages. Examples of indirect benefits and costs are: (1) public accidents in the production cycle and during transportation; (2) accidents related to the production process; (3) environmental benefits of energy wood procurement; and for example (4) positive effects that decreasing unemployment may have upon health.

Further, atmospheric emissions damage crops, forests, human health and materials. Most of the damage, however, is related to human health. For instance, 98% of the local and regional damages associated with natural gas are related to human health, where the largest contribution is to chronic mortality caused mainly by emissions of nitrates (Nielsen and Schleisner, 1998).

In addition to the important factor of security of supply of raw materials, the increasing awareness of environmental issues and sustainable consumption has brought environmental arguments into decision making. To give an example, a study carried out in Finland shows that Finnish consumers would like to increase the share of renewables in primary energy production. The consumers particularly emphasized security and safety, environmental friendliness and a secure supply as the most important factors when choosing primary energy sources in Finland.

Many rural areas especially in the northern parts are frequently facing severe winters with long cold periods. Heavy snow loads are breaking power lines, causing interruptions to the energy supply. These sorts of situations can be critical, especially for animal husbandry and related industries. Local energy production (fuelled with local resources) would increase security in remote communities and provide preconditions for sustainable development and entrepreneurship in rural areas.

2.3.4   Environmental and Ethical Questions

Sustainable development is linked to the utilization of renewable resources. The Bruntland Commission (1987) stated that the aim of sustainability is to attain 'development that meets the needs of the present without compromising the ability of future generations to meet their own needs'.

The goal of attaining economic welfare and, at the same time, protecting the environment and meeting regional social needs is a challenging task. The importance of the responsible use of renewable resources cannot be exaggerated when it comes to sustainable development, especially when taking into account the forecasted world population growth. The international community can push in the direction of sustainable production, for example by promoting the development of new technologies and methods for improved biomass use, both in industrialized and in developing countries.

The production of biomass for versatile non-food purposes and current industrial uses will require fertile land – which is already used for food production in most cases. Thus, one of the main concerns is the availability of land for food and non-food production. Furthermore, when changing the land use from traditional agricultural production to other purposes, the resulting consequences for ecosystems and intangible commodities should be taken into account. Demand for increased productivity may lead to more intensive crop systems, genetic modifications, monocultures and varying effects from ecological to visual quality. The land use planning should be based on a multi-objective approach taking into account ecological, economical, social, ethical and cultural aspects of biomass production and use. In most cases, comprehensive analyses of these aspects have not yet been carried out. Such comprehensive analyses would considerably help to integrate biomass systems into agriculture, forest, energy and other policies and strategies.

The ethical point of view has to be emphasized in the global context of land use, especially in developing countries where food security deserves the highest priority. This is especially the case when considering allocation of arable land for non-food production instead of food production, or when considering production of food crops for alternative uses. In future, consumer behaviour needs to be increasingly connected to ethical codes of production. Business ethics will also have a strong influence on all areas of future production.

2.4   Policy Towards Renewable Resources

In order to stimulate society and industry to use more renewable resources, a general policy must be developed. Although most people are convinced that an increased use of renewable resources would be advantageous for the environment and may stimulate the agricultural community and forestry, the manner in which the policy is developed in different countries may vary enormously. In this context, two important examples of how the drive towards renewable resources has been translated into a policy will be discussed – the European policy and the policy in the United States. From the completely different topics discussed in these two regions, it is immediately apparent that attitudes towards renewables are entirely inconsistent and focus on very disparate aspects.

2.4.1 The European Policy

Regulations

The focus of the EU Agricultural policy concerning industrial crops is on environmental factors and on the question of how to deal with the large agricultural surpluses (butter mountain, milk sea, cereal surplus) experienced in the 1980s. The costs of storing surpluses were too high and in addition some agricultural regions in Europe needed to be subsidized. In order to solve the problem of surpluses, the concept of compulsory set-aside land (arable land that was not allowed to be used) was introduced in 1992 and farmers were paid per acre for the loss of income from this land (set-aside premium). At the outset, the set-aside concept was introduced as a management tool in order to bring balance into the cereal market.

At a later date, the use of compulsory set-aside land to grow crops for non-food applications was permitted (e.g. rapeseed, short rotation coppice). In this way the farmer was encouraged to grow renewable resources for industry since he was allowed to keep the set-aside premium whilst growing industrial crops. The non-food set-aside concept proved to be a success and in 1995–1996 1×10^6 ha were classed as set-aside land. Later it reduced to 400 000 ha because the set-aside obligations declined from 15 to 5%.

In 1998, 2×10^6 ha of arable land (4% of the total arable land in Europe) were utilized for industrial crops (Versteijlen, 1998), e.g. textile fibres (650 000 ha), starches (650 000 ha) and oilseeds (350 000 ha). Recently, the most important crops are oilseed crops (for biodiesel production) and the energy crops.

The recent reforming of agricultural policy in the EU based on an acreage subsidy system, instead of the formerly existing production-bound system, has the advantage of offering more freedom to farmers in decision making and production planning according to the changing market demand. On the other hand, this means that society needs to find alternative measures for production control, for example towards environmentally sound production and to enhance secure production of commodities that have a high intangible value for society. However, nowadays subsidy systems are being minimized in order to keep the agricultural budget under control (Blair Hous agreement). The intervention price for arable crops has been cut, for example payments for oilseeds and non-textile linseeds are to be progressively reduced in three stages. Compulsory set-aside has been retained and voluntary set-aside is still allowed. This approach should be more effective and should have a more positive impact on the environment. The level of compensation for set-aside, both compulsory and voluntary, will be the same as that for cereals.

Following the 1997 publication of the first policy document of the EU on Renewable Energy, discussion regarding the potential use of biomass in the energy sector has increased. The first version of the Green Paper, 'Energy: let us overcome our dependence', was published in 2000. The subsequent encompassing discussions led to an update in 2002. The main elements of the White Paper were included in the new report with more emphasis placed on achieving the goals. The report has clearly indicated that in current conditions the market share of renewables will stagnate at around 7% within 10 years. Strong financial and R&D measures are needed in order to fulfil the original target of doubling the share of renewable resources in energy productions, from 6 to 12%.

The Union has to improve the preconditions in order to increase the use of biomass as an energy source. According to the Green Paper the obstacles are great. Apart from costs, the access conditions to the market in most countries are unfavourable. It has been recommended that the fiscal framework must be adapted in order to stimulate renewable biomass energy. It must enable RES to benefit from preferential conditions in order to be competitive with traditional energy sources on the market. The fate of biofuels, for example, is totally conditioned by tax exemptions in order to promote their use (http://europa.eu.int/comm/energy_transport/livrevert/brochure/dep_en.pdf). Renewables are far from replacing other sources of energy and the future policy of the EU will be based on diversified production opportunities.

Due to the turbulence in the energy sector the most discussed area concerning renewables has been related to energy policy. However, there are great opportunities to use biomass as sustainable raw material for different industries. Besides the type of raw material, harvesting and conversion technologies are the fields that are offering a great number of marketing opportunities for products.

Marketing, Push and Pull

Over the years the definition of biomass has been changing. During the time that the EU had problems with surpluses, the term biomass included vegetables and fruits, so that these were subsidized when they were used in renewable energy processes. Today however, the EU struggles with animal by-products such as bone meal (which is not allowed to be used as feed anymore) and manure and these materials are now included as renewable resources since the aim is to reach a share of 12% of energy production coming from renewable resources.

The actual definition states that biomass is the biologically degradable fraction of products, waste products and residues from agriculture (from plant or animal origin), forestry and related businesses, and also the biologically degradable fraction of industrial and domestic waste.

Since this definition is not always assumed in the national laws and policies in the various European countries and federal states, a confusing situation exists which should be eliminated as soon as possible, for example, animal waste is not considered as biomass by the Flemish Government (Van Outryve, 2003).

In 2003, the area in the EU being used for industrial crops (including cotton, flax, hemp) is estimated at 2.1×10^6 ha, of which approximately half is set-aside land. In the year 2000, 5.69×10^6 ha were set-aside, proving that the set-aside concept is not working efficiently.

The Common Agricultural Policy (CAP) has now embarked upon an approach based on a combination of lowering institutional prices and making compensatory payments. The aim of the new CAP reform is to deepen and widen the 1992 reform by replacing price support measures with direct aid payments in line with a consistent rural policy. In this way the aim is to stimulate European competitiveness, take care of environmental problems, guarantee an effective income for farmers and to simplify the legislation.

However, in the newest regulations of the CAP, it is proposed that industrial crops should not be grown on set-aside land due to the alleged loss of biodiversity. The production of biomass will have to use regular arable land. The European Commission is planning to stimulate the production of energy crops with a subsidy of €45/ha.

Applications Leading to Immediate Environmental Disposal

Although there are still not many obligations to use renewable materials for certain applications in Europe, under the pressure of the green movement in some European countries national laws are being passed regarding uses where products will inevitably enter into nature. One key example is the obligatory use of biodegradable lubricants for chain saws in forestry in Germany. Further, biodegradable lubricants and hydraulic fluids also need to be used for construction machinery (leakage) and heavy duty material in forestry, although these lubricants are more expensive than the synthetic equivalent products which are less biodegradable. This is because the oil is inevitably being released into the environment. Fatty acid esters (green solvents) are also being promoted for the cleaning of metal surfaces instead of organic solvents. Biodegradable materials are promoted for less obvious applications as well. Now, body bags also need to be biodegradable.

In the area of sports, the tee (golf) can be made based on starch so that the player does not need to pick it up (or in case they are forgotten) because they degrade after some weeks in a moist atmosphere. Clay-pigeons are now being made from bioplastics or starchy products because the pieces cannot be collected after clay-pigeon shooting. These are just a few examples to demonstrate that renewable resource-based products do have interesting applications in niche markets, even if they are more expensive than the fossil fuel-based substitutes.

2.4.2 Policy and Environmental Concerns in the United States

The world uses 382 quadrillion British Thermal Units (BTUs) or Quads of energy, of which almost 40% is derived from petroleum and 85% from fossil fuels. Only 8% of the world's energy comes from renewable resources such as hydroelectric, wind, solar and biomass sources.

The United States (US) consumes 97 Quads of energy, over a quarter of the world usage. Per capita energy use in the US is 355 million BTU compared to the world average of 65 million BTU. The US derives its energy from a similar mix of sources but uses a slightly lower percentage of renewable energy. The renewable energy sources used in the US are ethanol, geothermal energy, solar energy, hydroelectric power, wood and wood wastes. Of these, ethanol, widely used in gasoline, is the largest energy source. However, all of these combined account for a very small percentage of the total energy requirements of the US. The wood products industry consumes most of the wood and wood waste burned for energy, which is not for domestic consumption. The wood industry already produces the majority of its own electricity and even exports some electricity to the power grid for use by other electric customers.

After declining in the early 1980s, imports of petroleum have increased since 1985 and now account for 62% of the US demand, along with 16% of the demand for natural gas. The US is becoming ever more dependent on foreign petroleum for both the energy needs and the basic chemicals that feed industry.

The use of transportation fuel increased at a rate of 1.7% per year during the last decade of the 20th century (www.DOEenergy.gov). The transportation sector alone consumed 13.4×10^6 bbl./day in 1999. A major portion of the domestic petroleum energy consumption is used by transportation, with diesel-fuelled engines consuming over 20% of the total transportation energy use (Davis, 1999).

The problem is likely to become even more severe in the future. As the standard of living rises in the developing nations of the world, their demand for energy will undoubtedly increase. Competition for energy and for the chemicals and products we currently derive from fossil resources will drive prices higher and make shortages severe.

Regulations

Although there are laws in some European countries concerning environmental regulatory issues, there are no federal regulations that mandate the use of bio-based or low-toxicity biodegradable lubricants in the US. Some Executive Orders (EOs) do however exist. In 1993, EO 12873 was issued to provide guidance related to federal acquisition, recycling and waste prevention of environmentally preferable fluids and lubricants.

EO 13101, issued in 1998, extends EO 12873 to call for preferential federal purchasing of biodegradable lubricants listed in the Federal Register. Multiple federal activities in the last few years include Executive Orders and legislation to put into effect what has become known as the federal 'Bio-based Products Initiative'. In September 1999, President Clinton issued EO 13134 calling for an increase in bio-based product research and a tripling of federal consumption of bio-based products by 2010. In June 2000, Congress passed the Biomass Research and Development Act of 2000 as part of the Agricultural Risk Protection Act, which mirrored the call of the EO. Both actions called for the United States Department of Agriculture (USDA) and the Department of Energy (DOE) to be the lead agencies in this research initiative and for the implementation of a national purchasing policy for bio-based lubricants.

The Farm Security and Rural Investment Act of 2002 and the Farm Bill of May 2002 established a programme of affirmative procurement for bio-based products. It is now mandatory that all federal agencies establish a programme to purchase bio-based products. The bill defines 'biobased' as biological or renewable domestic products from agricultural or forestry products (Martin, 2002). Federal agencies are given some flexibility in that the bio-based products must meet performance requirements and be reasonably priced.

Implementation of the bill by USDA is expected to establish a designated bio-based product list and guidelines for its use. Each year the federal government purchases over $5 billion in lubricants alone. The Farm Bill of 2002 provides new incentives for the use of bio-based products that are stronger than the previous executive orders, but laws similar to those in Europe are not expected for several years.

Marketing, Push and Pull

The USDA has turned some attention to the necessary marketing efforts that will be required for the host of new bio-based products already in the market to gain market share. In some cases, the USDA has applied 'push' marketing with some success, such as the subsidy of biofuels, ethanol and biodiesel. By subsidizing the cost of production they help to push the product into the market by lowering the price.

Another approach being tried is to apply 'pull' marketing through harnessing the vast buying power of the federal government. Federal, state and local governments buy some $800 billion in goods and services, exclusive of categories such as weapon systems. These are real everyday products that consumers might buy. To that end, EO 13101 instructs all federal agencies to purchase environmentally preferable products including bio-based industrial products.

EO 13101 also instructs the USDA to prepare a list of such bio-based industrial products for procurement consideration by all federal agencies. Progress on that list has been slow due to the legal considerations of federal rule-making. To gain a place in the list of approved vendors one must obtain a series of forms. If approved, the documents ensure that the vendor's products will be supplied at a set price. The process results in a contract that allows the vendor to accept orders directly from users, which can easily take three to six months, and places the vendor on a list of approved suppliers. Government agencies have some latitude in that any product must meet performance requirements and be of a reasonable cost.

Electronic Source Book

A small level of appropriations have been supplied for the development of a website to provide a source book of bio-based products available for federal agency use. The draft website is available at www.usda-biobasedproduct.net and the project is slowly growing and evolving, as funds are available and with input from industry.

Waste Management

The Federal Resource Conservation and Recovery Act (RCRA) requires that used oil is treated as hazardous waste and either disposed of accordingly or recycled. Vegetable oils in their unused state are not considered hazardous, but when they are contaminated by use, that is contain more than a specified level of heavy metals for instance, then they become hazardous wastes.

Engine Emissions – Heavy-duty Engines

The Environmental Protection Agency (EPA) has published new guidelines for Particulate Material (PM), Nitric Oxides (NO_x) and other emissions allowable from heavy-duty engines and vehicles running either on gasoline or diesel. These come into effect with the 2004 model year and engine manufacturers have probably already decided on a way to address these new standards. Some diesel OEMs must meet stringent requirements, starting with the 2003 engines as a result of a negotiated consent decree. Fuel is a significant factor in the reduction of diesel emissions. Concerns focus on durability and wear as the levels continue to be reduced. Fuel economy is the most important driving force for passenger cars.

Additional concerns for OEMs arise when the Phase 2/Tier 3 plan is implemented to reduce tailpipe emissions of particulates beginning with the 2007 model year and phasing in through 2010. After-treatment systems will most likely be required. To meet these requirements, the EPA has also published standards to reduce the level of sulphur in diesel fuel to 15 ppm, which will have an effect on the lubricity of the fuel itself. In addition, reduction of lubricant sulphur and phosphorus may be required to improve catalyst efficiency (Perez, 2000). This poses the question of whether vegetable oil-based lubricants and lubricity additives can address the need for lower exhaust emissions and lubricity in these types of engines. In order to play a role, it would be necessary to work with the equipment manufacturers at the testing level to develop equivalent performance lubricants that meet specifications for use in these engines.

There are also new regulations for light-duty trucks and automobiles. These are primarily aimed at reducing NO_x emissions, but do set lower limits for other emissions such as particulates and non-methane hydrocarbons. The second phase of this plan comes into effect with the 2004 model year and also lowers the allowable sulphur content of gasoline. California has a more stringent plan that will be phased in between 2004 and 2007; other states are reviewing those standards and may adopt them. Sulphur and phosphorus levels are also concerns.

Motor Vehicle Economy

There is pressure on automakers to improve the Corporate Average Fuel Economy (CAFE) standards, especially for light trucks. CAFE requirements are currently at 27 mpg. It is not clear how legislation will affect CAFE in terms of increased market penetration of popular sports utility vehicles (SUVs) and family-size vans. If SUVs were to be included in the requirements, vehicle manufacturers would need to make adjustments to the size and number of vehicles sold. The House energy bill did not set a specific new standard but charged the Department of Transportation with setting a new standard for light-duty trucks that would save the equivalent of 5×10^9 gal. of oil.

Edible Oil Regulatory Reform Act (EORRA)

The EORRA, passed in 1995, directs agencies to differentiate between petroleum oils, and vegetable oils and animal fats. However, it has not resulted in any significant change from a regulatory perspective. There has been some accommodation for producers of vegetable oils and vegetable oil products in spill prevention planning, but there has been no real change in spill reporting or clean-up requirements. The EPA denied a petition by the vegetable oil producers for further reconsideration based on lower toxicity and bio-degradability, stating that vegetable oils could still cause significant environmental damage and that the products of biodegradation could be more toxic or harmful than the oil itself.

Environmental Marketing Claims

The Federal Trade Commission has established strict standards on what is required for an environmental marketing claim to be made. There are specific standards for making broad or unqualified claims for biodegradability, toxicity and recyclability. A broad claim saying the product is 'non-toxic' requires data to substantiate that there are zero health effects and zero environmental effects. Making a qualified claim such as a product is 'less toxic' than some other common material requires supporting data.

Definitions of Environmentally Acceptable Fluids

There still remains a lack of understanding of what is required to consider a fluid or lubricant as a marketable 'environmentally acceptable' or 'environmentally friendly fluid'. Renewable basestocks blended with selected petroleum basestocks result in improved low-temperature performance, better oxidative stability and higher biodegradability. However, properties and exhaust emissions of the used oil look similar to petroleum-based products. Selected synthetics are an improvement on the petroleum stocks but are more expensive. There is considerable activity in various technical organizations in the US and Europe to resolve the issues on terminology and testing methods needed in this area.

2.5   Conclusions

Environmental consciousness among consumers has increased in society during the past 15–20 years. Therefore, it can be assumed that this will also be reflected in the consumers' attitudes and interests towards bioenergy and other environmentally friendly products. Considering not only all the benefits of renewables on employment, on fossil fuel import, on an increased security of supply, on local and regional development but also the awareness of the finite nature of fossil fuel resources, it could be concluded that the development of renewables could be left to the market demand (pull effect). However, even though the utilization and the development of renewable resources are of major importance, the process still needs political action to proceed (push effect). If the supply and use of biomass and organic wastes are pursued seriously and will be enhanced by subsidy systems, the additional subsidies required will be withdrawn from other economic sectors or from other tasks, which requires wide political and social consensus (also Leibe, 1998).

All of the potential consequences of alternative production for societies have not yet been properly analyzed and understood. On the other hand, it is not only a question of development in Europe and the US, the global consequences should also be taken into account. The expected growth in energy consumption in many developing countries offers promising business opportunities for Western industries. However, a significant decrease in oil and coal production could weaken the economical welfare remarkably in some areas.

One example of consumer attitude is a survey carried out in 1997 in Finland to discover the opinions of Finnish consumers regarding energy, bioenergy and especially wood-energy-related opinions, beliefs and intentions of Finnish consumers. According to the results of the study, consumers' attitudes towards energy seemed to be composed of several dimensions, characterized by the willingness to compromise between the standard of living, economic growth and impacts on the environment. However, the results of the study indicated that price is a very important criterion of the consumers' energy choices. On the other hand, the results also suggested that the consumers would like to see bio-energy and other alternative energy sources increasingly used in primary energy production in Finland. It can be expected that environmental factors will become more important as one of the selection criteria of the energy supplier during the next few years.

Several measures need to be taken to aid development in the utilization of renewable resources. First, education and dissemination of information on environmental impacts of renewable and non-renewable products is certainly needed in order to influence consumer attitudes. This should then result in an increased awareness of the products they purchase. Secondly, a wider choice of environmentally friendly products should be made available to consumers. Finally, the competitiveness of alternative products should be continuously improved.

References

Energi från skogen. SLU Kontakt 9. Uppsala (1999). Swedish University of agricultural sciences. p. 132.

Davis, S.C. (1999). US Department of Energy Transportation Energy Data Book, Oak Ridge National Laboratory, Oak Ridge, TN.

Grassi, G. (2000). Overview on industrial perspectives in bioenergy. In: *1st World Conference on Biomass for Energy and Industry*. Eds Kyritsis, S., Beenackers, A.A.C.M., Helm, P., Grassi, A. and Chiaramonti, D. pp. 11–14. James & James Ltd. London.

Leibe, L. (1998). Use of biomass and organic waste – a contribution of agriculture to sustainable development? In: *Biomass for Energy and Industry*. Proceedings of the International Conference. Würzburg. Eds Kopetz, H., Weber, T., Palz, W., Chartier, P. and Ferrero, G.L. pp. 1187–1190. Germany. C.A.R.M.E.N.

Martin, J. (2002) (ed.). Proceedings of the United Soybean Board Soy Lubricants Advisory Panel. Omni Tech International, Ltd, Midland, MI, September 17, 2002.

Nielsen, P. and Schleisner, L. (1998). Environmental externalities related to power production on biogas and natural gas based on the EU externe methodology. In: *Biomass for Energy and Industry*. Proceedings of the International Conference. Würzburg. Eds Kopetz, H., Weber, T., Palz, W., Chartier, P. and Ferrero, G.L. pp. 1235–12348. Germany. C.A.R.M.E.N.

Perez, J.M. (2000). Exploring Low Emission Engine Oils. USDOE/NREL Workshop, Special Report No. NREL/SR-570-28521, NREL, 1617 Cole Boulevard, Golden, CO, USA 80401–3393.

Tahvanainen, L., Asikainen, A. and Puhakka, A. (1998). Bioenergy strategy for North Karelia. (Pohjois-Karjalan Bioenergiastrategia). Regional Council of North Karelia, Joensuu. p. 46 (in Finnish with English summary).

The European Commission (1997). Energy for the future – renewable sources of energy. White paper. COM (97) 599. p. 55.

Trossero, M.A. (2000). Food and biofuel production: New challenges ahead. In: *1st World Conference on Biomass for Energy and Industry*. Eds Kyritsis, S., Beenackers, A.A.C.M., Helm, P., Grassi, A. and Chiaramonti, D. pp. 17–21. James & James Ltd. London.

Van Outryve, J. (2003). Land- en tuinbouw: meer dan voeding? Europees en mondiaal perspectief. Studiedag Land- en tuinbouw: meer dan voeding? p. 9, April 2003, Brussel, Belgium.

Versteijlen, H. (1998). Material available from agriculture for non-food applications. In: *The First ERMA Conference. Agriculture: Source of Raw Materials for Industry*. pp. 13–16. March 1998, Brussels, Belgium.

Wisniewski, G. (1998). Institutional and social framework for the development and implementation of bioenergy strategy in Poland. In: *Biomass for Energy and Industry*. Eds Kopetz, H., Weber, T., Palz, W., Chartier, P. and Ferrero, G.L. Proceedings of the International Conference. Würzburg. pp. 1247–1250. Germany. C.A.R.M.E.N.

3

Integral Valorization

Werner Praznik and Christian V. Stevens

In co-authorship with *Anton Huber, Elias T. Nerantzis, Waldemar Rymowicz, Monica Kordowska-Wiater and Piotr Janas*

3.1 Introduction

The concept of the use of renewable resources is implicitly connected to the idea of integral valorization, meaning that it is important to be able to use the product delivered by nature or by agriculture completely.

Of course, this concept also holds for other types of production and needs to be one of the major concepts in designing new applications and new products. This concept is certainly not new and has been applied from ancient times when tribes had to survive under stringent conditions. All material delivered by nature and early agricultural production was utilized in some way.

Over the centuries, however, certainly in the Western World, society developed a consumer attitude in which resources were believed to be abundant. Nowadays, without being pessimistic, people realize that natural resources are very valuable and one needs to take care of them. Atom efficiency, integral valorization and waste minimization all refer to the same concept of being careful with resources. Integral valorization, however, is more connected to the concept of renewable resources and aims at a complete use of industrial crops.

Up to now companies have too frequently looked only at producing those products for which they already have a market. Many times they have preferred to pay to get rid of the side/waste products instead of trying to develop new applications for these materials. With a growing awareness of sustainable development, integral valorization has to be one of the starting points in designing new processes with optimal utilization of side products and metabolites. Further, next to maximizing the utility of the agricultural products,

products and side products form a closed resource cycle which is a prerequisite for sustainability.

Simply speaking, the plant produces raw materials from water and atmospheric carbon dioxide by photosynthesis. These materials can be extracted in a more or less pure state and can be treated, refined and further processed resulting in a great variety of finished products, auxiliary substances and additives. The cycle is closed when carbon dioxide and water are released after usage through biodegradation.

W. Umbach, Henkel/FRG

In this chapter, several points connected to integral valorization will be discussed:

- Specific aspects of the use of agricultural raw materials
- Approaches to integral valorization
- Several examples of valorization of by-products
- The importance of the concept for developing countries
- Ways to stimulate integral valorization.

3.2   Specific Aspects of Agricultural Production

This change of attitude in organizing production processes, involving integral valorization and sustainable development also leads to other accents in the market organization. A few examples of how people are dealing with this other way of thinking are given in Table 3.1.

However, using agricultural or natural raw materials in production processes has certain disadvantages compared to the production of goods from fossil fuels. One of the major disadvantages is the changing quality of the resources, depending on factors that are typically connected to agricultural production (soil type, weather conditions, time of harvesting, seed quality, etc.). This aspect cannot be ruled out completely by design of the production process and can only be partially controlled by a good understanding of crop production. Therefore, quality control of the products isolated from crops or animals is essential. This has been illustrated in a dramatic way by the problems associated with mad cow disease, polychlorobiphenyls in fats, dioxins, etc.

Using renewables for non-food applications will also need rigid quality control in order to prevent batches from going into products which would normally have been discarded because they failed to meet specifications. Quality control also implies trying to produce a constant quality. Here, mixing of resources can be a solution to guarantee a certain quality level, for example as is being done to produce champagne. The juices of different types of grape and of different years are blended to obtain a certain, more or less constant, taste. This is also being performed for non-food uses of agricultural products. Flax fibres of different years and regions are being cut and mixed to obtain a standardized flax mixture.

Next to the changing quality, the crop production itself is very dependent on, for example the weather conditions during the growth and harvesting seasons. A disappointing production can immediately lead to a serious price increase, while an overproduction can lead to bulk-purchase prices. These aspects might be disadvantageous to convincing some

Table 3.1 Comparison of a liberal/free market economy and an ecological economy

	Liberal/free market economy	Ecological economy
Principles	Profit/turnover	Improving life quality/sustainability; LCA; optimum energy utilization
Driving force	Shareholder value (advertisement induced); demand and supply	Preserve/keep resources (for the coming generations)
Limitations	Capital general legislation subsidiaries/toll	Actual needs/resources
Distances	Existing distribution (transportation and storage) infrastructure: 'just in time'; transportation for 'free'	Technology/downstream processes Problem of (localization of) primary processing/stabilization of raw materials
Positioning of enterprises/ processing technologies	Global; existing and established downstream processes; minimized waste for petrochemical raw materials	Focused on regional solutions, Adapted processing technology only; Valorization of by-products
Consumer	Passive 'extra-economical'	Participant actor in any processing/ consuming/delivering step
Quality	Applied state of the art technology	Known origin; known processing technology; minor waste
Price	De-coupled from product and resources 'total value'; major argument for/against any product	Connected with 'total value' (energy) of resources → products; one out of many arguments for / against any product
Acceptance	Price/quality ratio	Quality; Regional gain
Know-how (e.g. processing)	Protected Shareholder property	Public community property: regional solution
Globalization	Products, production facilities, waste	Know-how
Characteristics of raw materials	Fossil; permanently available; few deposits; limited	Biologically active; periodically available; harvesting/storage problem; ubiquitaires available; renewable

industries to turn to renewable resources in preference to products derived from fossil fuels. The reliable availability is certainly an important issue during the development of renewable resource-based processes.

Another drawback is that most of the agricultural production (certainly industrial crops) is limited to a certain period of the year, so that the yearly consumption has to be produced and downstream processed in a short period. This leads to the need for downstream processing units with a bigger capacity than in the case of fossil substrates which can be produced 365 days a year. Essential to a lot of processes using renewable resources is the transformation of the renewables to an intermediate which is stable. The stability of the isolated products towards bacterial or fungal attack, chemical oxidation or hydrolysis, storage problems, etc. is of major importance in developing a sustainable business. Examples are abundant in the food industry as well as in non-food applications. In the food industry, sugar production from sugar beet can be taken as an example. Sugar beets are extracted immediately in the presence of calcium hydroxide in order to prevent the formation of invert sugar, and the sugar syrups need to be concentrated in order to prevent fungal growth. In the field of non-food applications, the use of anti-oxidants is well known in products containing lipids or fatty acids. Therefore, the search for stable intermediates from renewable resources, which can serve as a basic product for applications, is very active. Ethanol, methanol, hydroxymethylfurfural, etc. are being promoted to mimic the position of ethylene in petrochemistry.

3.3 Approaches to Integral Utilization of Renewable Materials

3.3.1 Ecological Economy: Regional Solutions

Integral utilization of renewable resources is a generally accepted goal which may be achieved by several approaches. Maybe the best-known historical approach is the formation of self-sustaining ecological societies in rural areas with cyclic utilization of locally available resources. The open society of today, combined with a price-devoted liberal economy, destroyed these local communities, however, similar approaches are being developed again. Cooperations of modern raw material suppliers (agronomy, forestry), decentralized energy producers (local caloric power plants, biogas facilities) and nearby communities which consume the provided energy reinstall the initial self-sustaining model, at least partially. The key to success for these approaches is optimum by-product and waste management, and docking the general supplier lines which take over production and supply in periods of shortage.

An example of this concept is the oilmills for biodiesel production, combined with caloric power plants using waste from biodiesel production, with wood residues and biogas also partially supplied by waste from biodiesel production. The fact that the major suppliers (farmers) and the people working in and for the facilities are the consumers of the goods produced (biodiesel and energy) increases acceptance and shifts the price from a primary to a secondary position. The Mureck-energy-cycle (Figure 3.1), a rural area in the south of Austria, illustrates this concept. Processing of locally grown rapeseeds provides biodiesel (380 l/1000 kg) and protein-rich rapeseed feedstock (620 kg/1000 kg) (Mittelbach, 1996). The feedstock is redistributed in aliquots to the rapeseed farmers on request. Any surplus is sold or utilized in the connected caloric energy power plant. To increase efficiency of the biodiesel production facility, residual nutrition oils (0.85 l biodiesel/1 kg oil) are collected with specifically developed collection logistics from private households, restaurants and communities within a radius of several 100 km. Thus, the nearby located public transportation of Graz (200 000 inhabitants) completely runs on Mureck-Biodiesel. Glycerine, which is a by-product of the biodiesel production, is utilized for energy production by the Mureck-community in the biodiesel-facility-connected caloric power plant (2×2 MW). Excess thermal energy is used for drying wood residues for the caloric power plant and for heating Mureck-public buildings and selected private buildings. Excess energy is provided/sold to the public energy supplier. Midrange future aspects of the Mureck-concept include the implementation of a biogas facility for still remaining biogenic residuals.

The advantages of processing renewable resources according to the local-solution concept of Mureck are: local and relatively small energy-production facilities; short distances from production site to consumer site; high acceptance as consumers are either

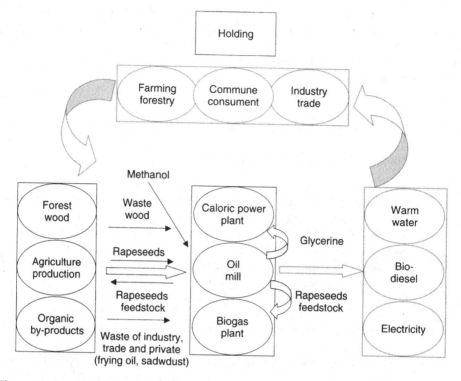

Figure 3.1   *Scheme of the 'Mureck' local solution approach*

suppliers and/or employees; optimum processing; and high pressure for improvements as the consuming community is the owner. Temporary surplus delivered to the super-regional energy provider generates credits for shortage periods or for new investments. Jobs, even midrange and high-qualified jobs, are generated in a basically rural area. And, although not yet studied in detail, life cycle analysis is expected to yield good values to this concept.

3.3.2   Ecological Economy: Globalization of Strategies to Improve the Efficiency of Downstream Processing of Bio-Materials

One of the possibilities to overcome some of the problems due to the use of renewable resources is vertical cooperation or vertical integration. Essential in the concept of vertical cooperation (partners do not belong to the same industrial holding) is that there are contracts between the farmers who produce the crops or animals and the industrial companies that are performing the downstream processing and the production of the final products. In this way, a lot of uncertainties regarding price stability, prices versus quality, volumes, etc. can be avoided. The farmers are sure that they will be able to sell their crops at a certain price and the industries get warranties that they will have crops at that price and of a certain quality. The vertical cooperation can go very far; from the companies that sell the crop seeds, over the farmer, to the downstream processing industry. This type of organization is essential for a sustainable renewable resources business and most of the firms using big quantities of renewable resources have their own network of farmers and production units. Many examples are known in the production of bio-ethanol for production of biofuels, the production of inulin for food and non-food applications, etc. Since the conversion of biological products to stable intermediates is crucial, the vertical cooperation helps to reduce the loss of material due to degradation or microbial spoilage.

In practice, the system remains complex since the prices in the world market will also influence the price negotiation between producers and industry. This, in turn, will be influenced by the prices of the final products, which have to compete with possible substitutes. The economic success of a company selling biodiesel will depend on the price of the oils (e.g. soya bean oil) and the price of glycerol as a side product. The profitability of biodiesel in turn will depend on crude oil prices and therefore on the international political situation.

3.4   Food versus Non-Food Applications

An important asset in the production of renewable resources is the fact that the products can be used either for food production or for some non-food applications. In this way more flexibility towards the market is built into the business.

One of the most popular crops, for food as well as for non-food utilization, is maize. As maize crops have a high yield, it may be regarded as a good crop for widespread integral utilization. Figure 3.2 illustrates different possibilities for main- and by-products (in small or large amounts) of equivalent importance from the point of view of total utilization. Wheat production can either go to food or feed application or can be used to produce starch for the paper industry. Changes in market opportunities can be partially levelled if different outlets are available. A very important aspect is that the technology for the use of the renewable resources for non-food applications needs to exist and preferably be the same as the technology used for the fossil-derived substitutes. If this is not the case, companies will be very reluctant to evaluate some of the new renewable resource-based alternatives. Biopolymers can be successful only if they can be transformed and utilized with the machinery (extrusion, moulding equipment, etc.) that is traditionally used in the chemical industry.

3.5   Integral Use of Bio-Materials

Using renewable resources in industrial processes has often been limited to the use of the root, the seeds, the fibres or the leaves while the other parts of the plant were discarded or were used as feedstock. A typical example is the inulin producing industry where the chicory root is extracted and the pulp is added to feed. However, the chicory pulp contains some interesting bitter compounds which could be valuable if a good isolation process was available.

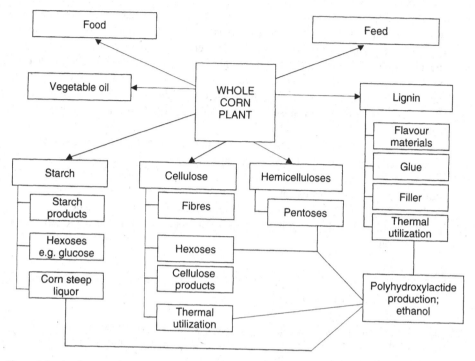

Figure 3.2   *Application of a whole corn plant*

A major evolution in the integral use of bio-materials is the attention that is currently being paid to the valorization of waste materials from biological processes. Increasingly, companies are realizing that developing a useful outlet for side products and waste streams can contribute to the sustainability and economical health of their process. Mostly, waste streams do contain valuable products but the effort to separate or process them has been neglected in the previous decades. Many examples of the upgrading of waste streams are known. In this section some selected examples will be discussed.

3.5.1   Valorization of Waste from the Wine and Brandy Industry

The grape and wine industry is well developed in many countries, as wine production is known from ancient times. People have been using grapes for the production of wine almost from the beginning of civilization. Wine-making passes through certain routes common to all varieties of wine production. The wine-making process can be separated into three distinct stages (Figure 3.3).

Each stage includes certain tasks necessary for the wine-making process and produces certain by-products, which can be further utilized.

The prefermentation stage involves the preparation of the vineyard, harvesting of the grapes and the processing of the must. Preparing the vineyard can start very early in the year. This process includes caring for the vine in order to obtain the highest productivity

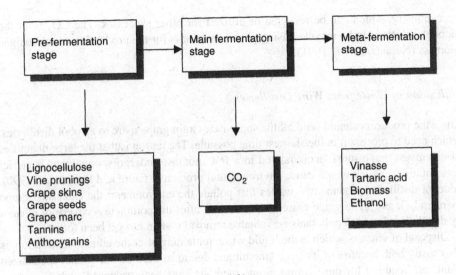

Figure 3.3 *The three stages summarizing the wine-making process*

and a better quality of grapes. At this stage the wastes are mainly lignocellulosic material with the dominant by-products, the prunings.

The main fermentation involves the fermentation of the must, with the main waste material being the CO_2 and the biomass produced simultaneously with the wine. The third stage, the meta-fermentation (or post-fermentation) stage, consists of the clarification and stabilization processes of the wine. In this stage various by-products are separated, such as the biomass produced during the fermentation and the tartaric acid which is separated during the stabilization process.

Utilization of the Vine Prunings as Wood Substitute for Particleboard Production

The results of environmental pressure and preservation of forests in some countries resulted in the prohibition of further forest harvesting, leading to wood shortage and unemployment in the wood industry (McNutt, Haegglom and Raemoe, 1992; McCleery, 1995).

However, the demand for wood products and especially for wood panels is increasing with the consequence of a continuous search for alternative sources. Lignocellulosic agricultural by-products have been considered as an appropriate source for wood fibre. Therefore, vine prunings have also been considered as an appropriate alternative for particleboard production, as vine prunings are available in large quantities in wine producing countries (Ntalos and Grigoriou, 2002).

Recycling of the CO_2

The production of CO_2 follows the fermentation processes, independent of the nature of the raw material. It is a by-product of the fermentation industry, particularly the wine and beer industry, which can be recycled or utilized for other processes. The CO_2 from the alcohol industry can be recycled using photobioreactors for the production of microalgae biomass (Nerantzis *et al.*, 1991).

Utilization of Waste from Wine Distilleries

The wine producers should send all the sugar-rich virgin grape marc to alcohol distilleries, which need to process it in the shortest time possible. The reason is that the large quantities of exhausted grape must, accumulated in a few months, undergo spontaneous bacterial fermentations, which would cause environmental problems (Faure and Dechamps, 1990). Alcohol distilleries produce two wastes that pollute the environment: the exhausted grape marc and vinasse. The grape marc is the remnant after the complete recovery of ethanol by distillation. For both pollutants no valuable removal system has yet been found.

Disposal of vinasse, which is the liquid waste remaining after the ethanol distillation, is also costly both because of its large amount and due to its dark colour so that specialized plants are required for purification. Some methods have been proposed such as vinasse concentration followed by burning or grape marc composting.

Grape marc for composting

In many areas, for example in Greece and in the south of Italy, the exhausted grape marc is generally utilized in distilleries as low-caloric fuel for steam production. However, problems arise because of the contaminated smoke production, mainly due to the combustion of seed oil which cannot be extracted in a profitable way. A possible solution could be composting for the production of biofertilizers or yeast production for animal feed or feed supplement.

Composting usually takes place on the field. The material is accumulated in piles and is mixed and agitated by mechanical means. However, this method is not controllable and the compost is not of high quality. Proper composting takes place only in well-controlled environments. The temperature, moisture content and good aeration are the main parameters for a good composting process. If the agitation and aeration are not sufficient, anaerobic processes occur with detrimental effects on the final product. The composting takes place at mesophilic temperatures. In countries where the climate is warm, thermophilic processes are applied as they are faster and more efficient. In some cases, vermicomposting has been applied with good results. The vermicomposting normally uses composting vessels, and earthworms like *Eisenia foetida* have been used successfully for different substrates, resulting in a high-quality fertilizer.

Biogas production from vinasse

Vinasse is characterized by a high concentration of organic acids and polyphenols. This determines its relatively high values of pollution indices like the chemical oxygen demand (COD). Vinasse is subjected to anaerobic fermentation under mesophilic and thermophilic conditions (Peres, Romero and Sales, 1998). The usual method of biomethanization of vinasse is a one-stage process. However, a two-stage system for methanization is sometimes more suitable (Shin *et al.*, 1992).

The methanogenic bacteria have low growth rates, therefore the fermentation cycle is long and the residence time of the substrates is high. This is the reason why the volumetric capacity of the bioreactors for methanogenesis should be large in order to be economical. The need to keep high levels of biomass inside the fermentors leads to the use of many techniques including immobilization on various supports and microencapsulation. The immobilization techniques using cheap materials are more preferable due to their low cost and ease of use. A polymeric material, used for the immobilization of methanogenic bacteria and for the production of biogas from vinasse, is the copolymer of acrylonitrile – acrylamide. The porous granules, with a diameter of 1.5 mm, are activated for 4 h in 12.5% of formaldehyde in a 0.1 M phosphate buffer at pH 7.5 and at a temperature of 45 °C. The ·biomass is adapted to vinasse and is added to the liquid in suspension with the activated granules. In this way a covalent immobilization is achieved. This system is ready to be used in bioreactors in a continuous mode (Lalov, Krysteva and Phelouzat, 2001).

Extraction of Tartaric Acid

Tartaric acid is the most abundant organic acid in grapes. In the must tartaric acid is found in the range of 3–7 g/l. Tartaric acid is the only acid that can be used by law for the increase of the total acidity of wines. The level at which this acid is present is related to the chemical and biological stability of wines.

The recovery of tartaric acid from grape wastes is of economic importance since its chemical synthesis is more expensive. Tartaric acid is usually recovered from wine-making by-products and mainly from grape seeds and red grape skins (Palma and Barroso, 2002). It has been found that grape seed and red grape skins contain high amounts of tartaric acid.

3.5.2 Production of Ethanol as Fuel

The production of alcohol is a process which involves separation of the alcohol, produced by fermentation of sugar or starchy raw materials by physical methods, usually by distillation or vacuum distillation. The scope of this case study is to present some uses of the main waste (by-products) for non-food applications. Ethanol can be used as a fuel for the formulation of gasohol and alternative fuel resources.

The production of ethanol from industrial and agricultural wastes has been a research topic for a long time. Despite the fact that ethanol can be produced for uses other than food, for example as car fuel, this is not yet permitted in countries like Greece and Italy. In the United States and Brazil gasohol has been used as an alternative energy source.

Lignocellulose-based ethanol production includes the following stages:

- feedstock acquisition and pre-processing;
- detoxification (elimination of toxic products of pre-treatment);
- enzymatic hydrolysis of the cellulolytic complex;
- fermentation; and
- alcohol recovery.

The first stage seems to appear as one of the limiting factors. The aim of the pre-treatment of the native feedstock is to hydrolyze hemicellulose, partially degrade lignin and increase the specific surface area of the substrate and reduce the crystallinity of the cellulose. Many methods of pre-treatment of cellulose have been developed including: alkaline and acid maceration, autohydrolysis (heating with water steam at a temperature of about 200 °C) and acid-catalyzed steam explosion. They appear economically feasible.

The economic feasibility of the production of ethanol as fuel depends on the price of oil. The technological advances of the fermentation industry, using high-productivity yeast strains developed by biotechnological approaches, are promising for a feasible production of fuel ethanol (see p. 63).

Since 1975 the Brazilian National Alcohol Program has used sucrose from sugar cane as the main substrate for the production of alcohol. Over 50×10^9 l of ethanol were produced during the first decade of the programme. In 1989 about 40% of Brazil's 14 million vehicles were operating on ethanol (95% ethanol, 5% water) and 60% on a blend of 78% gasoline and 22% ethanol. This process was heavily subsidized by the government in Brazil. However, with the subsidy now being reduced, the bio-ethanol project is ready to die. In the US during the year 1988 about 840×10^6 gal. of ethanol from corn and other starch-rich grains were used in gasohol blends (10% alcohol to 90% gasoline) (Rosillo-Calle, 1989). Figure 3.4 presents the conversion of various feedstocks to alcohol.

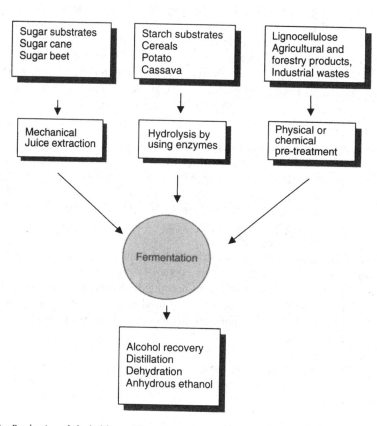

Figure 3.4   *Production of alcohol from different sources*

There are different processes for the production of ethanol to form these by-products. An important parameter is the nature of the carbon source. The sugar industry, for instance, provides molasses which contain 50% of fermentable carbohydrates by weight. When the substrate is a starchy or lignocellulosic material the process involves the hydrolysis of the substrate to fermentable sugars. The hydrolysis of starch usually involves enzymatic or acid treatment of the substrates.

In the United States the major feedstock for the production of fuel alcohol is corn starch. For the production of fermentable sugars from corn starch the substrate passes through several stages. The dry corn is milled, then water is added and the suspension produced (the slurry) is hydrolyzed by heating to solubilize the starch. After cooling the enzymes, α-amylases and amyloglucosidases are added to convert the starch to glucose. In the main fermentation stage, the simple sugars produced from the polysaccharides pass to the fermentors for conversion to alcohol.

The glucose used for the production of ethanol is only a part of the total amount used by the yeast cell. The growth and maintenance of yeast during the fermentation requires energy. Consequently, the maximum yield of ethanol is expected to be approximately 86%, allowing some carbon expenditure for growth. In fermentations the production of ethanol is reaching 90–95% of the theoretical value. The faster the growth, the smaller the proportion of energy going to cell maintenance. In yeast fermentations the higher the ethanol content in the medium, the higher the maintenance energy requirement of the cells and the lower the ethanol yield will be. In this way, during the fermentation the ethanol concentration is increasing, leading to an increased use of ATP by the yeast for cell maintenance at the expense of reproduction. Therefore, the fraction of the carbon source in the bioreactor allocated to yeast cells decreases from 14 to 5% or less of the starting glucose – with a corresponding increase in the yield of ethanol (Lynd *et al.*, 1991).

One of the key factors affecting the economics of ethanol production is the ethanol concentration in the fermentation broth. The higher the concentration in the fermentor, the lower the distillation cost per litre of the product, since the separation of ethanol from water requires very high amounts of energy. On the other hand, ethanol is toxic to yeast cells at concentrations ranging between 8 and 18% by weight, depending on the yeast strain and fermentation conditions, so that the concentration in the medium is limited. Genetic modification of yeast can give strains with a higher alcohol tolerance.

Ethanol can also be produced with the bacteria *Zymomonas mobilis* and *Clostridia* species from cellulosic and hemicellulosic by-products. Another important aspect to consider is the waste of the alcohol industry. Particularly in Brazil, the wastewater is poorly treated (if at all) and leads to severe reduction of the water quality in rivers and streams, causing health problems in the area. Therefore, the high biological oxygen demand (BOD) of the effluent from the distilleries will need to be used for the production of methane, reducing the problem of pollution and adding to the economic feasibility of the project by creation of energy for heating distillation units.

3.5.3   Biomass Production from Agricultural Waste

Next to the agricultural production of plants and cattle for the food industry, an important source of valuable material is formed by the waste products of the food and non-food industry, which can be utilized favourably for a lot of applications (even in human food and animal feed). In this part, the use of these side streams of raw materials will be highlighted and their value will be illustrated by several examples.

Yeast Production

Different kinds of microorganisms can be cultured for the production of biomass. These microorganisms are yeasts, bacteria, fungi and algae. They are considered protein sources. The cell biomass as a protein source for consumption by animals and humans is referred to as single cell protein (SCP, the content of protein is more than 65%). If the protein content is less than 65% the term 'single cell biomass' (SCB) will be applied. The definition of 'food yeast' is as follows: 'Food yeast is a yeast that has been killed and dried; it should have no diastase activity and has not been submitted to extraction processes nor received any additives.' Utilization of yeast for food applications is possible only if the RNA content of the SCB is lower than 2%.

The industry is the oldest in the field of biotechnology. It is a high-tech industry which has benefited from many scientific advances. The culture processes have improved considerably because of the increased knowledge of biology and cell physiology. The culturing of yeast for its nutritional value started at the end of the First World War in Germany. After the Second World War, the production of fodder yeasts was developed in the USA and in European countries.

The advantages of yeast SCB over plant and animal proteins are the following: due to the short generation time (yeasts have a generation time of 1–5 h) yeasts can utilize wastes or raw materials which are available locally in adequate quantities, such as lignocellulosic and starchy materials, whey, molasses, sulphite waste liquor, animal wastes, forest wastes and lipid by-products; the production of SCB can be carried out in a continuous culture with a limited land area and water requirement, and is independent of climate changes; yeasts are high in protein (30–65%, dry weight) as compared to other natural sources.

The nutritional value of yeast biomass is directly related to its protein and amino acid composition and its lipid, vitamin, mineral and nucleic acid content. Depending on the specific organism, the protein content of SCB ranges from approximately 50% for yeasts to over 70% for bacteria. The nutritional value of yeast has been tested on a large number of animals including rats, chickens and pigs. The amino acid composition of protein determines its biological value as a nitrogen source. Most of the plant proteins are often deficient in lysine, methionine and tryptophan, while yeast protein has an amino acid profile very similar to that of animal protein such as milk or meat (Table 3.2).

In general, the nutritional application of SCB yeast products with added methionine (0.1–0.2%) can replace soya cake protein and fish meal. All the essential amino acids, except sulphuric amino acids, are present in the yeast products at levels exceeding the requirements per 100 g of protein (FAO/WHO, 1989).

Various yeast species are used for biomass production (Table 3.3). The yeast for biomass production are obtained by natural selection from the environment, or by mutation or gene manipulation. There are only a few food yeast species permitted for human consumption. According to the Food and Drug Administration only the dried cells of

Table 3.2　Comparison of Essential Amino Acids in biomass of yeasts (g/100 g protein)

Amino acid	Soya bean meal	Fish meal	Yeast	Yarrowia lipolytica*	Candida Intermedia*	FAO reference
Lysine	2.8	4.9	7.8	9.0	6.4	4.2
Leucine	3.5	5.0	7.8	7.4	8.0	4.8
Isoleucine	2.2	3.2	5.3	3.7	2.5	4.2
Valine	2.3	3.7	5.8	4.9	3.3	4.2
Methionine/cystine	1.3	2.6	2.5	2.3	1.3	4.2
Tryptophan	0.6	0.9	1.3	1.5	1.0	1.0
Threonine	1.9	3.0	5.4	5.1	7.5	2.8
Phenylalanine/tyrosine	5.9	7.4	8.8	11.7	9.4	5.6

* Yeast cultured on media containing rapeseed oil as a carbon source.

Table 3.3　Yeasts used according to conventional and non-conventional carbon sources

Microorganisms	Substrate
Conventional carbon sources for Single-Cell-Biomass production	
Kluyveromyces fragilis, Kluyveromyces lactis	Lactose whey
Candida utilis, Saccharomyces cerevisiae, Saccharomyces carlsbergensis	Beet or cane molasses
Candida utilis, Torula utilis	Sulphite waste liquors
Non-conventional carbon sources for Single-Cell-Biomass production	
Candida tropicalis, Candida lipolytica, Candida maltosa	
Candida quilliermondii	n-alkane
Candida utilis	Ethanol
Pichia pastoris, Candida utilis	Methanol
Yarrowia lipolytica, Trichosporon cutaneum, Candida rugosa	Lipids
Schwaniomyces castelii	
Schwaniomyces alluvins and Lipomyces kononenkoe*	Starch
Hansenula polymorpha, Candida utilis	Cellulose hydrolysate
Candida tropicalis and Aspergillus niger*	Fruit processing wastes
Saccharomycopsis lipolytica and Trichoderma reesei*	
Geotrichum candidum and Trichoderma viride*	
Pichia pinus	

* Yeasts and fungi were employed in symbiosis.

Saccharomyces cerevisiae, Candida utilis and *Kluyveromyces fragilis* can be used. Recombinant DNA technology has led to the possibility of constructing new strains of both baker's and fodder yeast. In the near future, baker's yeast will be grown using new raw materials such as starch, cellulose wastes or cheese whey. In addition, yeast biotechnology will offer a profitable use of these resources, of which some are potential pollutants. However, consumer resistance to genetically modified organisms is the main barrier (Glazer and Nikaido, 1995).

In recent years, a lot of companies have produced baker's or fodder yeast with a production capacity ranging from 4 to 100 t/day. Nearly 40% of European yeast production is exported outside the European Union. The European yeast business has a turnover of about €800 million and generates over 8000 jobs (Crueger and Crueger, 1994).

Substrates for yeast production

Sugar beet and sugar cane molasses   Molasses are a cheap by-product of sugar beet and sugar cane refining. Beet molasses have all the advantages of cane molasses, besides the benefit of having extra protein. In addition to a large amount of sugar, molasses contain nitrogen compounds, vitamins and trace elements (Table 3.4). Therefore, they require little treatment before use as a culture medium. Preparation is frequently limited to clarification, adjustment of the pH and the sucrose concentration (to the level required) and addition of nutrients, e.g. nitrogen, phosphorus, trace elements and vitamins. Baker's yeast, *Saccharomyces cerevisiae*, is the main biomass product from molasses.

Whey   Whey is a by-product of cheese and casein production. Annually, about $60-80 \times 10^6$ t of whey are produced worldwide. The whey composition varies depending on its origin. Sweet whey is a by-product from the manufacture of various hard and soft cheeses, whereas acid whey is generated from the production of cottage cheese. Whey contains about 4–5% of lactose which can be utilized as a good carbon and energy source by yeasts for the production of SCB. A mixed continuous culture of the yeasts *K. marxianus* and *C. utilis* will be able to utilize all of the substrates present in food wastes, at the same time reducing the level of pollution of the dairy wastes as measured by their BOD. It permits reduction of the pollutant load from 50 000 to 3000 mg/l. In this case, several organic constituents of the wastes, mainly lactose, lactic acid and proteins, are converted to yeast biomass by aerobic cultivation. The final biomass concentration of *K. fragilis* cultured on whey can reach up to 90 g/l.

Table 3.4   Composition of sugar beet and sugar cane molasses (Average values for components at 75% dry matter)

Component		Sugar beet molasses	Sugar cane molasses
Total sugars	(%)	48–52	48–56
Sucrose	(%)	ca. 48.5	ca. 33.5
Invert sugar	(%)	ca. 1.0	ca. 21.2
Crude ash	(%)	12–17	9–12
Crude Protein, i.e. (N×6.25)	(%)	6–10	2–4
Phosphorous	(%)	0.02–0.07	0.6–2
Sodium	(%)	0.3–0.7	0.1–0.4
Potassium	(%)	2–7	1.5–5
Magnesium	(mg/g)	ca. 0.09	ca. 0.06
Calcium	(mg/g)	0.1–0.5	0.4–0.8
Copper	(µg/g)	9–11	7–15
Biotin	(µg/g)	0.02–0.15	1.2–3.2
Panthothenic acid	(µg/g)	50–110	15–55
Thiamine	(µg/g)	ca. 1.3	ca. 1.8
Inositol	(µg/g)	5000–8000	2500–6000
PH		4–5	4–5

Lipids   Lipids (fats, oils and waxes) are insoluble in water. Various types of lipid substrates such as vegetable oils (soya, rapeseed, sunflower, maize, olive and palm oil) can be used for biomass production. Animal fats, however, are not easy to use because of their solid form, since it is easier to work with liquid oil than with a solid fatty substrate. The most exciting idea is to promote non-conventional protein production in the oil seed-producing developing countries using oils and by-products as carbon sources for yeasts. These can be used as energy and carbon sources by the genera *Candida*, *Yarrowia*, *Torulopsis*, *Trichosporon* and *Cryptococcus*. Several yeast strains possess adequate lipase activity for growth on di- and triglycerides.

Wastewater produced from olive oil processing is a serious pollution problem in the mediterranean area (the level of COD is very high; 200 000 mg/l). For example, in Italy approximately $2 \times 10^6 \, m^3$ of wastewater is produced each year by the olive oil industry. The wastewater is dark and contains fats, sugars, phosphates, phenols and metals. *Yarrowia lipolytica* was found to be the best yeast for growth in this medium. The production of biomass by *Y. lipolytica* on crude rapeseed oil and waste fatty acids was studied in Poland. The biomass obtained on lipids has a high biological value ranging from 73 to 78%. The amount of essential amino acids is also in agreement with the FAO standards for fodder yeasts. The biomass contained 31.3–50.9% of protein and 11.9–28.1% of lipid, depending on the yeast strain. The optimization of the medium also plays a fundamental role in preventing metabolic deviations that directly effect the yield (Montet *et al.*, 1983). In tropical countries such as Malaysia, microbial utilization of plant waste lipids was performed using *C. curvata*, *C. rugosa* and *C. lipolytica*.

Lignocellulosic and starchy substrates   Lignocellulosic and starchy substrates are present in fruit processing wastes, corn stover, bagasse, wood pulp, sawdust, wheat bran, paper mill wastes, etc. The effective production cost of SCB is related to the raw materials and the efficiency of conversion of these raw materials into the desired biomass. The use of waste by-products as substrates for SCB production becomes an efficient tool in environmental pollution control and a means of very cheap product recovery.

Various types of raw material from fruit processing industries are used for SCB production, including mango peel extract, virgin grape marc, apple pomace and orange peel (Lo Curto and Tripodo, 2001). Apple pomace, a solid waste from the apple processing industries, is used as a raw material for the production of protein-enriched products. The maximum amount of protein was obtained using *Saccharomycopsis lipolytica*, with a growth yield of 0.45 g protein produced per gram of sugar consumed after 35 h culture. The protein content of the final product was 13%, which is suitable for cattle feeding.

Mango peel extract, a polysaccharide hydrolyzate containing monosaccharides like glucose, galactose, arabinose, xylose and unknown sugars, is used for SCB production. The protein content of the yeast ranged from 66 to 70.4%. The amino acid composition of *Pichia pinus* shows that essential amino acids are present in adequate amounts according to the FAO standards, with cysteine and methionine present in higher than standard amounts. Therefore, the biomass of *P. pinus* can be used as a good nutritive material for animals.

Agricultural residues, such as stalks, bran, straw, bagasse, wood and industrial wastes, are potentially unlimited, cheap and renewable substrates, which can easily be converted into SCB products. The main problems encountered with the biological conversion of lignocellulose arise from its inaccessible structure. The lignocellulose is good particularly for the growth of fungi or mixed cultures, e.g. fungi and yeasts. The pre-treatment and hydrolysis of lignocellulose can be carried out physically (e.g. steam treatment), chemically (by acid or alkaline hydrolysis) and enzymatically (cellulases, hemicellulases and ligninases from various fungi), or using a combination of these methods. After hydrolysis of lignocellulose, fermentable sugars such as glucose and xylose are obtained. The direct fermentation of pre-treated lignocellulosic materials using a mixed culture can be used for SCB production.

The production of SCP from rice hulls pre-treated with sodium hydroxide and saccharified with *T. viride* cellulase gave the highest glucose yield and yeast protein production (ranging from 26 to 44 mg/100 ml culture medium). Wheat straw and corn stover can be enriched in protein for animal feed by Solid State Fermentation (SSF) with a mixed culture of *Chaetomium celluloliticum* and *C. lipolytica*. Under optimal pre-treatment and fermentation conditions, the maximum cellulose conversion of 33–40% and maximum final protein content of 15–17% are reached. The fermentation media prepared from wood hydrolysates are suitable as carbon sources for various yeasts.

Various types of renewable starchy raw materials (wheat, maize, cassava, potatoes) are used in the production of fodder yeast. Large amounts of starch are also in wastewater from the potato industry. Two strains of yeast, *C. utilis* and *Endomycopsis fibuligera*, are employed in symbiosis for SCB production from wastewater and wastes from a potato processing factory. Other yeasts used for the production of SCB from potato starch are *S. cerevisiae* and *Cephalosporium eichhorniae*.

3.5.4 Valorization of Waste Materials from Wood Technology in the Field of Composites

The wood industry produces a lot of sawdust, wood chips, etc. that are not always easy to upgrade towards valuable consumer products. However, because of the continuous pressure for biodegradability, many product designers try to incorporate nature-based components into their products even as filling material. The polymer industry certainly tries to use wood-based products as filling material to produce composites. Lots of research is being done on the processing of thermoplastics reinforced with wood fillers (Bledzky, Reihmane and Gassan, 1998; Bledzky and Gassan, 1999; Li, Ng and Li, 2001). Because of the different nature of the materials, several technical problems arise which require a great deal of fine-tuning before the material can be produced. These problems are associated with the high level of moisture absorption of the filler, the poor wetability and the poor adhesion of the filler and the matrix. The compatibility can be improved by physical (e.g. steam explosion) and chemical methods (acetylation of cellulose, grafting of cellulose or by the use of coupling reagents). One of the best coupling agents for wood-thermoplastics is polymethylenepolyphenyl isocyanate.

In this way not only can waste material be utilized but also the biodegradability of the end products ameliorate. In this area, a lot of research is currently being done to improve the compatibility between the natural waste (wood or fibre materials) and the organic matrix that is being used in composite materials.

3.5.5 Hydrolysis of Wood and Plant Components to Industrial Feedstock

Hydrolysis of Cellulosic Material to Glucose

Cellulose is the main component of the ligninocellulosic complex that constitutes plant biomass. Individual cellulose chains adhere to each other by hydrogen bonding and van der Waals forces to form microfibrils. Microfibrils of cellulose consist of a crystalline core with a perfect 3-D array. Surrounding this core is a region of cellulose molecules, which are less-ordered, called the paracrystalline or amorphous region (see also Chapter 7). Between the cellulose molecules of the amorphous region of the microfibril are matrix polysaccharides.

Enzymatic hydrolysis of cellulose is influenced by many factors which include: pre-treatment of the substrate; activity; thermostability; concentration and adsorption of enzymes on the cellulose; time of hydrolysis; substrate concentration; and inhibition of cellulase activity by the end products of the hydrolysis. Bioconversion of this substrate requires a pre-treatment stage. As mentioned before, many methods of pre-treatment of cellulose have been developed including: alkaline and acid maceration; autohydrolysis; and acid-catalyzed steam explosion.

Cellulolytic enzymes are produced by both bacteria and fungi. The cellulolytic bacteria include species such as *Bacillus*, *Cellulomonas* and *Clostridium*, but these microorganisms produce low yields of cellulases. Fungi are the best producers of cellulases. The most active strains with regard to cellulolysis are species like *Aspergillus fumigatus*, *Chaetomium thermofile*, *Mucor pussilus* and fungi from the genera of *Trichoderma*, *Stachybotris* and *Hypocera*.

Due to the resistant character of the substrate, a high concentration of enzyme is needed in cellulose hydrolysis. The production of cellulolytic enzymes is the most expensive part of the process and accounts for up to 50% of the total cost of hydrolysis. Efforts have been devoted to optimize the growth conditions as well as to search for cheap, soluble sources of carbon for the biosynthesis of these enzymes. Many cellulolytic microorganisms are known but only a few of them are used on an industrial scale. This is due to the fact that the production of the full cellulase complex is needed for the complete degradation of native cellulose to glucose. Cellulases are multicomponent enzymatic complexes which contain three basic types of enzymes that act synergistically. They include cellobiohydrolase, *endo*-β-1,4-glucanase, and β-glucosidase. According to a hypothesis proposed by Enari and Niku-Paavola (1987), cellobiohydrolase hydrolyzes insoluble cellulose into soluble cellodextrin and removes cellobiose units from the non-reducing ends of the cellulose chains. *Endo*-β-glucanase acts on the internal 1,4-β-D-glycosidic linkages in cellodextrin and hydrolyzes them into cellobiose. The cellobiose is then split into two glucose units by β-glucosidase. This enzyme also removes glucose from non-reducing ends of cellodextrin.

At present, one of the favoured organisms for production of these enzymes is the wild strain of *Trichoderma reesei* QM 6a. Since then, many multistage mutagenization programmes of this strain have been performed in laboratories in Europe and in the US. Some mutants produce many times the cellulolytic activity of the wild strain. Some of them are resistant to catabolic repression of glucose. The application of genetic engineering to *Trichoderma* has made it possible to modulate cellulase production in such a way that the new strains produce novel cellulase profiles of commercial potential.

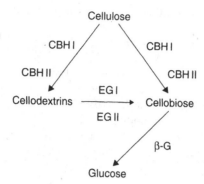

Figure 3.5 *Alternate route to enzymatic cellulose degradation. CBH I, CBH II – cellobiohydrolase; EG I, EG II – endoglucanase; β-G – glucosidase*

An alternative route to cellulose degradation has been presented in a paper of Teeri (Teeri, 1997) (Figure 3.5). According to this scheme, cellulose chains are initially attacked in the amorphous region by endoglucanase, resulting in a rapid decrease in the degree of polymerization and generating free chain ends on the cellulose surface for the exoglucanases to act upon (endo–exo synergism). Cellobiohydrolases initiate their action from the ends in crystalline parts of the cellulose chains. They produce primarily cellobiose and decrease the substrate very slowly. All cellobiohydrolases have been thought to release cellobiose from the non-reducing end of the cellulose chains. Now, there is evidence that different cellobiohydrolases have opposite chain-end specificities. For example, CBH I from *Trichoderma reesei* as well as *Th. fusca* exoglucanase E4 and E6, prefer reducing ends, while CBH II of *T. reesei* and endoglucanase E3 of *Th. fusca* act at the non-reducing ends.

Cellulases are also used to increase the digestibility of fodder with a high content of fibre, such as seeds of rice, wheat and barley. The hydrolysis of cellulose results in an increased assimilation of other compounds. Cellulose hydrolyzates can also be applied for assimilation of other high-value biotechnological products like organic acids (fumaric, citric and lactic acid).

Cellulases for textile applications became the third largest application (after their use in the detergent and starch industry) and one of the fastest-growing markets for industrial enzymes. Cellulase preparations are applied for quality improvement of cellulosic textile fabrics. They are used in biopolishing of cellulose-made fabrics which includes desizing, scouring, bleaching, dyeing and finishing. The action of cellulases consists of the removal of impurities and small, loose fibre ends from the fabric surface. Cellulases can also control the fibrillation of artificial cellulosic fibres like lyocell.

Hemicellulosic Material from Wood and Crops

Utilization of hemicellulosic materials from wood (sulphite pulp), sugar beet, chicory pulp and cereals for fermentation to ethanol and for the production of yeast has already been mentioned.

Hemicellulose, an amorphous heterogeneous group of branched polysaccharides, surrounds the cellulose fibres and intrudes into the cellulose through pores. D-xylose, L-arabinose, D-mannose, D-glucose, D-galactose, glucuronic and galacturonic acids are the major sugar residues, which create respectively xylan, galactan, glucan, mannan, arabinan and pectin as their heteropolymers.

Xylan is a complex polysaccharide comprising a backbone of xylose residues linked by β-1,4-glycosidic bonds. Xylan is the most common hemicellulose polymer in the cell walls of plants, representing up to 30–35% of the total dry weight. It is the major hemicellulose in hardwood from angiosperms, but is less abundant in softwood from gymnosperms; it accounts for approximately 15–30% and 7–15% of the total dry weight respectively. Xylan from hardwood consists of at least 70% of β-xylopyranose residues. Hardwood xylan are highly acetylated, while softwood xylan are not acetylated but are linked with α-L-arabinofuranose units.

The main component of xylan is D-xylose, a five-carbon sugar that can be converted to SCP and chemical fuels or other substances. But xylan must first be hydrolyzed by mineral acids (e.g. sulphuric, hydrochloric) or a complex of several hydrolytic enzymes: β-1,4-endoxylanase, β-xylosidase, which acts on the main chain of xylan and α-L-arabinosidase, α-glucuronidase, acetyl xylan esterase and phenolic acid (ferulic and p-coumaric acid) esterase hydrolyzing side branches. The most important ones are endo-β-1,4-xylanases, which depolymerize almost completely xylan to xylobiose, xylotriose, xylooligosaccharides and β-xylosidase, which hydrolyze xylooligosaccharides to xylose from the end of the chain. These xylanolytic enzymes are produced by moulds (e.g. *Aspergillus* sp., *Aureobasidium* sp., *Fusarium* sp., *Geotrichum* sp., *Trichoderma* sp., *Thermomyces* sp.), yeasts (*Cryptococcus* sp., *Trichosporon* sp., *Pichia* sp.), actinomycetes (*Streptomyces* sp.) and bacteria (*Bacillus* spp., *Cellulomonas* sp.). Recombinant strains are also constructed to degrade xylan and utilize it immediately, for example yeasts *S. cerevisiae* and *S. pombe* have been hosts for genes from *T. reesei* responsible for xylan hydrolysis.

Products of xylan hydrolysis such as xylose, xylooligosaccharides and arabinose can be used in chemical and biotechnological processes to obtain sweeteners (xylitol, arabitol), ethanol, xylulose, triglycerides, SCPs, furfural and organic acids, as well as in xylanolytic enzyme production (Beg *et al.*, 2001).

Enzymatic hydrolysis of pentoses, xylose and arabinose yields valuable raw materials for the food and non-food industry if transformed to furfurals. Maybe the most efficient feedstock for furfural production is corn-cobs or corn stover. Huge quantities of this material are currently left on the fields as waste.

Another application is microbial fermentation to lactic acid which is already used in biomedical applications as intravenous and dialysis solutions (Litchfield, 1996). In cosmetics, lactic acid is added as moisturizer, buffer and acidifier.

Xylitol Production

An interesting example of a sweetener is xylitol, produced as a result of the biotransformation of xylose by selected yeast strains. It has a sweetening power like sucrose, is metabolized independently from insulin, and can be used by diabetics and persons who are deficient in glucose-6-phosphate dehydrogenase. The raw materials needed for the industrial process are wood pulp from paper processing, lignocellulosic materials (hardwood chips), plant biomass and other agricultural by-products. The existing chemical method for xylitol production involves the hydrolysis of the xylan-hemicellulose portion of the plant biomass to obtain D-xylose for the catalytic hydrogenation to xylitol and a subsequent purification by crystallization. This conventional method of xylitol production poses a significant technical problem generated by the poor selectivity, leading to a complex mixture of polyols and sugars that are subjected to expensive refining treatments. The chemical process is also performed at high temperature and high pressure, which increases the cost of the production. It seems that a better method is the biotransformation of xylose to xylitol by microorganisms belonging to the yeast family (strains from *Pichia*, *Candida*, *Debaryomyces* genera and genetically modified *Saccharomyces cerevisiae*). The biotechnological synthesis of xylitol is becoming more attractive due to its cheapness; it does not necessarily require a pure D-xylose solution and does not produce toxic residues compared to the chemical method.

The whole process involves pre-treatment of the raw material by hydrolysis (acid, enzymatic) followed by electrodialysis and reverse osmosis for the demineralization and concentration of the hydrolysates respectively, prior to biotransformation. The biological method requires that some conditions are fulfilled in order not to limit the production: a high oxygen-transfer rate; a high initial xylose concentration; the presence of a nitrogen source; and an appropriate temperature and pH according to the microorganism used.

The immobilization of yeast cells may be successfully used, e.g. in Japan. After transformation, the yeast biomass is separated either by centrifugation or filtration techniques. The cleaned broth is subsequently processed to recover purified xylitol by chromatographic techniques and crystallization. The problems encountered when using lignocellulosic hydrolysates in xylitol production include their complex nature and the presence of sugars acting as catabolic repressors, as well as the presence of antimicrobiological agents such as acetic acid and furfural. Although this method is not yet optimized, it is interesting to study in depth (Udeh, Kordowska-Wiater and Targonski, 1997).

3.5.6   Use of By-products (e.g. Molasses) to Produce Bacterial Exopolysaccharides

Transformation of by-products like molasses by means of enzymatic catalysis can lead to a new class of biocompatible polysaccharides. Many biopolymers are now being developed and important applications are emerging in the areas of packaging, food production and medicine. Possible applications range from thermoplastics that are truly biodegradable to novel medical materials that are biocompatible and to water-treatment compounds that prevent mineral build-up and corrosion. Some biopolymers can directly replace synthetically derived materials in traditional applications, whereas others possess unique properties that could open up a range of new commercial opportunities. Novel biopolymer compounds are being investigated by established agricultural and chemical firms, as well as small biotechnology enterprises. Even if some biopolymers have better environmental characteristics than conventional polymers, much work needs to be done to bring down the costs of production. Commercially available biopolymers are typically two to five times more expensive than synthetic resins. There are only a few specialized applications such as biomedicine, where the relatively high costs of the biopolymer materials are not likely to impede market growth.

Xanthan Gum

Xanthan gum is a copolymer of five sugars produced by the bacterium *Xanthomonas campestris*. It was one of the first commercially successful bacterial polysaccharides to be produced by fermentation (Rocks, 1971). Genetic modification of *Xanthomonas* using recombinant DNA technology has increased the rate of xanthan production by more than 50%. Recombinant DNA technology may enable entirely new xanthan biosynthetic pathways to be created in host organisms (Becker *et al.*, 1998).

Xanthan gum is produced by large-scale fermentation of *X. campestris* using a number of different feedstocks including molasses and corn syrup. The gum is excreted from the bacterium during the polymerization process and can be recovered by alcohol precipitation following removal of the bacterial cells.

For some applications such as enhanced oil recovery, the crude culture broth can be used directly following sterilization. Probably, the most significant technical problem in the production of xanthan is the fact that as the polymer is produced the fermentation medium becomes increasingly viscous. This increases the energy required for the mixing process that feeds oxygen to the bacterial cells.

The unusual physical and mechanical properties make xanthan gum an attractive polymer for non-food and food applications. One of the most important non-food applications for xanthan is increasing recovery in mineral oil production (provides viscosity control in drilling mud fluids) and mineral ore processing (used as a biocide). Xanthan is used for paper manufacturing (as a modifier), in agriculture (acts as plant growth stimulator), in pharmacy (being evaluated for sustained drug release) and in cosmetics (controls dust release). In the area of food technology, xanthan is applied as a gelling agent in cheese spreads, ice creams, puddings and desserts, and as an ingredient of packet soups and many of the fat-free products. Further, xanthan is a very interesting component of clear-gel toothpastes. The worldwide production of xanthan is currently in the range of 10 000–20 000 t and is increasing. About 60% of the xanthan produced is used in foods, with the remaining 40% used for industrial applications. Food-grade xanthan costs about €16–20/kg, while non-food grades sell for about €10/kg.

Dextran

The exopolysaccharide dextran is commercially produced from the microorganism *Leuconostoc mesenteroides* (Kenne and Lindberg, 1983). The dextran can be synthesized using either large-scale industrial fermentors or enzymatic filtration methods. The latter approach is generally favoured since it results in an enhanced dextran yield and a uniform product quality, which allows the product to be readily purified. Both of these production methods permit system conditions to be adjusted in order to control the molecular weight range of the products. This feature is an integral requirement for polysaccharide biosynthesis.

Dextran polymers have a number of medical applications (Sandford and Baird, 1983). Dextrans have been used for wound coverings, in surgical sutures, as blood-volume expanders, to improve blood flow in capillaries in the treatment of vascular occlusion, and in the treatment of iron-deficiency anaemia in both humans and animals. Dextran-haemoglobin compounds may be used as blood substitutes that have oxygen delivery potential and can also function as plasma expanders. Chemically modified dextrans such as dextran sulphate have both anti-ulcer and anticoagulant properties. Other modified dextran products are used as the separation material for biological compounds. Further, dextrans are being incorporated into X-ray and other photographic emulsions. This results in a more economical usage of silver compounds and at the same time reduces surface gloss on photographic positives. Dextrans are used extensively in oil-drilling muds to improve the ease and efficiency of oil recovery. They also have potential in agriculture as seed coating and soil conditioner. The protective polysaccharide coatings are found to improve germination efficiencies under suboptimal conditions. Today, although many applications have been proposed for dextrans, only a small number of these have been realized and developed on a large scale.

There is considerable potential for low-molecular-weight dextrans in the biomedical industry surgery and drug delivery systems. However, low-molecular-weight dextrans sell for about €160/kg. As new and higher-volume applications for these materials are developed, large-scale production of dextrans may represent a major new market for the sugar cane and sugar beet industries.

Hyaluronic Acid (HA)

Hyaluronic acid is composed of a linear polysaccharide of D-glucuronic acid (β-1,3) N-acetyl-D-glucosamine disaccharide units. The range of molecular weights is 70 000 to $2–4\times10^6$ Da in a highly polymerized preparation. Because of the high density of negative charges along the polymer chain, HA is very hydrophilic and forms gels even at very low concentrations. It is extremely flexible and has a high viscosity. In the last two decades, the therapeutic and esthetic uses of HA have been extended to a number of areas including treatment of joint pain, use in fertility clinics, and tissue augmentation. In the past few years, biotechnology has been used to develop HA derivatives with tailor-made molecular sizes, which will further increase the potential applications of this remarkable molecule.

HA plays an important role in tissue hydration, lubrication and cellular function, and is able to hold more water than any other natural substance. Its unmatched hydrating properties result in increased smoothness, softening and decreased wrinkles. It is a popular biomaterial in orthopaedics as a result of its proven biocompatibility, biodegradability and fixation capacity to laminar bone. In addition, it has also been applied to the development of a number of artificial skin products (Radice *et al.*, 2000).

In recent years, the properties of the HA biopolymer have undergone significant modification for use in tissue engineering applications. Varying levels of esterification can reduce the water solubility of the polymer (extensive esterification resulting in water-insoluble films). HA esters have been developed, for tissue engineering, that retain the biological properties of HA and continue to exhibit cell-receptor recognition and interaction with other extracellular matrix molecules. Through a number of defined chemical modifications, HA-derived viscoelastic biopolymers have now been commercially produced, ranging in composition from soluble to solid hydrogels.

The unique physiochemical and structural characteristics of HA make it an excellent candidate for applications that require biocompatibility. The prices of HA are extremely high, at more than €100 000/kg, and give the chance for a high price product.

3.6 Integral Use of Renewables, an Interesting Concept for Developing Countries?

During the last century, the Western countries turned away progressively from agriculture and moved to petrochemistry as a source for materials and industrial products. This tendency is one reason for the economical and social gap between North and South. From this point, the increased awareness in the Western World of the industrial utilization of renewable materials should be an important signal to sustainable development in developing countries. Of course food production must be the basic concern, but utilization of 'current waste materials' could provide a stimulus for regional activities in developed as well as in developing countries.

As is described in Chapter 2 and in this chapter (Mureck model), stimulating the use of renewable resources can be extremely valuable for small communities. It can increase the number, maybe even the qualification, of jobs, and societies can become less dependent on imported goods. However, any prediction is difficult since other parameters such as local acceptance, regional identification or politics (legislation, subsidiaries, protective tolls) are major factors that influence the success of the concept.

There is no doubt that all activity which increases the number of jobs, and generates jobs with midrange and high qualification, reduces the tendency to move to the cities and reduces the need for import of expensive Western products.

3.7 How to Stimulate the Idea of Integral Use of Renewables?

Research initiatives for integral utilization of renewable resources need international networks, in particular cooperation of scientific institutions with ecological and politically active lobbies from agronomy/forestry, trade, industry and public communities. Such interdisciplinary cooperation is expected to provide a climate to accept new ideas, in particular if a flexible schedule of networking, workshops, conferences and symposia is run via lean-structurized projects. Participation of industrial partners in such networks should increase acceptance of obtained results for upscaling to pilot or even production scales. Participation of public and governmental partners is supposed to affect legislation and to provide channels for widespread public information, and hence, identification.

Strategies to establish integral utilization of renewable resources include:

- Increased utilization of local crops and other renewable resources such as algae, marine organisms and products based on animals.
- Development and application of downstream processes focus not only on products, but also on most efficient total utilization of locally available renewable resources.

- Development/modification of downstream processes to manage local requirements and, as far as possible in a modular way, for potential application at different locations.
- Development/modification of processes so that they support local cyclic management of renewable resources.
- Increased implementation of biotechnologically supported processes on any processing level.
- Development/modification of processes as local public property which is a source of income for local communities and which strongly supports local identification and self-respect.
- Increase the awareness of people using eco-boni.

3.8 Conclusion

Integral utilization of renewable resources is capable of adding value to traditional agronomy and forestry. Sustainability emerges from regional focus and will be supported by biotechnologically processing modules wherever applicable. Valorization and added value arises from local jobs, increased qualification of local jobs and relatively short distances between resources and consumers – a large proportion of whom are also suppliers. Focus on total utilization will minimize waste and in many cases may increase income due to speciality products from former waste and development of processing know-how. Therefore, the development of processes needs to focus more on the integral use of the renewable raw materials. The research needs to be intensified in order to cope with some specific problems of the production and downstream processing of the renewable raw materials.

References

Becker, A., Katzen, F., Pühler, A. and Lelpi, L. (1998). Xanthan gum biosynthesis and application: a biochemical-genetic perspective. *Appl. Microbiol. Biotechnol.*, **50**, 145–152.

Beg, Q.K., Kappor, M., Mahajan, L. and Hoonda, G.S. (2001). Microbial xylanases and their industrial applications: a review. *Appl. Microbiol. Biotechnol.*, **56**, 326–338.

Bledzky, A.K., Reihmane, S. and Gassan, J. (1998). Thermoplastics reinforced with wood fillers: A literature review. *Polymer-Plastics Technol. Engg.*, **37**, 451–468.

Bledzky, A.K. and Gassan, J. (1999). Composites reinforced with cellulose based fibres. *Progress in Polymer Science*, **24**, 221–274.

Crueger, W. and Crueger, A. (1994). In: *Biotechnology, A textbook of Industrial Microbiology* (ed. E.T. Brock), pp. 1–316.

Enari, T.M. and Niku-Paavola, M.L. (1987). Enzymatic hydrolysis of cellulose: is the current theory of the mechanism of hydrolysis valid? *CRC Crit. Revs. Biotechnol.*, **5**, 67–87.

FAO/WHO (1989). Protein Quality Evaluation, Report of a Joint FAO/WHO Expert consultation, Rome, FAO.

Faure, D. and Dechamps, A.M. (1990). Physico-chemical and microbiological aspects in composting of grape pulps. *Biol. Wastes*, **34**, 251–258.

Glazer, A.N. and Nikaido, H. (1995). *Microbial Biotechnology*, pp. 365. W.H. Freeman & Co, New York.

Kenne, L. and Lindberg, B. (1983). Bacterial polysaccharides. In: *The Polysaccharides* (ed. G.O. Aspinall), pp. 287–363. Academic press, Orlando.

Lalov, G., Krysteva, A.M. and Phelouzat, J.L. (2001). Improvement of biogas production from vinasse via covalently immobilzed methanogens. *Bioresource Technol.*, **79**, 83–85.

Li, T.Q., Ng, N. and Li, R.K.Y. (2001). Impact behavior of sawdust/recycled-PP composites. *J. Appl. Pol. Sci.*, **81**, 1420–1428.

Litchfield, J.H. (1996). Microbial production of lactic acid. *Adv. Appl. Microbiol.*, **42**, 45–95.

Lo Curto, R.B. and Tripodo, M.M. (2001). Yeast production from virgin grape marc. *Bioresource Technol.*, **78**, 5–9.

Lynd, L.R., Cushman, J.H., Nichols, R.J. and Wyman, C.E. (1991). Thermophilic ethanol production – Investigation of ethanol yield and tolerance in continuous culture. *Appl. Biotechnol.*, **28**, 549–553.

McCleery, R. (1995). Resiliency: the trade mark of American forests. *For. Prod. J.*, **45**, 19–28.

McNutt, J.A., Haegglom, R. and Raemoe, K. (1992). The global fibre resource picture. In: *Wood Product Demand and the Environment*, A Forest Products Research Society International conference Proceedings, pp. 39–53.

Mittelbach, M. (1996). *The High Flexibility of Small Scale Biodiesel Plants*. Proceedings of the 2nd European Motor Biofuels Forum, Graz, Austria.

Montet, D., Ratomahenina, R., Ba, A., Pina, M., Graille, J. and Gabzy, P. (1983). Production of single cell protein from vegetable oils. *J. Ferment. Technol.*, **61**, 417–420.

Nerantzis, E.T., Stamatiadis, S., Giannakopoulou, L. and Maniatis, L. (1991). The production of microalgae in a hellicoidal photobioreactor for feed in vermiculturing. *Forum Appl. Biotechnol.*, Gent, Belgium, pp. 91–92.

Ntalos, G.A. and Grigoriou, A.H. (2002). Characterization and utilization of vine prunings as a wood substitute for particleboard production. *Ind. Crops Prod.*, **16**, 59–68.

Palma, M. and Barroso, C.G. (2002). Ultrasound-assisted extraction and determination of tartaric and malic acids from grapes and winemaking by-products. *Analytica Chimica Acta*, **458**, 119–130.

Peres, M., Romero, L. and Sales, D. (1998). Comparative performance of high rate anaerobic thermophilic technologies treating industrial wastewater. *Water Res.*, **32**, 559–564.

Radice, M., Bnrun, P., Cortivo, R., Scapinelli., R., Battaliard, C. and Abatangelo, G. (2000). Hyaluronan-based biopolymers as delivery vehicles for bone-marrow-derived mesenchymal progenitors. *J. Biomed. Mater. Res.*, **50**, 101–109.

Rocks, J.K. (1971). Xanthan gum. *Food Technol.*, **25**, 22–31.

Rosillo-Calle, F. (1989). A reassessment of the Brazilian national alcohol programme (PNA). In: *Resources and Applications of biotechnology: The New Wave* (ed. R. Greenshields), pp. 332–345. Stockton Press.

Sandford, P.A. and Baird, J. (1983). Industrial utilization of polysaccharides. In: *The Polysaccharides* (ed. G.O. Aspinall), pp. 411–490. Academic press, Orlando.

Shin, H.S., Bae, B.U., Lee, J.J. and Paic, B.C. (1992). Anaerobic digestion of distillery wastewater in a two phase USAB system. *Water Sci. Technol.*, **25**, 361–371.

Teeri, T.T. (1997). Crystalline cellulose degradation: new insight into the function of cellobio-hydrolases, *Tibtech.*, **15**, 160–166.

Udeh, K.O., Kordowska-Wiater, M. and Targonski, Z. (1997). Xylitol: metabolism, microbial production and perspective in polymer substrates utilization. *Biotechnology*, **3**, 27–37.

4

Primary Production of Raw Materials

Waldemar Rymowicz

In co-authorship with *Bernd Honermeier, Simone Siebenborn, Tanja Schaefer, Wieslaw Kopec and Christian V. Stevens*

4.1 Introduction

When agricultural raw materials need to be used to produce consumer products and industrial products in a sustainable way, the production and the breeding of plants and animals will be of utmost importance in order to select and develop the best plants and grow them under the most optimal conditions and in the most economical way. In this chapter, the production and several aspects of the downstream processing of the raw materials will be highlighted. The chapter is divided into four parts:

1. raw materials from plant origin;
2. raw materials from animal origin;
3. raw materials from marine origin; and
4. materials from microbial origin, produced from agricultural waste products.

Not only will plants, animals or marine organisms be produced for non-food applications, but also the use of side and waste products will be optimized to develop products with increased added value.

Renewable Bioresources: Scope and Modification for Non-food Applications. Edited by C.V. Stevens and R. Verhé
© 2004 John Wiley & Sons, Ltd ISBNs: 0-470-85446-4 (HB); 0-470-85447-2 (PB)

4.2 Raw Materials of Plant Origin

4.2.1 Fibre Plants

Economic Importance

Plant fibres have been used since ancient times. Archaeologists found knotted and woven textile materials in ancient graves of Central America, India and Peru, bearing proof of the craftsmanship of the people of those times. In former times, plant fibres were mainly used to produce clothing, twines and nets (Bòcsa, Karus and Lohmeier, 2000). Nowadays, they are mainly applied in the textile industry. Further areas of application are the building and construction industry (insulation material and filling material), paper production (writing, cigarette and photographic paper), the furniture industry (upholstery material for furniture and mattresses), automobile production (dashboards), landscape designing (securing of embankments) and horticulture (planting vessels).

In the year 2000, the world's production of plant fibres amounted to approximately 25×10^6 t, cotton being the most important fibre plant accounting for 77%. Further important fibre plants are jute (and fibres similar to jute) accounting for 16%, flax for 2%, sisal for 1.5% and hemp for 0.3% of the world's fibre production (FAO, 2002). Plant fibres hold a share of 65% (46% of which is cotton) of the overall fibre production, whereas synthetic fibres account for 30% and animal fibres for only 5% on the world market.

Taxonomy of Fibre Plants

The group of fibre plants consists of a wide range of plant species belonging to different taxonomic classes. They are spread all over the world and are well adapted to the prevalent climatic conditions. In humid and moderate regions (e.g. Europe) flax is the main cultivar. Cultivation of fibre nettles (great stinging nettle) is equally possible in these regions. Fibre hemp, however, can be found in moderate and warmer regions (e.g. Asia). In the tropics and subtropics a large variety of fibre plants are cultivated (e.g. cotton, jute, sisal, palm tree species and agave species). All these plants contain cellulose, which is a common characteristic of them, but their morphogenesis varies largely. The fibres in the plant are either present in the form of seed hairs, or they can be produced from the shoot axis, leaves or the pericarp (Table 4.1) (Berger, 1969).

Table 4.1 shows the classification of fibre plants. Cotton belongs to the group of fibre plants producing fibres out of seed hairs. In botanical terms, cotton is part of the genus of *Gossypium* and the family of *Malvaceae*. Various species can be found within the genus *Gossypium*, which are well adapted to the prevalent site conditions (temperature, supply of water and nutrients). Cotton fibres are produced by the pericarp and each consists of only one single cell. When separated from the seed grain, they are referred to as 'lint'. Since cotton needs high temperatures and is characterized by a long period of growth, it is mainly cultivated in tropical and subtropical regions. However, cotton fields also exist in the south of Europe (e.g. Greece, Italy).

The seed hairs of the kapok tree are located in the epidermis of the pericarp. This tree belongs to the family of *Bombaceae* and is also known as the silk-cotton tree, owing to its

Table 4.1   Classification of fibre plants

Fibre name	Exact location	Example	Latin name
Seed hairs	One cell fibre	Cotton	*Gossypium* spp.
	Fruit hairs	Kapok tree	*Ceiba pentandra*
Bast fibres	Inside the stem	Flax	*Linum usitatissimum*
		Hemp	*Cannabis sativa*
		Nettle	*Urtica dioica*
		Ramie	*Boehmeria nivea*
		Jute	*Corchorus capsularis* (white jute)
			Corochus olitorius (Tossa jute)
		Roselle hemp	*Hibiscus sabdariffa*
		Kenaf	*Hibiscus cannabinus*
Leaf fibres	Under leaf	Abaca (Manila hemp)	*Musa textilis*
	Petiole	Palm species	*Palmae* spp.
	Leaf blade	*Agave* sp. (sisal)	*Agave* spp. *Agave sisalana*
Fruit fibres	In mesocarp	Coconut	*Coco nucifera*
		Sponge gouge	*Luffa cylindrica*
Wood fibres		Spruce	*Picea abies*
		Pine	*Pinus* spp.

short fibres which are silky, glossy, downy and very soft. Because of their elasticity and springiness, these fibres are suitable for filling materials. Kapok trees grow exclusively in tropical climates and require a deep sandy loam. Precipitation during vegetative growth and dry weather during the flowering phase and at the harvesting time are ideal preconditions for the cultivation of kapok. Harvesting may start when the trees are 3 or 4 years old. Until 10 years of age, the fibre yield increases from approximately 1.5 kg to a maximum of 10 kg per plant. In older trees even higher yields can be achieved. Kapok trees can be harvested until they are 30 years old (Berger, 1969).

The group of bast fibre plants includes a variety of plant species. In all of these plants, the fibres are located underneath the shoot axis and extend over its whole length. In contrast to fibres produced out of seed hairs, these fibres are not present in the form of individual elementary fibres. Fibre cells are interconnected by the so-called 'binding substances' to fibre bundles. As a result, these plants, which can reach a height of 2–5 m, are characterized by a good stability. Jute is the most important bast fibre plant (genus *Corchorus*, family *Tiliaceae*), especially used when a rough and sturdy fibre structure is desired in the end product. Approximately 75% of jute is destined for the production of sacks and bags. It is also in use in the production of twines, chair covers and backs of linoleum or carpeting. The colour of the fibres is determined by the jute species, e.g. those of *C. olitorius* are yellow, reddish or grey, while those of *C. capsularis* are white. A slightly higher price is paid for *C. olitorius*, owing to the more favourable properties of its fibres (fine, soft and sturdy structure). Under the climatic conditions in North and Central Europe, only fibre flax (family of *Linaceae*), fibre hemp (family of *Cannabaceae*) and fibre nettle (family of *Urticaceae*) are cultivated. Flax and hemp are annual plants, whereas fibre nettle may be cultivated perennially. A finer thread is obtained from flax, traditionally used for tablecloths, linen and high-quality textiles, whereas hemp fibres are used in the production of working clothes and harvesting bags (Berger, 1969).

In addition, hemp fibres are renowned for their suitability for the production of high-quality paper and moisture-resistant and tear-proof materials (e.g. ropes, cords, canvas). As a result of the introduction of synthetic fibres at the beginning of the 20th century, the importance of hemp and flax as fibre plants has decreased considerably. In some countries, cultivation of hemp has been generally forbidden for a long time. Since 1996, hemp may be cultivated in some countries for fibre production under sharp restrictions (because of the presence of narcotic secondary metabolites). In addition, new areas of application have been developed over the last few years, especially in the field of building and construction (mats for thermal insulation of buildings) and in the automobile industry (inside trimmings).

The bast fibre plant ramie belongs taxonomically to the family of *Urticaceae*, but takes on economic relevance only in China and Japan. Length and stability of the fibres are major attributes of this plant. Ramie can be applied in the same branches as fibre flax and fibre hemp. The main reason for low worldwide distribution of ramie may be seen as high costs of handpicking of the fibres. Ramie can be grown in tropical climates and in the warmer areas of moderate climatic zones. Profitable cultivation needs sites with fertile soils, high air temperature and sufficient precipitation. *Hibiscus* species (kenaf and rosella) belong to the family of *Malvaceae*. The properties of their fibres are similar to those of jute. However, *Hibiscus* fibres are coarser and less pliable. As a result, they cannot always be a substitute for jute fibres.

Hibiscus species have the advantage of being better adapted to various climatic conditions. For this reason, they can produce high yields not only in tropical but also in subtropical climates. Kenaf (*Hibiscus cannabinus*) is used in a number of countries in the production of ropes, twines and fishingnets. Occasionally, the fibres are mixed with jute and used in paper production. Paper made from a fibre mixture consisting of kenaf and jute has a quality similar to that of paper made from wood fibres. Kenaf is a short-day plant and can be harvested as early as four or five months after planting. Rosella demands fertile soil and sufficient precipitation and needs a growing period of up to 8 months. Therefore, rosella is not as widespread as kenaf (Berger, 1969).

Leaf fibres, very often referred to as hard fibres, can be obtained from the leaf stem or the shoot axis. Abaca, like manila hemp, is a banana species (*Musaceae*). This perennial plant forms rhizomes out of which pseudo-stems grow. The pseudo-stem is formed out of the widened leaf base and the surrounding leaves. Fibres, present in the pseudo-stem, can be extracted by peeling the rind. Abaca is cultivated by planting the rhizomes with at least three eyes. The first harvest may be carried out as soon as the flower appears. Later, every four to six months, two to four pseudo-stems can be harvested. The plant can grow for a period of 12–15 years. Annual precipitation of at least 2500 mm, air temperatures of 27–29 °C and extremely fertile soil are necessary preconditions for successful cultivation of abaca. Its fibres are mainly used in the production of ropes, cords and rigging. They are resistant to moisture and salt. Sisal and henequen (equally known as Mexican or Yucatan sisal) are important agave species cultivated for fibre production. These plants have a fibre content of 2.5%, the fibres being present in the leaf blade. Each species is perennial, forms rhizomes and grows under tropical and subtropical conditions. The plants can be harvested until they enter the generative phase, after which they die. Fibres of agave species are not only applied in the production of cords, ropes and twines, but they can as well be used in the production of sacks, carpet backs and chair covers.

The fibres of the coconut palm growing under tropical conditions are present in the mesocarp of the fruit and are extracted by stripping off. These fibres are usually a waste product in the production of coconut milk. Production of coconut mats or carpets is very common. The network of vascular bundles of the vegetable sponge is used for bath sponges, insoles, upholstery material and in the production of wallpaper. It is cultivated in Japan, India and Egypt (Berger, 1969). Wood fibres are primarily used in the production of paper, mainly wrapping material, writing paper and newsprint.

Production of Fibres from the Plant

While the seed hairs can be spun directly after cleaning the seeds, in the production of other fibres it is necessary to separate the fibres from the remaining parts of the plant. Fibres usually do not occur in an isolated form but are connected with each other and the surrounding tissue by binding substances (pectines, hemicellulose). A retting process (dew retting) is usually carried out after harvesting the bast fibres in order to dissolve the connections. The plants are left lying on the field for approximately 2–4 weeks. During this period, pectines and hemicellulose are microbially degraded by bacteria and fungi under the influence of changes in temperature and moisture conditions. Another method of producing fibres is water retting. In this process, harvested plants are put into a retting basin filled with water (in former times rivers or ponds), and left soaking for 5 and 10 days for hemp and flax respectively, and up to 30 days for jute. If the water in the basin is heated (warm water retting) the process is shorter. In this process, the degradation of binding substances by microbes takes place much faster than during dew retting. A high content of nutrients in the retting water may be a problem. If the process is carried out in natural waters, it may lead to eutrophication. This is the reason why it is no longer applied in Europe. The fibres produced in this manner, however, are much finer and cleaner than those produced by dew retting. The length of the retting time depends on the physiological age of the plant and the temperature during the retting process. In dew retting, the number of sunny and rainy days is of additional importance. The course and the result of the retting process are checked by collecting the samples of straw. The colour of straw, fungal infestation and separation of rind (bast) from the wooden parts of the shoot axis are assessed. Isolation of the fibres is commonly carried out mechanically by crushing rolls (contrarotating pairs of rolls) or hammer mills. Chemical pulping by lyes (NaOH, soda), or pulping by steam pressure may be carried out after the first stage of separation if the fibres are to be further processed into textiles. Mechanical pulping is generally sufficient if the fibres are processed into insulation mats, fleece or compression moulded parts. If this method of processing is applied, bast fibres and wooden parts (shive) are obtained as a waste product which can be used as litter, material for thermal insulation or in the production of light fibre-boards. Fibres of leaves are not processed by retting. In sisal, the green leaf substance surrounding the fibres is removed by crushing or scraping. Immediately afterwards, the fibres are dried and pressed into bales (Berger, 1969).

The area of application of plant fibres depends on their morphological properties (e.g. length and radius of the individual fibre and also fibre bundles). Table 4.2 shows some morphological characteristics of the most important plant fibres.

Table 4.2 *Length of individual fibre, fibre bundle and shoot axis and fibre content of various fibre plants*

Fibre plant	Length of individual fibre (mm)	Length of fibre bundle (cm)	Length of shoot axis (cm)	Fibre content (% DM*)
Cotton	25–28	does not exist	–	100% in the lint
Flax	20–100	50–70	up to 100	25
Hemp	10–50	up to 200	up to 350	25–35
Nettle	50–100	60–120	up to 200	5–10
Ramie	150–180	up to 200	>200	12–15
Jute	2–5	100–300	up to 400	4–8
Kenaf	2	100–300	up to 400	
Roselle	2	100–300	up to 400	

* DM – dry matter.

In order to assess the possibilities of application of plant fibres, knowledge about their morphological and technological properties, such as tensile strength, breaking elongation, heat resistance, colour and density, is crucial (Scheer-Triebel and Léon, 2000). The necessary data are obtained in laboratory trials by comparing plant fibres with mineral fibres. In general, plant fibres exhibit a lower density ($1.2–1.4\,g/cm^3$) than glass fibres ($2.6\,g/cm^3$). This can be favourable when applying prefabricated moulded parts in automobile production in order to reduce weight. High flexural strength, on the other hand, is crucial in elastic sandwich material. Heat resistance of plant fibres ($230\,°C$), however, is relatively low as compared to glass fibres ($700\,°C$). It is worth noting that plant fibres are of great importance, since they are environmentally friendly and biologically degradable.

4.2.2 Medicinal Plants

Introduction

The earliest indications on the use of medicinal plants can be found in the writings of the ancient advanced civilizations, for example on Egyptian papyri, Assyrian cuneiform scripts, in the Indian Vedas and also in the 'Pen-t'ao knag mu', an extensive compendium of Chinese medicine. In Europe, the ancient art of healing was continued in the monastic schools, and the medicinal plants were cultivated in monastery gardens. Many modern medications have been developed from the old recipes, and since the last decades of the 20th century the interest in phytopharmaceuticals on the drug market as a whole has been steadily rising. Whereas in earlier times, raw materials were mostly obtained from wild collections, cultivated goods are increasingly coming into demand nowadays. This development can not only be attributed to high quality standards of the pharmaceutical industry, but also to the protection of endangered species. Complete control from the nursery bed to harvest and subsequent processing is possible only when cultivating medicinal plants. Thus, falsifications and unwanted pollutant entries can be avoided and necessary preconditions for optimal production of ingredients be created. The market demands are determined by the regulations of the pharmaceutical industry. These, in turn, are based on the regulations of the national pharmacopoeias in which vegetable raw materials and quality demands are described in the so-called drug monographs. In order to standardize these regulations on a European level, the European pharmacopoeia, which is increasingly gaining importance, has been introduced.

Pharmaceutically Effective Ingredients

The effective ingredients of medicinal plants, like those of spice plants and dye plants, are mostly secondary metabolism products. In contrast to primary metabolism products, like carbohydrates, proteins and fat, only small amounts of these products are present in plants, and they are normally characteristic of certain groups of plants. They are often found only in certain cells or cell groups and sometimes only during certain phases of differentiation. The content of active substances within a plant can vary considerably, depending on the genotype and environment (climate, soil, season and time of day). Pharmaceutical terminology refers to the chopped up and dried vegetable material as a drug. Since the drug is the basic material for further industrial processing, it should contain all the active substances relevant to the pharmaceutical product.

Basic materials of phytopharmaceuticals can be categorized in different ways, depending on their properties. They can be categorized from a pharmaceutical point of view according to their effectiveness or according to special characteristics such as colour, taste, smell, solubility, basicity, and effectiveness. Group designations for pharmacologically effective content materials, which are still applied nowadays, have been the result of these categorizations (Bisset and Wichtl, 2001; Wagner, 1999).

- essential oils
- alkaloids
- flavonoids
- cardio-effective glycosides
- saponins
- bitter substances
- slimy substances
- tannic acids
- anthraglycosides.

Essential oils are widespread in the vegetable kingdom and represent the most important active substance group in medicinal plants. They are volatile and lipid-soluble. The characteristic smell and taste of a plant variety or a chemotype (variation within a variety) result from the varied composition of the essential oil. The largest portion of the essential oil constituents is composed of terpene compounds (see Chapter 10). Furthermore, phenylpropane derivatives, simple phenols and other aliphatic and aromatic compounds are present. Essential oils take effect in a multitude of ways. They are antibacterial, antiphlogistic and antispasmodic and frequently used for colds and flu.

Alkaloids are plant bases of a mixed structural principle, all containing nitrogen. Very often they have a distinctive effect on various areas of the central nervous system. The group of alkaloids covers a wide range of natural substances which very often have a complicated chemical structure.

Flavonoids are ubiquitous among higher plant species and can be found in various parts of the plant in the form of pigment soluble in cell sap. The pigments located in the flowers are especially attractive to insects. One of the characteristics of flavonoids is a C-15 carbon skeleton along with phenolic OH-groups which can be present as free, methylated or as a glycoside compound. Owing to their structural diversity, flavonoids have a variety of pharmacological effects, the most important one being the stabilizing effect on the blood vessels. Furthermore, they exhibit an antiphlogistic and antispasmodic effect.

Cardio-effective glycosides have been found so far in plant families. They are *steroid-glycosides*, with a specific effect on the dynamics and rhythm of the insufficient heart muscle. One of their characteristics is a five- or six-membered lactone ring and a sugar bond (glycoside bond) at the C-3-hydroxyl. All cardio-effective glycosides are qualitatively equal in effect, but differ in quantity. Depending on their structural peculiarity (degree of polarization, lipid solubility, sugar bond), they show differences in the latent period, resorption and subsiding rate and, therefore, in the time span over which they are effective.

Saponin drugs contain triterpene glycosides and steroid glycosides as their main active substances. Their name comes from their most important capability of forming durable foam in combination with water (Latin sapo=soap). Owing to their ability to reduce interfacial tension between two phases, saponins are now frequently used as emulsifiers or dispersion agents. As drugs they have antibiotic and diuretic effects and help to bring up phlegm.

Bitter substance drugs are used for stimulating salivation and secretion of the glands of the stomach and the gall bladder because of their intensely bitter taste. From a chemical point of view, they do not belong to a uniform class of substances. Bitter substance plants very often contain terpenoids; non-terpenoid bitter substances are flavonoglycosides, phloroglucine derivatives and alkaloids. Slimy substances consist of heteropolysaccharides, and owing to their typical hydrophilic groups have a strong swelling effect. They are applied externally as antiphlogistics and internally as light laxatives.

Tannic acids are high molecular compounds without nitrogen, widely spread in the flora and used in former times for tanning leather. They have an astringent effect which helps in the healing of wounds.

Anthraglycosides or anthraquinones belong to the most frequently used laxatives. The compounds of this group of drugs are derived from anthracene and are for the most part present as glycosides. Glycosides, which are slightly soluble in lipids, are their transport form. Aglycones having a laxative effect are then released as a result of the activity of the glycoside-hydrolyzing bacterial flora in the colon. The dianthrones of St John's Wort (Hypercines), which are violet-red in colour, do not have a laxative effect, but are supposed to be partly responsible for the antidepressive effect.

Raw Material Production for the Pharmaceutical Industry

Supply shortages on the open world market, as well as occasional offers of poor-quality raw material, have encouraged a number of companies to sign long-term cultivation contracts in order to meet the quality and quantity requirements. In this way, falsifications found in the collections of the wild flora can be avoided. Farmers specializing in the cultivation of medicinal plants and providing the pharmaceutical industry with raw materials can realize considerable profit margins.

However, cultivation is not without risk and cultivation of new varieties of medicinal plants in particular requires experience. Many medicinal plants still show the characteristics of wild plants, they do not germinate well, do not mature evenly or burst easily. Special equipment is necessary to harvest root, herb and flower drugs. Planting is also a problem; not all varieties can be sown directly onto the field, but have to be raised in nursery beds and then transplanted. This again requires specific technical facilities (greenhouses, planting machines), investment in the purchase of seeds and high labour costs. Weed control is another major problem. According to the new European herbicide regulation, herbicides can exclusively be used for those plants that have been tested and have received certification. Generally, this does not apply to medicinal plants. For this reason, pharmaceutical companies, cooperatives and state institutions try to attain supplementary regulations for the use of herbicides on medicinal plants. Only after the regulations in question have been passed can herbicides be implemented in this area. Otherwise weed control has to be conducted mechanically, and very frequently even manually. In crops with slow and uneven germination and lack of ground cover, this can increase the costs and working time.

Since medicinal plants rarely come fresh onto the market, they have to be dried after harvesting. Air-drying is possible only in regions with high temperatures and low precipitation. The purchase of dryers is expensive, and therefore practised in the regions where cultivation of medicinal plants represents a considerable proportion of the overall production of agricultural or gardening businesses.

The decision to cultivate certain medicinal plants is therefore strongly dependent not only on the financial means of a business, but also on on-site conditions, which should meet the requirements of the plants in terms of warmth, precipitation, light and soil condition. Thus, many warmth-loving essential oil plants such as *Lavandula*, *Thymus*, *Rosmarinus* and *Citrus* are exclusively cultivated in southern Europe (e.g. in the south of France, Italy, Spain). Although some of the Mediterranean plants can be cultivated in cooler hemispheres, they produce far less essential oil under such conditions, and for this reason cannot compete on the market. Many medicinal plants are indigenous to the tropics and can be cultivated only in these regions (e.g. *Cinnamomum*, *Capsicum*, *Chinchona*, *Strychnos*, *Harpagophytum*). A number of species are perennial plants and require extra care in order to prevent outlaying; in colder climates winter losses are of special concern. Many perennial species (*Mentha*, *Melissa*, *Hypericum*) can be used as early as in the first year of cultivation, whereas others have to grow 5 years or longer before harvesting can begin.

Infrastructure is also an important factor in the choice of the site. Production in the vicinity of the processing industry helps to keep costs of transportation and storage low, costs for drying may not occur at all. Regional cooperatives can be a profitable option with regard to the communal usage of machinery and exchange of know-how. Moreover, a better basis for negotiations on the market is possible.

Production technology, especially harvesting technology, for medicinal plants is dependent on the part of the plant to be harvested, i.e. leaves, herbs, flowers, seeds or roots. Occasionally resin, bark or the whole plant can be utilized. Therefore, the most important species of medicinal plants are listed in Table 4.3 with regard to the parts of the plants used (Dachler and Pelzmann, 1999; Marquard and Kroth, 2001).

Table 4.3 *Important medicinal plants used as leaf and herb drugs*

Plant species	English name	Family	Plant part for use	Ingredients
Mentha piperita *Mentha crispa*	Peppermint, spearmint	*Lamiaceae*	Leaves	Essential oil (menthol, menthon, menthofuran)
Melissa officinalis	Common balm	*Lamiaceae*	Leaves	Essential oil (citronellal, citral)
Salvia officinalis	Sage	*Lamiaceae*	Leaves	Essential oil (cineol, thujon)
Thymus vulgaris	Thyme	*Lamiaceae*	Herb	Essential oil (thymol, carvacrol)
Origanum vulgare	Wild majoram	*Lamiaceae*	Herb	Essential oil (thymol, carvacrol)
Hypericum perforatum	St John's wort	*Hypericaceae*	Flowering herb	Dianthrones (hypericin), flavonoids, hyperforin
Fagopyrum esculentum	Buckwheat	*Polygonaceae*	Flowering herb	Flavonoids
Cynara scolymus	Globe artichoke	*Asteraceae*	Leaves	Bitter substances, flavonoids
Solidago virgaurea	Golden rod	*Asteraceae*	Flowering herb	Flavonoids, essential oil, saponins
Digitalis lanata *Digitalis purpurea*	Grecian foxglove Purple foxglove	*Scrophulariaceae*	Leaves	Heart glycosides

Leaf and Herb Drugs

Generally, the technical effort for the cultivation of leaf and herb drugs (Table 4.3) is comparatively low. The crop has to be cut several times a year. This can be done using an ordinary cutter loader, common on numerous farms. Crop management aims at the production of vegetative mass. In an intensive production of *Mentha* or *Melissa*, nitrogen rates of up to 200 kg/ha/yr are frequently applied when the plants are cut several times a year.

Plants such as *Thymus* and *Origanum*, which naturally tend to grow on poor soils, should be fertilized moderately. If the crop is harvested only once a year, this should be carried out when the optimum accumulation of active substances is reached. In essential oil plants, for example, this is usually the case shortly before or during flowering. In some plants such as *Hypericum* or *Fagopyrun*, only the flowering horizon in the upper part of the plant is harvested.

It is essential that the material harvested should reach the next stage of processing, usually drying, as soon as possible. Essential oil plants must not be dried at temperatures above 40 °C, otherwise volatile active substances will be lost. Further stages of processing such as separation of leaves from stems (leaf drugs) or cutting of the drug require special technical equipment. It is advisable to produce essential oil directly on the field or on the farm. In order to do this, the fresh material is processed using a mobile or stationary distilling apparatus. Distillation residues remain on site, and only pure essential oil is marketed – the complicated process of drying can be dropped.

Table 4.4 *Important medicinal plants used as flower drugs*

Plant species	English name	Family	Ingredients
Camomilla recutita	Camomile	Asteraceae	Essential oil (bisabolol, chamazulen), flavonoids
Chamaemelum nobile	Roman camomile	Asteraceae	Essential oil, bitter substances, flavonoids
Arnica montana	Arnica	Asteraceae	Essential oil, flavonoids
Calendula officinalis	Marigold	Asteraceae	Flavonoids, saponines, caratinoids, essential oil
Lavandula angustifolia	Lavender	Lamiaceae	Essential oil (linalool, linalylacetat)

Flower Drugs

In this group of plants (Table 4.4), the active substances are mainly present in the flowers. Since other parts of the plant are not needed in the drug, one has to make sure during harvest that the proportion of stems in the drug is kept at the lowest level possible. Handpicking is the best method to obtain high-quality material, but this is possible only in countries with low labour costs. Large quantities of high-quality camomile drug, for example, come from Egypt. It is mainly sold in pharmacies as loose-tea, since the presence of whole flowers without stems is an essential quality characteristic of these goods. If the flowers are used for the production of an extract, a certain proportion of stems can be tolerated, as long as the content of active substances meets the minimum quality standards. In this case, harvesting can be carried out by special flower-picking machines. Along with a couple of other machines on offer, there are very often modified combine harvesters picking the flowers by rotating picking-combs. The picking-combs, which make handpicking easier and faster, are a kind of compromise between the use of harvesting machines and hand-picking. However, the material harvested by the machine is of much lower quality than that handpicked.

As for fertilizer use, it is important to know that nitrogen improves the formation of herbage to a higher extent than the formation of flowers. In turn, too much herbage mass makes mechanical harvesting more difficult and has a negative influence on the quality of the material, owing to a high proportion of stems. For this reason, a crop with a large number of flowering plants, little branching and little herbage mass is to be aimed at.

Seed Drugs

Plants in which the active substances are mainly contained in the mature multiple fruit or seeds are usually harvested using a combine harvester (Table 4.5). Combine harvesting is possible provided that it takes place in due time and the plants are resistant to damage. Strong branching has a negative effect on the ripening process, since the secondary branches are very often still green while the main branches are mature. Therefore, the time of harvesting frequently involves a compromise between the risk of loss and the harvesting of unripe fruit, with consequent obstructions during threshing resulting from the presence of green foliage. The tendency towards a strong formation of secondary branches can be avoided by higher crop density. In extremely problematic crops, it is possible to compensate for uneven ripening by occasional handpicking. In regions with long

Table 4.5 *Important medicinal plants used as seed drugs*

Plant species	English name	Family	Ingredients
Carum carvi	Caraway	Apiaceae	Essential oil (carvon)
Foeniculum vulgare	Fennel	Apiaceae	Essential oil (trans-anethol, fenchon, methylchavicol)
Silybum marianum	Milk thistle	Asteraceae	Flavonoids (silybin, silydianin, silychristin)
Oenothera biennis	Evening primrose	Onagraceae	Fatty oil with γ-linolenic acid
Borago officinalis	Borage	Boraginaceae	Fatty oil with γ-linolenic acid
Papaver somniferum	Poppy	Papaveraceae	Alkaloids (morphin)
Linum usitatissimum	Linseed	Linaceae	Mucilage
Sinapis alba	White mustard	Brassicaceae	Mustard oil, essential oil

periods of good weather, the crop can be cut and dried on the field for a couple of days and threshed afterwards.

Root Drugs

Plants cultivated for the use of roots (Table 4.6) demand specific soil properties for optimum growth of the roots and easy harvesting. For this reason, light soils are good for the cultivation of root drugs. Soils containing a high proportion of sand are better than more fertile but heavier clay or loamy soils. The most convenient time for harvesting root drugs is at the end of the growing season, in late autumn. Depending on the weather, harvesting can be carried out in spring, before sprouting sets in. Harvesting during the growing season usually results in insufficient content of active substances, owing to substance conversion during the phases of growing and maturing. Time-consuming harvesting by hand with a digging fork is nowadays carried out on smaller sites only. In extensive cultivation, first the green matter is chopped off (or sometimes killed by chemical agents) and then the roots are harvested by extractors. Some species require highly specific harvesting machines which have to be adjusted or even modified according to need. Because of the loss of active substances and risk of rotting, any injury to the roots can cause a reduction in quality.

Table 4.6 *Important medicinal plants used as root drugs*

Plant species	English name	Family	Ingredients
Valeriana officinalis	Valerian	Valerianaceae	Valeopotriates, essential oil
Gentiana lutea	Yellow gentian	Gentianaceae	Bitter substances
Inula helenium	Elecampane	Asteraceae	Bitter substances, essential oil, inulin
Angelica archangelica	Angelica	Apiaceae	Cumarines, essential oil
Armoracia rusticana	Horse radish	Brassicaceae	Mustard oil
Echinacea purpurea	Purple Coneflower	Asteraceae	Echinacosid, polysaccharides, essential oil
Echinaceae pallida			
Echinaceae angustifolia			
Taraxacum officinalis	Dandelion	Asteraceae	Bitter substances, flavonoids

The harvested roots have to be cleaned. This is usually done by means of vegetable washers, normally applied in root vegetable cleaning. It is advisable to divide larger rhizomes into smaller pieces in order to facilitate subsequent cleaning and drying.

4.2.3 Dye Plants

Introduction

Cultivation and use of dye plants was very common in Europe up to the end of the 19th century. Pigments from leaves, fruits, seeds, wood and roots were used as dye stuffs for textiles and as paint in art and craft. The discovery of synthetic dyes led to a breakdown of the natural dye market and, as a result, cultivation of dye plants came to a standstill. However, over the last few years synthetic dyes have been losing good reputation because of the risk of toxicity, negative influence on the environment and high allergic potential. Consequently, an increasing demand for natural dyes has developed. In nutrition, the application of natural dyes such as chlorophylls, carotenoids or fruit juices is already firmly established. In the dying of textiles, some problems occur because of the requirements concerning colour stability and non-fading properties. Nevertheless, companies specializing in the production of natural clothing show a particularly large interest in dying technologies based on natural raw materials.

The mechanisms of the dying process prove to be fairly complex. Chemical reactivity and histological structure of the fibres are essential in addition to certain properties of the dyes, such as acid, alkaline and zwitterion content, along with the size of molecules. The capacity of natural fibres to take on dyes depends to a large extent on whether they are of animal or plant origin.

Animal fibres consist of proteins, present at the isoelectric point in the form of zwitterions. When using acid dying agents, the dissociation of the carboxyl groups occurs, resulting in a surplus of charged amino-groups that bind the acid dye molecules via the salt bond. On the other hand, in the environment of low alkalinity, the dissociation of the carboxyl group is enhanced, and therefore binding of basic dyes is possible.

Vegetable fibres consist of cellulose molecules, which neither release molecules nor are able to take them up. Therefore they have no affinity to acid or basic dying agents. Thus, chemical modification of the cellulose prior to the dying process, such as treatment with tannic acids, becomes necessary.

Apart from a few exceptions, most natural dyes are rather ill-suited for direct dying of natural fibres. Also in the dying of protein fibres, a treatment of the textiles with metal salt mordants is advisable in order to bind soluble dyes as an insoluble colour layer on the surface of the fibres. Mordants are chemicals which enable binding of fibres and dye. They are usually metal salts (Al-, Fe-, Cu- and Cr-salts) forming water-soluble hydroxides which bind with the amino acid groups of the protein fibres and, in the majority of cases, acid natural dyes. Thus, protein fibres of animal origin (wool, silk) can be dyed without much problem. The most important mordant is alum, especially potassic alum, which exists in nature ($KAl(SO_4)_2$ $12H_2O$) and has been produced in alum slate mines since ancient times. Whereas alum and chromium salts result in unchanged, light shades of colour, addition of copper or iron salts is responsible for colour changes. The use of chromium, copper and tin salts, however, is not recommended for ecological and toxicological reasons, and therefore an environmentally friendly mordant, such as aluminium sulphate or iron sulphate, should be used preferentially. As mentioned above, adhesion of the dyes on vegetable cellulose fibres (cotton, linen) is much more difficult, since they require pre-treatment with soda and tannins. Dying of the famous 'Turkish red', in which madder was used, required as many as 20 stages. Apart from the so-called mordant dyes, water-insoluble vat dyes are another important class of natural dyes used in the textile industry. Before dying, they are converted to a water-soluble form by reduction. The dyestuffs start adhering to the textile material in the vat and, by means of oxidation in the presence of air, the original non-water-soluble colour is produced again. Indigo belongs to the class of pigments which can be found in various plants giving an off-blue dye (Marquard and Siebenborn, 1999; Siebenborn, 2001).

Dye Plants and Dye Drugs

Dye plants, parts of these plants or animals giving off dye (e.g. cochineal *Poryphyrophora* sp., kermes *Kermococcus vermilio* Planchon), conserved by drying are referred to as dye drugs. In the vegetable kingdom, names such as 'tinctorus, -a, um' (dyer) and 'coriarius, -a, -um' (tanner) refer to dye plants. Table 4.7 contains a list of the most popular dye plants. On the whole, the range of plants giving an off-yellow dye is far wider than the species giving an off-red or a blue dye. Green shades of colour are usually produced by mixing various dye drugs, since dying using the green plant pigments containing chlorophyll and xanthophyll (spinach, stinging nettle) does not give long-lasting colours. The following species have proved to be especially suitable for cultivation under Central European conditions: *Polygonum tinctorium* for blue; *Rubia tinctorium* for red; and *Reseda luteola*, *Anthemis tinctoria* and *Solidago canadensis* for yellow.

4.3   Raw Materials of Animal Origin

4.3.1   Selection of Edible and Non-Edible By-Products

Meat is the primary product of the livestock sector but it is only 35–55% of the live weight of slaughtered animals, depending on the species. The remaining part of the body of slaughtered animals is referred to as meat by-products and the main categories are: bones (16–18% of live weight); organs and glands (7–16%); skin or hide with attached fat and hairs (e.g. bristle or feathers) (6–15%); blood (3–4%); and horns, hooves, feet and skull (5–7%). Generally, by-products are considered to be interior protein sources as compared to skeletal muscles.

By-products are divided into edible and non-edible products on the basis of regulatory requirements, hygiene, tradition or religion. Traditionally, edible by-products such as livers, tongues, stomachs, hearts, lard and brains are called 'variety meats' (Goldstrand, 1988).

Generally, edible by-products are suitable for food applications in the form of cooked, uncooked or cured further products or their parts (i.e. casings). They are consumed directly, as processed meat, sausage or food ingredients. In some cases, products for animal feed

Table 4.7 *Important dye plants*

Plant species	Plant parts for use	Dye components	Colour
Reseda luteola L.	Flowers, leaves, stems	Flavonoids (Luteolin)	Yellow
Solidago canadensis L.	Flowers	Flavonoids	Yellow
Genista tinctoria L.	Herb	Flavonoids (luteolin, genistein)	Yellow
Anthemis tinctoria L.	Flowers	Flavonoids	Yellow
Serratula tinctoria L.	Herb before flowering	Serratulin	Yellow
Berberis vulgaris L.	Roots, stems	Alkaloid berberin	Yellow
Crocus sativus L.	Stigma	Carotinoids (crocetin)	Yellow
Agrimonia eupatoria L.	Leaves, stems	Flavonoids, tanning agents	Yellow
Alchemilla vulgaris L.	Leaves	Flavonoids	Yellow
Vitex agnus castus L.	Whole plant	Flavonoids (vitexin)	Yellow
Chlorophora tinctoria	Wood chips	Flavonoids (morin)	Yellow
Cotinus coggygria	Wood	Flavonoids, tanning agents (fisetin)	Yellow
Carthamus tinctorius L.	Flowers	Flavonoids (carthamin)	Yellow
Lawsonia inermis L.	Leaves	Naphtoquinones	Red
Caesalpinia brasiliensis	Wood chips	Pyran derivates (brasilin)	Red
Alkanna tinctoria L.	Roots	Hydroxyanthraquinones	Red
Galium sp.	Roots	Hydroxyanthraquinones	Red
Rubia tinctorum L.	Roots	Hydroxyanthraquinones	Red
Asperula tinctoria L.	Roots	Hydroxyanthraquinones	Red
Phytolacca americana L.	Fruits	Anthocyanes	Red
Pterocarpus santalinus L.	Wood chips	Flavonoids (santalin)	Pink
Haematoxylon campechianum L.	Kernel wood	Pyran derivates (haematoxylin)	Blue
Polygonum tinctorium LOUR.	Leaves	Indigo	Blue
Isatis tinctoria L.	Flowering herb	Indigo	Blue
Indigofera tinctoria L.	Herb	Indigo	Blue
Quercus robur L.	Bark	Tanning agents, Flavonoids (quercitin)	Brown
Juglans regia L.	Roots, bark leaves	Naphtoquinones, tanning agents	Brown

production are also classified in this group. The most important edible by-products are the following: variety meats for human consumption; blood and its fractions (mainly plasma) used in sausage manufacturing; pork and poultry skins in sausage production; edible rendered fats for shortening; and the collagen derivative – gelatine. The group of products for animal feed consists of not only meat, bone, blood, hoof, horn and feather meals, usually obtained after drying of animal by-products, but also dried plasma and gelatine. A separate group of products is pet food (Goldstrand, 1988).

The following can be included in non-edible by-products: hides and skins for leather goods production or gelatine manufacturing; non-edible fats used as lubricants; hair, wool and feathers for brushes; felts, insulation and clothing; glands and some organs (stomach, liver, lungs) for pharmaceuticals or human implants manufacturing (especially hearts and their parts); blood fraction (especially serum and albumin) and bones for gelatine production.

The application of both edible and non-edible by-products depends on legislation preventing the risk of spreading TSE diseases (transmissible spongiform encephalopathy – e.g. scrapie and BSE). The aim of the European legislation in this field is exclusion of specified risk material (SRM) from any feed chain. SRM includes mainly: skull with brain and eyes; tonsils and spinal cord; ileum; spleen; intestine; vertebral column or thymus of ruminants (cattle, sheep, goats) aged over six months. SRM can be processed only by incineration. In addition, feeding of farm livestock on animal-origin feed meals has been prohibited in the EU since the year 2000 (Council Dec. 2000/766). The by-products, for purposes other than animal or human consumption, which can be referred to as renewable resources, are by no means high-risk materials and they are included in category 3, i.e. low-risk materials according to European legislation (Prop. Europ. Parliament, 2000).

4.3.2   Utilization of Collagen- and Keratin-rich Raw Materials (Skins with Hair, Bones)

Collagen contains about 30% of animal body protein. Collagen is a large extracellular protein composed of three polypeptide chains (α-chains, each containing about 1000 amino acids) with a sequence of repeating amino acids: glycine – proline – Y (in which Y is often hydroxyproline). Three α-chains form a triple helix (about 300 nm), with a shorter globular domain (telopeptides) at the ends of the molecule, called tropocollagen. Collagen is not a single type of protein but a family of related molecules coded in DNA by different genes. Up to date, 19 different forms of collagen (types) have been identified, with different molecular mass (around 300 kDa) and amino acid sequence. Most collagen types self-assemble and create fibres of high mechanical strength by reducible cross-links formed by intra- (aldol) and inter-molecular (aldimine and oxo-imine) lysine cross-links (Bailey and Light, 1989).

The triple helix of collagen is also resistant to most proteinases except collagenases (e.g. metalloendopeptidase). Other proteinases like trypsin, papain or pepsin do not act on the triple helix, but are able to cleave telopeptides, especially in acid conditions (pepsin) when collagen is partly solubilized. Thermally denatured collagen (gelatine) is easily digested by most common proteases. Hair is composed almost entirely of keratin, which consists of insoluble, elastic, fibrous proteins rich in cystine (25%), proline (10%), arginine (10%) and glycine. There are two main types of keratin: hard keratin in nails, hooves and cortex of hairs; and soft keratin. Hard keratin is built by very densely packed keratin filaments and is higher in sulphur content than the soft keratin (horny matrix) filling cells of the inner root sheath of hair follicles. Hard keratin is resistant to most proteolytic enzymes including pepsin and trypsin, but can be thermohydrolyzed with water under pressure.

Leather Production from Hides and Hair Utilization

Hides and skins are recognized as the most valuable animal by-products (their commercial value is about 60% of the by-product price) (Ockerman and Hansen, 2000). The skin protects the animal body against solar radiation and microbial infection. It comprises a surface epidermis, an underlying connective tissue of corium (dermis) and a deep subcutis which is a loose network of collagen fibres with subcutaneous fat. Skin (epidermis) produces hair, feathers, nails, hooves and horns.

Corium, which is the main raw material for tannery, consists of two main layers. The upper layer (called grain – about 10% of skin) is built by thin collagen fibres containing sebaceous glands. The deeper layer of the corium is composed of coarse collagen fibres which contribute to the inherent strength of the leather. Hair roots extend into the corium (beef hide) or subcutis (pigskin). Hides, after removal from the carcasses (mainly using a pulling technique), are cooled and have to be preserved because of their high microbial contamination and high water content. Pigskins are usually scalded in hot water (at about 60 °C) which denatures proteins in the hair follicle and consequently loosens the hair. When the temperature during scalding is below 60 °C, pigskins (after dehairing) can be used for leather production. The hair removed from scalded carcasses, i.e. bristle, can be used for the production of brushes (if the length of hair is above 20 mm) after washing in a 2% Na_2CO_3 solution and drying. Shorter bristle is utilized as animal feed only after thermo-hydrolysis, or as fertilizer, but this by-product is rather resistant to biodegradation processes.

Keratin-rich raw materials can also be utilized as feed after depolymerization, during thermohydrolysis with acids, alkali and urea. Acid thermohydrolysis causes tryptophane damage and partial deamination of all amino acids. On the other hand, alkali treatment causes a significant loss of sulphur-containing amino acids, and the biological value of the obtained hydrolyzates is lower than that obtained after acid hydrolysis.

'Tanning' is a general name for the multi-stage chemical procedure in which hides undergo processes changing them to strong, flexible leather resistant to mechanical deformation, thermal treatment, water and spoilage.

There are different possibilities for conducting preliminary processing of hides or skins destined for leather production. They can be sent to a tannery after cooling, or after operations such as fleshing (removal of the subcutis), curing, trimming and even chrome tanning (the so-called 'wet blue' technique), sometimes performed at a slaughter facility.

The subcutis is removed using a fleshing machine (the main part is a rotating cylinder equipped with spiral knives cutting off the flesh layer). The removed layer (about 15% of 'green' hide mass) is the main constituent of fleshings, which are utilized for glue stock production. The thick fleshing of pigskin with a high content of fat enables salt diffusion to the corium during the next process, i.e. curing (or any other method of hide preservation).

Curing can be done using salt (NaCl with preservatives), followed by piling up many hides or skins together. They are then stored for one month to enable the diffusion of salt and the decrease of the water content. Due to the high water content in hides, 30–50% of salt in relation to 'green' hide mass has to be added to prevent the growth of micro-organisms. To enhance the lethal effect, an addition of 2% Na_2CO_3 (which also prevents the occurrence of the so-called 'salt stains') or fluorides and naphthalene (which prevent 'red and violet stains' in leather caused by microbial growth) is commonly done.

Brine curing is a faster (10–24 h) and easier method and includes skin leaching (washing) to remove grease, blood, etc. In some cases, a combined method of preservation is applied, e.g. pickling, which consists of curing in brine of 10% NaCl and in 1% H_2SO_4 or other acids.

Another technique of skin preservation is drying in order to reduce the moisture content to 12–15%. Drying needs temperatures below 35 °C because gelatinization of collagen can occur when the temperatures are higher.

Because only the matrix of corium (dermis) can bind tanning agents, that part of the hides or skins is the raw material for the leather industry. Other skin constituents, i.e. epidermis, hair and subcutis, should be removed. Preparation of the skin for tanning is usually conducted in a beamhouse and consists of the following operations: trimming, soaking, fleshing and unhairing (also called liming). After that treatment, the so-called limed pelt, which is a semi-product in the tanning process, is obtained in the beamhouse. During the soaking process, salt adsorbed by the hide during curing is removed, and rehydration of the corium occurs during stirring of the hides in water with the addition of detergents, disinfectant or 0.35% alkali salts in the case of hides, or 0.4–1.5% Na_2CO_3 for pigskin which is higher in fat. The aim of soaking is also the removal of non-collagenous proteins soluble in water. During this process, dirt and blood are also removed.

The aim of the next process – liming or unhairing – is the loosening of the connection between corium and epidermis, removal of hair, loosening of hair roots and elastin fibres, extraction of non-collagenous proteins (albumins, globulins) and also partial saponification of fat without damaging collagen (Ockerman and Hansen, 2000).

During the unhairing process the hair is loosened by the use of chemicals. Most often $Ca(OH)_2$ solution (lime liquor) is applied as a depilatory agent, but other strong hydroxides can also be used. The effect of lime on epidermis and hair is the chemical modification of the keratin structure (keratinolysis) mainly due to cleavage of cross-linking ionic bonds. To accelerate hair loosening or dissolving, Na_2S is applied as a lime liquor component. The impact of sulphides on keratin results in the reduction of disulphide bonds.

Generally, sulphides highly accelerate the liming process. Due to high pH (>11.5), hides swell during the process. The increase in their volume during liming is 20–50%, these values are lower in the case of Na_2S and higher only when $Ca(OH)_2$ is used for lime liquor preparation. Calcium is bound to limed pelt in a physical way, i.e. dissolved and bound in capillaries between collagen fibres, chemically bound to collagen molecules or present in calcium soap (lime soap) which is insoluble in water.

Deliming, being the next operation during the preparation of hides for tanning, enables the removal of calcium hydroxide and Na_2S or other alkali substances used in previous operations. The result of the deliming procedure is also the reduction in swelling.

Strong mineral acids could cause acid swelling of collagen (also dissolution of fibres), therefore weaker acids (especially organic acids such as lactic acid) are preferred. However, the best method is deliming of pelts with ammonium salts (sulphate, formate or butyrate), resulting in a fast reduction of the pH.

Bating, pickling, splitting and shaving operations are referred to as the 'tanyard process'. The aim of bating, which is the process of enzymatic treatment of delimed pelts, is the digestion of the remaining non-collagenous components (remaining keratin of hair follicles, sweat and sebaceous glands, protein constituents, residues of triglycerides). Common preparations (bates) contain mainly alkaline proteinases and also lipases of different origin: from pig pancreas, bacteria, fungi and plants (Ockerman and Hansen, 2000). Pelts for upper parts of shoes undergo proteolysis for a shorter time, while leather for gloves manufacturing takes a longer time.

The next process is pickling, in which the pH of the pelts is changed to an acidic pH (about 3) to enable the reaction with tanning agents (mainly chrome salts). Usually, the pickling procedure is performed in rotating drums using solutions containing NaCl (preventing acid swelling) and H_2SO_4.

Hides and skin collagen have the ability to react with chemical substances called tannins, which consequently affect durability and resistance to water and proteolytic enzymes. Due to the mechanism of the tanning reactions, the methods can be divided into three main groups:

1. creation of covalent bonds between the tanning agents and the collagen molecules;
2. creation of coordinative bonds; and
3. tanning with creation of different kinds of bonds.

The most common tanning agents are:

- mineral salts of chromium, aluminium and zirconium;
- phenolic compounds present in plants (vegetable tannins) or synthetic compounds (syntans); and
- aliphatic agents: natural fat tanning, aldehydes.

The most popular method of tanning uses chromium salts (mainly Cr III). The proposed mechanism consists of the creation of coordinative bonds with free carboxylic groups in the collagen side chains. The resultant bluestock can be called leather.

Today, vegetable tanning is only an additional process performed after the reaction with chromium, which prevents it from leaching and improves the leather softness. The active tanning agent in vegetable extracts is tannic acid (tannin). Vegetable tannins are polyphenolic compounds of two types: they are derivatives of pyrogallol or catechol. The mechanism of the tanning reaction is probably based on hydrogen bonding between phenolic groups and peptide bonds of protein chains. Syntans, produced by condensation of phenols or aromatics, sulphonic acids and formaldehyde, are also applied in the tanning process (Ockerman and Hansen, 2000).

The next operation is splitting, in which the hide is cut (split) into two parts, the grain split (with grain layer) and the flesh split. When the latter is thin it can be utilized as a by-product for glue or collagen casing production or for suede type leathers. The final operations during the tanning procedure are colouring, fatliquoring (which is lubricating with oils or emulsions for the softness of the leather), massaging, buffing (smoothing the surface by mechanical sanding) and finishing (covering with impregnating materials like acrylic polymers, blood or egg albumin, vinyl polymers, polyurethanes and waxes).

The tanning process generates a lot of wastes because of the high water consumption (a 15-fold increase in volume) and because of the large volume of offal in processing. Many tanning by-products are used for gelatine manufacturing.

Gelatine Production from Hides and Bones

Gelatine is a heat denatured and partly degraded product of collagen, forming cold-setting gels. During heat treatment, a two-stage process of collagen triple helix denaturation occurs: separation of polypeptides and denaturation of the helical form of peptides. Collagen molecules denature in the narrow range of temperatures with a mid-point T_D (melting point), and create 'random-coil' configurations due to the breaking of hydrogen bonds (Bailey and Light, 1989).

As a result of the denaturation, a mixture of polypeptides consisting of one, two or three connected peptide chains is created. So gelatine is not characterized by a defined molecular mass, as this can vary from 10 to 250 kDa.

The technical product of quite high purity contains not only gelatin, but also some inorganic material and a proportion of macromolecular impurities. In its crude form it is glue. Gelatine and glue are produced from collagen-rich by-products such as hides, or parts not used for leather production (or by-products of that process), and bones.

Bones contain 40–45% moisture, 20–30% protein (mainly collagen), 12–22% fat, and on average 20% minerals (with 35% calcium and about 15% phosphorus content).

Ossein is produced from bones after removing the mineral constituents (mainly calcium phosphate) by acid (HCl) treatment. The raw materials for ossein manufacturing are crushed bones (particles sized 10–20 mm) obtained after pre-treatment. The next steps are milling and extraction of the remaining fat with organic solvents (hexane). Crushed bones are mixed with an HCl solution (~1 M) in acidproof vessels for 24 h. Fresh ossein is characterized by a high water content and has to be dried to a 10% moisture level.

The technology of gelatine and glue production depends on the degree of their inter-molecular cross-linking, related to the 'age' of collagen. If gelatine is produced from hides of older animals or ossein containing more cross-linked collagen, pre-treatment of raw materials with alkali-lime for several weeks or with NaOH for a few days is necessary. The alkaline method is the most popular in gelatine manufacturing. The technology includes washing (hydration) of the collagen raw material in cold water. Then, the process of liming is conducted using a saturated solution of calcium hydroxide in large quantities. A long period of $Ca(OH)_2$ action (from seven days to three months – longer period for ossein) causes hydrolytic reactions in collagen, but no solubilization. After liming, collagen is partly depolymerized, which enables easy melting of collagen fibres and molecules during heating.

Processing of hides of relatively young animals (e.g. pigs) or bone materials (ossein) is conducted with mild acid treatment for collagen extraction, mainly due to the breaking of aldimine intermolecular cross-links. The yield of gelatine is 20–30% from pigskins and 12–13% from bones (ossein).

After the extraction process, a 5% gelatine solution is purified from fat and fibre residues using a centrifuge and filtration with diatomaceous earth. In the case of gelatine production for pharmaceuticals or photographic materials, an additional purification step including ion-exchange chromatography for removing calcium and sodium ions is conducted.

The thin gelatine solution is concentrated in a multi-stage evaporator at 50–100 °C to 'honey' consistency. Concentrated gelatine solution of high viscosity is filtered with a cellulose-plate filter and sterilized using an UHT method at 140 °C. Then, solidified and extruded gelatine bands are dried in horizontal continuous drying stores.

After the alkali treatment, the final product is almost fully deaminated and characterized by an isoelectric point near pH 5.0 (class B gelatine). Class A gelatine is produced using an acid method and has an isoelectric point in the pH range of 6.0–9.0. Gelatine is mainly characterized by gel strength values (expressed in °Bloom) determined for gels containing 6.7% protein and compressed in standard conditions (Bloom apparatus).

Gelatine is still very important for the photographic industry (accounts for 20% of gelatine production) as a carrier of silver iodide in photographic plates and for the making of silver halide emulsions. The photographic plates can be composed of 20 different gelatine layers in order to have the correct photographic characteristics.

The highest amount of gelatine in the pharmaceutical industry (10% of production) is used for the production of capsules, enabling storage of pharmaceutical substances and shielding them from light and oxidation. Depending on the kind of encapsulated substance, hard or soft capsules are utilized. Soft capsules are manufactured, formed and usually filled with liquid and pasty drugs. All these operations are performed in a one-stage process. They have thicker walls (about 300–400 μm) than the hard ones (about 100 μm). Many kinds of gelatine can be applied for soft capsules: gelatine B with gelling strength of 150–175° Bloom, and also gelatine A from bones or pigskin with gelling ability of 150–210° Bloom. Hard capsules (two-piece) are formed on steel pins immersed in 25–30% gelatine solution without any additives at temperatures from 45 to 50 °C. Gelatine clotted on pins is air-dried at 25 °C for 30 min and at 40–50% humidity. Gelatine is also widely used to coat pills and tablets (Ockerman and Hansen, 2000).

Vitamins A and E, soluble in oils, or carotenoids are confectioned as powders processed by spray drying of emulsion containing gelatine, sugars and sometimes starch. Low-Bloom gelatine types A and B are used for emulsion preparation. Concentration of vitamins and carotenoids in the mass is about 1%.

Blood plasma substitutes or expanders applied in medicine can be based on gelatine because of its ability to degrade gradually. Gelatine is not accumulated in the body and does not cause allergic reactions. Plasma expanders are manufactured from bone B gelatine of high Bloom value, usually chemically modified (succinylated).

In surgery and dental operations, gelatine sponges are used for controlling bleeding and for blood absorption. Sponges are produced in the process of foaming of pigskin gelatine solutions. It is also known that gelatine consumption has a positive effect on joints. Recent clinical results have shown that gelatine hydrolyzate has a positive effect on the joint mobility even with advanced arthritis, and a reduction of pain is observed. In order to obtain good results, a regular intake of about 10 g of gelatine hydrolyzate per day for a few weeks is recommended.

Cosmetics contain gelatine as a thickener and carrier of active substances. In shampoos, shower gels and washing liquids, gelatine hydrolyzates reduce allergic reactions of detergents on human skin.

Hide glue has a higher pH (6.5–7.4) than glues produced from bones, which absorb less water and set slower. The main application of animal glue is in plywood making, polygraphy, electrometallurgy for smoothing surfaces and for gummed tapes manufacturing.

4.3.3   Blood as a Source of Protein Isolates and Medical Preparations

Blood is an internal medium of the organism consisting of intercellular plasma and cellular elements: erythrocytes (red cells), leukocytes (white cells), and thrombocytes (blood plates). Blood is composed of 16.5 (sheep) to 18% (cattle and pigs) protein, which is well balanced in amino acids. Blood is rich in lysine (about 10% of the total amino acids in the protein), but deficient in isoleucine. The main kinds of blood proteins are: albumins (about 3.5%), globulins (3%), fibrinogen (0.4–0.6%) – the main plasma proteins – and haemoglobin (about 10%) present in erythrocytes (30% of their mass).

About 50% of blood is collected from animals during the process of bleeding and the yields are: about 10 dm^3 from a cow; 2.5 dm^3 from a pig; and 1.5 dm^3 from a sheep. Blood clots as early as 5 min after bleeding because of the conversion of soluble fibrinogen into insoluble fibrin, which is catalyzed by the thrombin enzyme and calcium ions. Therefore, chemical binding of calcium prevents blood coagulation. Most often, sodium citrate

(0.2% solution) or a mixture (1% in relation to blood) of NaCl (40%) and phosphates (60%) are used as anticoagulants (Ockerman and Hansen, 2000).

Centrifuge separation of blood collected at slaughter lines into plasma and (red) cell concentrate is one of the most common unit operations in the meat industry. Blood suitable for food and pharmaceutical applications has to be collected aseptically using knives in a closed system and needs to be mixed immediately with anticoagulants (sodium citrate, phosphate). In most common techniques, a solution of anticoagulant is supplied directly to a pipe in the knife handle. Blood is then stored at 2–4° before separation. This should be carried out within 20 min of the blood collection from the animal, in conditions preventing both haemolysis of red cells and contamination with microorganisms. Haemolysis results in the liberation of haemoglobin from red cells, which decreases the plasma quality. From 100 kg of blood, 65 kg of plasma containing 8% protein and 35 kg red cells sediment rich in haemoglobin (containing 32% protein) can be obtained. Both plasma and red cells can be incorporated into food products (mainly meat products), but only plasma proteins are widely used for increasing the protein level or the water-holding capacity. Plasma proteins (7–8% concentration) consist of about 50% albumin, 25% globulins (including macroglobulin and immunoglobulins) and 20% fibrinogen (Knipe, 1988).

Blood serum (fibrin-free plasma) is most often obtained by sterile bleeding of animals, it is then chilled and left to be clotted. The clots are cut into cubes to increase the surface area, which shortens serum release during cold storage. Next, serum is centrifuged, pasteurized (55 °C) and sterilized with microbial filters. Finally, it is frozen or freeze-dried and used in a laboratory for preparation of bacteriological media, and also used for the production of virus vaccines or standard solutions. Peptone prepared from blood fibrin is a well-known nutrient used for the growth of microorganisms (Ockerman and Hansen, 2000).

Blood albumin is serum from raw materials collected in non-sterile conditions, without anticoagulants. In the case of non-food applications serum can be chemically preserved (with 0.2% phenol). Albumin can also be produced from blood containing anticoagulants, using CaCl$_2$ to obtain a 1% solution in plasma after cell centrifugation. With this technology, a higher yield of albumin (about 5% in relation to blood weight) is obtained, as compared to the production from blood without anticoagulant. Selective fractionation, especially of blood plasma, yields extracts such as albumin, transferrine, fibrinogen and immunoglobulins. Bovine serum albumin is extensively used in research and clinical medicine. Pork blood fibrin is a source of amino acids in parenteral nourishing of some surgical patients. Hog blood plasmin is used in patients after heart attack for fibrin digestion in blood clots (Ockerman and Hansen, 2000).

Cell sediment (red cell paste) is a raw material for hemin and amino acids production. Many food or potential non-food uses of red cell concentrates depend on the possibility to remove the haem pigment from the haemoglobin moiety. The decolourization process consists of two main steps: reduction to choleglobin with ascorbic acid and haem removal with acidified acetone, to form 'globin'.

Some blood products can be used for the manufacturing of foam compounds applied in extinguishers with organic solvents and additives (oils, hexane, fats). Foam compounds are produced from blood, spray-dried blood or blood meal (dried coagulated blood) after hydrolysis in a 2% NaOH solution and heating at 90 °C (Ockerman and Hansen, 2000).

4.3.4   Animal Organs and Glands – Utilization in Biopreparations Production

The glands and organs for the production of bioactive substances are: pancreas, pituitary, adrenal, thyroid, stomach, heart, liver, lungs and intestines.

Endocrine glands have to be chilled or quickly frozen because of autolysis and destructive bacterial growth. Some glands can be preserved using chemical substances (acetone, phenol or formaline). Preservation has to be done within one hour after removal from the carcass. Frozen glands should be placed in covered containers for storage. Many of the frozen glands are not allowed to thaw and are processed frozen using chopping equipment and necessary solvents (e.g. acidulated alcohol for insulin production) or are dried under vacuum.

The pituitary is one of the most valuable animal by-products because of the possibility of obtaining many important hormones regulating the activity of other endocrine glands. The most important are growth promoting (GH), adrenal-cortex stimulating (ACTH) and thyroid stimulating (TSH) hormones. The pituitary (after drying as a crude product) can be used in therapy, but most often hormones are extracted directly from the fresh gland. ACTH is the most commonly produced hormone used in the therapy of arthritis, leukaemia and skin disorders. Adrenal glands contain adrenaline and noradrenaline, which can be used in surgery to shrink blood vessels, stimulate heart action or reduce peripheral blood flow (Ockerman and Hansen, 2000).

Insulin production is one of the best-known applications of animal glands. Insulin controls the level of sugar in the blood and is produced by β cells of the internal part of the pancreas. Insulin is first extracted from the frozen gland with acidified ethanol, and then salted out from the extract and dissolved again with alcohol. Although today insulin is produced with biotechnological methods using genetically modified bacteria, pig insulin (which is chemically very close to the human hormone) is still important in the therapy of diabetes.

Heparin, a mucopolysaccharide, prevents blood clotting or dissolves blood clots and is also used in the treatment of burns. It is extracted from pig and bovine intestines (inner lining), lungs or livers in salt solution and is then precipitated using acetone. Bovine arteries for implants are treated with ficin, an enzyme, to remove parenhymatous proteins and they are then tanned with dialdehyde starch to strengthen collagen.

Pig hearts are a source of heart valves used in surgical implantation. Hearts, deeply cooled immediately after slaughter, are sent quickly to the facility producing the final product, where implants are prepared. Aortic valves are excised from the heart and processed in glutaraldehyde solution, which cross-links collagen and strengthens the tissue (Ockerman and Hansen, 2000).

A lot of edible and non-edible by-products, including organs and glands, are sources of proteolytic enzymes and inhibitors. Pancreases from chicken and turkey are especially rich in proteolytic enzymes. The process of isolation and purification of trypsin, chymotrypsin and elastase includes extraction of the pancreatic zymogens with $CaCl_2$, purification of enzymes by serial affinity column chromatography with respective inhibitors bound to Sepharose 4B followed by ion-exchange chromatography (Wilimowska-Pelc *et al.*, 1999).

Pancreatin is a mixture of amylolytic, lipolytic and proteolytic enzymes obtained from pig pancreas, and is used in flavourings production and milk peptonization and as a bating agent in leather manufacturing.

Two enzymes of great importance for cheese making can be extracted from pig (pepsin) or calf (chymosin – rennin) stomachs. Pepsin (acidic protease), extracted from stomach lining, has a lot of applications. It is used, for example, for extraction of native collagen from skins and hides, to clot milk, etc. Rennin is the main constituent of the enzyme mixture used for milk coagulation (often with swine pepsin).

Serine inhibitors isolated from bovine and swine pancreas (Kazal type) are also active against cysteine proteinases of *Porphyromonas gingivali* – gingipains, responsible for the development of paradentosis. Because the gingipains are not inhibited by cystatins, Kazal inhibitors from pancreas of slaughtered animals, after modification of their molecule, can effectively prevent the development of paradentosis. Chicken liver is a raw material for extraction of low-molecular-mass serine proteinase inhibitors, which are active against trypsin, human plasmin and human cathepsin G (Wilimowska-Pelc et al., 1999).

4.4 Raw Materials of Marine Origin

For a long time, the marine environment has been utilized as an almost endless source of food products that only needed minor attention. However, over-fishing of certain areas and the fear of the disappearance of certain fish species led to severe legislation and international fishing rules. The policies on fishing grounds have become an important aspect within the European Agricultural Policy, which initiated several strikes by the fishermen. The concern for certain fish populations led to the development of fish farms for the food industry using high-tech screening methods for the species and high-tech production techniques. The marine environment can also be seen as a source of raw materials for non-food applications, especially if we consider by-products of the food industry. Several products of marine origin are being used and developed for the industry. One example of the use of raw materials of marine origin is the utilization of chitin, which can be obtained from crab shells. The isolation and use of chitin and its derivatives (chitosans) are discussed in more detail in Chapter 7.

Another example which is well known is the use of agar, a strongly gelling hydrocolloid extracted from seaweed, which is extremely important in microbiology as a medium to grow bacterial and fungal cultures. Agar is the only hydrocolloid that gives gels and that can stand sterilization temperatures, and is completely soluble in boiling water. It also has good synergies with sugars and different hydrocolloids. The sources of agar are limited to a group of red algae (Class Rhodophyta, Family *Phillophoraceae, Gelidiaceae, Gracilariaceae*), although the alkali treatment of seaweeds has made the species *Gracilaria* the most competitive agar producing seaweed. The red-purple seaweeds from which agar is obtained grow in nearly all seas of the world. Agar is characterized by repeating units of D-galactose and 3,6-anhydro-L-galactose, but an exact structural composition cannot be given. Its non-food applications are situated not only in the domain of microbiology but also in the medical (cell culture, immunology) and chromatographic fields (electrophoresis, affinity chromatography) (Matsuhashi, 1998). Agar is also used for the micropropagation of *in vitro* plants and plant tissue culture. The absence of salts, acids and pathogenous agents make the agar an excellent product for botanical use. Further technical uses are in dental applications for dental impression, and in cosmetics and shampoos as thickeners.

Because of some high-tech applications agar can also be seen as a raw material with a high added value.

Another important gelation product for non-food applications is carrageenan. It is a polysaccharide found in *Chondracanthus exasperatus* and is used extensively in food and non-food applications. It is estimated that its production involves the use of 250 000 t of algae per year and generates more than 100×10^6/yr. Although its gelling ability is not as strong as that for agar, it is used in the pharamaceutical industry and in the cosmetics industry. Carrageenan is also used to provide the viscosity necessary for marbling on fabric.

Carrageenan comprises two components, the kappa and the lambda carrageenan, and forms a gel when heated or cooled in the presence of potassium ions. Nowadays much research is being done to grow the seaweed on rocks and other substrates in order to explore the aquaculture of this *C. esperatus*. The large-scale cultivation has been quite successful and it should be possible to have two harvests in one growing season. Further, efforts have been performed to grow the seaweed in tanks filled with seawater and this technique also seems to be useful.

A major advantage of the cultivation of the seaweed is that the harvest can be performed in the same phase of the growth, which makes it possible to have a more uniform end product. After harvesting, the weed is dried and washed with fresh water. The algae are then boiled in water (carrageenan is soluble in hot water) and filtered. Then the remaining water is evaporated, leaving behind the powdered form.

The materials of marine origin are now also intensively studied in order to find new lead structures for the development of new pharmaceutical or biologically active compounds. Hundreds of rare sponges and other marine organisms are being extracted in order to find elusive compounds with a specific pharmacological effect.

4.5   Materials of Microbial Origin Produced from Agricultural Waste Products

4.5.1   Processes

The last decade has witnessed an unprecedented increase in interest both in submerged (SmF) and in solid state fermentation (SSF) for the development of bioprocesses such as remediation and biodegradation of hazardous compounds, biological detoxification of agro-industrial residues, biotransformation of crops and crop residues for nutritional enrichment, production of valuable products such as biologically active secondary meta-bolites, including antibiotics, alkaloids, plant growth factors, etc., enzymes, organic acids, biopesticides, including mycopesticides and bioherbicides, biosurfactants, biofuel and aroma compounds (Hesseltine, 1972).

Solid state fermentation is largely involved in the production of various secondary metabolites of plants and microorganisms. Microbial secondary metabolites are high-value products which are not only produced by a liquid culture, but can also be produced by SSF. Recent data in the literature have shown that different kinds of secondary metabolites can be produced by SSF, e.g. antibiotics, food grade pigments, mycotoxins and alkaloids. Other products are, e.g. various kinds of enzymes, organic acids, flavour compounds and biopesticides. SSF systems exhibit certain advantages over SmF (higher product yields, easier product recovery and reduced energy requirements). Due to the lack of free water, smaller bioreactors can be used for SSF and less effort is required for downstream processing. The quality and properties of the products synthesized by SSF and SmF are sometimes different, for example fungal glucosidase is more thermotolerant and fungal spores have a better quality and viability after storage than those obtained by SmF (Hardin, Mitchell and Howes, 2000).

Mycelial organisms which can grow at high nutrient concentrations near a solid surface are the most important group of microorganisms for SSF. These microorganisms include a large number of fungi and a few actinomycetes. Yeasts and some strains of bacteria are also used in SSF. Filamentous fungi are ideal for SSF due to their hyphal mode of growth and capabilities. The genera of fungi used in SSF processes are: *Aspergillus, Rhizopus, Trichoderma, Penicillium, Chaetomium* and others. Filamentous fungi grow well at low water availability and pH values. These conditions are not suitable for growth of most bacteria and yeast contaminants. Many filamentous fungi can also produce a range of hydrolytic enzymes to degrade macromolecules (starch, cellulose, hemicellulose, lignin) present in solid particles. Amylases and cellulases are the most important enzymes for growth on solid substrates of agricultural origin; however, proteases and lipases are helpful for degradation and penetration. SSF is usually carried out non-aseptically because many fungi grow rapidly and are competitive with other contaminants. In some cases, the process requires sterile conditions, e.g. for slow-growing microorganisms, genetically modified microorganisms and for the production of pharmaceutical compounds and organic acids (Mudgett, 1986; Pandey, Soccol and Mitchell, 2000).

Two types of SSF systems can be distinguished, depending on the nature of the solid phase:

- The first and most commonly used system involves cultivation of microorganisms on a natural solid substrate from agriculture.
- The second system involves cultivation of microorganisms on an inert support impregnated with a defined liquid medium.

SSF on inert supports offer numerous advantages over natural solid substrates:

- improved process control and monitoring;
- enhanced process consistency;
- more reproducible fermentation process (e.g. spores of fungi can be readily extracted without destroying the particles); and
- the recovery process is simpler and the particles can be reused.

4.5.2 Substrates

Solid state fermentation can be carried out on a variety of agricultural residues and can be divided into three main groups, according to the main carbon source: starch, cellulose or hemicellulose and soluble sugars.

Starch is a major reserve polysaccharide in the plant kingdom and is mostly found in the seeds of maize, wheat, barley, oats, rice and sorghum, potato tubers and roots, arrowroot and cassava and in the pith of the sago palm. Starch has great potential as a raw material for the fermentation industry and as an energy source, since glucose can be further converted to other products, such as antibiotics, organic acids, ethanol and others.

Starchy substrates used in SSF processes are: whole rice kernels, pearled barley, oats, cassava, wheat bran, rice bran, cassava meal, corn meal, sweet potato residue, and others. Some of these substrates provide a nutritionally complete medium for growth and product formation, whereas others need nutrient supplements (e.g. peptone, glucose, asparagines). Products from starchy substrates include protein-enrichment feed for animals, organic acids such as citric acid, gluconic acid and γ-linolenic acid, secondary metabolites such as antibiotics, fruity aroma compounds, pigments, etc. Lignocellulosic substrates used in SSF processes are: wheat straw, cracked corn and rice stover, wood, hemp, sugar beet bagasse, sugar cane bagasse, sorghum, and others. Solid substrates with soluble sugars are: grape pomace, sweet sorghum, sugar beets, pineapple, fig, coffee pulp and kiwi. Starchy and lignocellulosic substrates generally require some kind of pre-treatment before use. The pre-treatment may be either mild or drastic, e.g. steaming, pearling, cracking, grinding, ball-milling, sieving and treatment with alkali or sodium chloride. After pre-treatment solid substrates are more susceptible to microbial penetration and modification. During the growth of the microorganism the solid medium is degraded by enzymes (glucoamylase, α-amylase, cellulase, hemicellulase, cellobiase), and as a result some changes will occur in the geometric and physical characteristics of the solid substrate (Gawande and Kamat, 1999).

4.5.3   Products

Citric Acid

Citric acid (CA) (2-hydroxy-1,2,3-propanetricarboxylic acid) is present in several fruits, such as lemon, orange, pineapple, plum, pear and peach. CA is widely used as an edible acidifier due to its high solubility, low toxicity, strong chelating power and pleasant taste. It is applied as a condiment, preservative (beverages and sweets), antioxidant with ascorbic acid (fruit freezing) and pH adjustor (preparation of sweets and fruit jelly). Sodium, potassium and calcium salts of CA are chelating agents and food dyes. CA also plays an important role in the pharmaceutical industry. In the chemical industry, triethyl acetocitrate and tributyl acetocitrate can be used as non-toxic plasticizers to make plastic film. CA is also more and more commonly used as a detergent ingredient, to replace sodium tripolyphosphate which causes eutrophication of surface waters, and therefore proves to be a green chemical derived from agricultural sources. CA is a major organic acid produced during fermentation. It is commercially produced by submerged fermentation of starch or sucrose-based media using the filamentous fungus, *Aspergillus niger*. The surface fermentation using *A. niger* with beet molasses as a raw material is still extensively employed by major manufacturers. Cane molasses are not used in the surface culture because of very low yields. In the surface culture, it is the CA-producing mycelium that is grown on molasses in shallow trays. These are kept in stacks in constant temperature rooms.

In 1952 the Miles company was the first to have adopted the deep-level fermentation method (submerged fermentation) to produce CA on a large scale. In recent years, many countries have invested a lot in research on CA production. As a result, the world production of CA increased rapidly. At the end of 1978, 35 countries were producing CA with a total output of 400 000 t, which in 1995 increased to 500 000 t. The unit price of CA varies in different countries from US$1.2 kg in Europe to US$ 1.9 kg in the United States and the Middle East. The use of the submerged fermentation gives a higher productivity and takes less space and labour. This method gives a possibility to utilize various raw materials, therefore, it has been adopted in many countries.

The raw materials that can be used for citric acid production can be divided into four categories:

1. dissolved pure sugar (sucrose, glucose), refined sugar;
2. starch or other raw materials containing starch;
3. sugar cane or beet molasses; and
4. a raw material containing cellulose, hemicellulose and pectin.

Pure sucrose or starch as a carbon source are easy to control in fermentation production. The other materials such as sugar cane or beet molasses are of different quality, depending on the season and also the refinery. It is therefore necessary to make a selection of molasses with regard to their performance.

Okara (soya-residue) and tofu (soya panner) cellulosic by-products of the soya milk industry have been used for the production of CA by SSF using a cellulosic *A. terreus* and CA-producing *A. niger*. Corn cobs are used as a substrate for CA production by *A. niger* NRLL 2001. This strain produces the highest level of CA (250 g/kg dry matter of cobs at 30 °C after 72 h in the presence of 3% methanol, which significantly effects fungal production of CA). The other solid substrates have been reported as carbon sources for CA production including kiwi fruit peel, apple pomace, wheat bran, fresh kumara (*Impomoea batatas*) and taro (*Colacasia esculenta*).

Citric acid is commercially produced by SSF in Indonesia and Thailand. The kumara and taro are excellent substrates for citrate production by SSF. For example, an overall reactor productivity of 0.46 g CA/kg wet weight kumara/h and a yield of 0.54 g citrate produced/g starch used, are obtained. Furthermore, methanol (3%) stimulated the CA productivity considerably. Citrate has also been produced in Japan by cultivating *A. niger* on cooked vegetable residues in a tray-fermentor. The solid-state method for citrate production is equal or superior to submerged fermentation (Hang and Woodams, 1998).

Gluconic Acid

Gluconic acid is produced in submerged culture with strains of *A. niger*. The raw material is normally glucose, supplemented with a cheap nitrogen source. The fermentation time is approximately 36 h, which is very short for a mould process. Continuous gluconic acid fermentation has been developed in Japan. The volumetric gluconic acid productivity is about twice as high as in the batch process. Gluconic acid can also be produced enzymatically using both free and immobilized glucose oxidase and catalase.

Gibberellins (Gibberellic Acid) and γ-linolenic Acid

Gibberellins are potent plant hormones, predominantly regulating plant growth. The gibberellins are produced by *Gibberella fujikuroi*. The most important product is gibberellin A3 (gibberellic acid), but gibberellins A4 and A7 have also found some commercial applications. The fermentation is carried out in a fed-batch culture. The gibberellic acid is produced in the stationary phase. The SSF technique has been used for gibberellic acid (GA) and γ-linolenic acid (GLA) production. GA is produced by some strains of *G. fujikuroi* or *Fusarium moniliforme*. GA is a very important organic acid, widely used in agriculture. It can be used for stopping seed dormancy for the acceleration of flowering and germination, e.g. in the brewery industry. Maximum GA concentrations of 300 mg GA/kg dry matter and 240 mg/kg dry matter are obtained in SSF using wheat bran or cassava as a substrate respectively. γ-linolenic acid is produced by oleaginous fungi of *Cunninghamella elegans*. The maximum amount of 14.2 mg of GLA/g of dry substrate is produced on a solid substrate composed of a mixture of barley, spent malt grains and peanut oil.

Antibiotics

Antibiotics are produced by various microorganisms in the stationary phase and are active against other microorganisms at low concentrations. Of the more than 6000 known natural microbial compounds with an antibiotic activity, about 150 are produced on a large scale and used in medicine and agriculture, mainly as antibacterial, antifungal or antiviral agents. All commercial antibiotics are produced by microorganisms in a conventional submerged culture. In a few cases, a natural microbial product can be chemically or enzymatically converted into a so-called semi-synthetic antibiotic with superior therapeutic properties. Presently, several antibiotic bioconversions involving single-step reactions are run with industrially immobilized enzymes or cells as biocatalysts (Robinson, Singh and Nigam, 2001).

As an industrial example of the use of enzymes or cells to convert peptide antibiotics into therapeutically useful derivatives, free and immobilized penicillin acylases, producing the penicillin nucleus 6-aminopenicillanic acid (6-APA), are able to synthesize semi-synthetic β-lactam antibiotics such as cephalosporins, nocardicins and monobactams.

Antibiotics are traditionally produced by submerged fermentation. Low yields of antibiotics are obtained in these cultivation systems and therefore extensive downstream processing is required. The production of these secondary metabolites by SSF allows a more concentrated product. Actinomycetes and some strains of bacteria are suitable for the production of such antibiotics as rifamycin B, cefamycin C, penicillin, cyclosporin A, tetracyclines, inturin and oxytetracyclines under SSF conditions.

Rifamycin B is produced from wheat bran by *Amycolatopsis mediterranei* VA18 under SSF conditions. After optimization, the maximum production (rifamycin B, 39 g/kg substrate) is obtained with 90% substrate moisture, which is almost 16-fold higher than that obtained in the submerged culture. Rifamycin B can be converted into rifamycin O or rifamycin S by chemical methods.

Cefamycin C is produced by a variety of strains, e.g. *Nocardia lactamdurans, Streptomyces catteya* or *S. clauverigerus*. Cefemicin C produced under SSF conditions reaches a maximum concentration after day 5 and is more stable than that produced by SmF. Wheat bran is used as a substrate. The addition of cotton seed and corn steep increases yields further.

Today, penicillin (the first β-lactam antibiotic) is produced in submerged culture by high-yielding strains of *Penicillium chrysogenum*. In the early days of fermentation, surface culture with *P. notatum* was employed. With the introduction of submerged culture, *P. chrysogenum* was found to be a better producer giving yields of 300 units/ml. Currently, yields of up to 25 000 units/ml have been reported and even higher yields can be expected. The fermentation product, penicillin G, is used for semi-synthetic penicillins. New strains and media are continually evaluated to increase the yield in the fermentation. For example, under SSF conditions a higher value of penicillin (about 13 mg/l) is produced, while under SmF conditions a maximum value of 9.8 mg/l is obtained. This illustrates the economic advantage of SSF technology.

Cyclosporin A (antifungal peptide) is produced by *F. solani* and *Tolypocladium inflatum*. Antifungal antibiotics are a very small but significant group of drugs and play an important role in the control of mycotic diseases. About 300 species of fungi have been reported to be potentially pathogenic to humans. Approximately 90% of human fungal infections are caused by *Aspergillus, Candida, Cladosporium* and *Trichophyton*. Cyclosporin A is produced from wheat bran by a high-yielding mutant of *T. inflatum* under SSF conditions, reaching a maximum of 12 mg cyclosporin A/kg of wheat bran.

Inturin (cyclic heptapeptide) is an antifungal antibiotic produced by *Bacillus subtilis*. Soya bean curd (known as okara) is a very good substrate for inturin production in SSF by *B. subtilis*. A total inturin production is superior in SSF as compared to that in SmF. Additionally, the inturin produced by SSF has a stronger antibiotic activity, the antibiotic extraction is much simpler and requires less solvent. Maximum inturin production in SSF occurs after day 2 as compared to day 5 in SmF. SSF was found to be 6–8 times more efficient.

Enzymes

Various kinds of enzymes are produced in submerged culture, which is now a favoured process. In fact, production of industrial enzymes in a fermentor is relatively simple, but the product recovery is the main obstacle to higher yields. The loss of 50% of the enzyme activity is often encountered during product recovery. Solid-state methods have been used for the commercial production of cellulase, amylase, pectinase, protease and lipase. The major industrial use of the Koji process is for the production of fungal enzymes. Such processes are widely used in Japan for enzyme manufacture. Commercial production of cellulases from wheat bran by *Trichoderma* sp. has been carried out in SSF. Wheat bran is one of the substrates yielding the highest activities in the solid-state process.

A lot of microorganisms are employed for enzyme production under SSF and SmF conditions such as *A. niger, Rhizopus oligosporus, Aspergillus oryzae, Aspergillus oryzae, T. harzianum, T. viride, P. citrinum, Mucor meihi, B. licheniformis, Rhizopus oryzae, S. cyaneus, Clostridium thermosulfurogenes,* etc. Many filamentous fungi, thermophilic bacteria and thermophilic actinomycetes can produce a wide range of hydrolytic enzymes, e.g. proteases, amylases (α-amylase, glucoamylase), lipases, cellulases, hemi-cellulases and other enzymes including *endo*-β-glucanases, pectinases, chitinases, various α-galactosidases, phenol oxidases (laccases) and pullanases (Viswanathan and Surlikar, 2001).

Microbial phytase is used in order to reduce the environmental loading of phosphorus from animal production facilities. The limiting factors in the use of this enzyme in animal feeds can be overcome by the production of amylase from rice by *A. oryzae*. This process was carried out in different kinds of bioreactors such as tray, packed bed, rotating drum, stirred and air-solid fluidized bioreactors, but high production levels of enzymes were obtained only in a spouted bed bioreactor with intermittent spouting with air.

SSF is a promising technology for commercial enzyme production on natural substrates containing high molecular weight biopolymers, e.g. starch, cellulose, hemicel-lulose, pectin, protein and lipid with lower production costs.

Aroma Compounds

Solid state fermentation has been mostly employed for bioconversion processes. The agro-industrial residues such as cassava bagasse and gigant palm bran are used for aroma compounds by *Kluyveromyces marxianus* in SSF. *T. harzianum* is employed for the production of 6-phenyl-α-pyrone (6-PP), a compound which has a strong coconut-like aroma. The sugar pith bagasse impregnated with liquid media is used in aroma compound production as a support in SSF.

The steam-treated coffee husk, containing some amount of caffeine and tannins, is an ideal substrate for the production of volatile fruity aromas by a fungal culture of *Cerato-cystic fimbrata*. Fungi from the genus *Ceratocystic* produce a wide variety of fruity-like aromas (peach, pineapple, banana, citrus and rose), depending on the strain and culture conditions. Compounds such as acetaldehyde, ethanol, isopropanol, ethyl acetate, ethyl isobutyrate, isoamyl acetate, isobutyl acetate and ethyl-3-hexanoate were identified in the headspace of the culture. A strong odour of banana and pineapple was detected in the culture enriched with glucose and leucine respectively (Soares *et al.*, 2000).

Biopesticides (Biological Control Agents)

Biological control agents (BCAs) are potential alternatives for the chemical fungicides presently used in agriculture in plant disease control. BCAs have been successfully introduced into the market. There are a number of fungal biocontrol products available on the markets in Europe, registered as biopesticides, e.g. Binab-T WP, Harzian 10, Trich-dermin and Bio-Fungus. The use of fungal pathogens and their toxins is now technically feasible and several agents have been tested. Spore production is essential if an effective fungal preparation is to be obtained. For this reason, surface rather than submerged culture is employed, the final product is usually an aqueous spore suspension.

Solid state fermentation has proven to be a useful technology for the production of fungal spores of fungi suitable as biocontrol fungi including *Beauveria*, *Fusarium*, *Metarhizium*, *Trichoderma* and *Coniothyrium*. *Coniothyrium minitans* is an example of a promising fungal BCA of the plant pathogen *Sclerotinia sclerotiorum*, which substantially decreases the yield of many crops. Non-mixed and mixed sterile packed bed reactors can be used for the production of large amounts of *C. minitans* spores (de Vrije *et al.*, 2001).

4.6 Conclusions

Producing industrial and consumer products for non-food applications from agricultural raw materials is in many cases dependent upon a lot of conditions that are more difficult to control than in petrochemical production of consumer goods. Therefore, a very good knowledge of plant species, plant–soil interactions, animals, downstream processing and the type of application is necessary in order to be able to develop an economical process. On the other hand, powerful enzyme systems in plants and animals are available to construct structures that have a high complexity. This tool needs to be exploited in every detail, as in the case of the production of complex pharmaceutical molecules. Further, next to the food application of plants and animals, side products and waste products can be valorized again using the powerful enzymatic systems of bacteria or fungi, or even by chemical modification.

Since agricultural production will always be connected to the use of arable land, and in competition with the production of food, options for production with a high added value will be an important aspect in the planning of industrial processes. The concept of integral valorization, which was treated in the previous chapter, is again well illustrated in the production of raw materials and the downstream processing.

References

Bailey, A.J. and Light, N.D. (1989). *Connective Tissue in Meat and Meat Products*, pp. 1–339. Elsevier Applied Science, London and New York.

Berger, J. (1969). *The World's Major Fibre Crops – their Cultivation and Manuring*, Centre d'etude de l'azote 6, Zürich.

Bisset, N.G. and Wichtl, M. (2001). Herbal drugs and phytopharmaceuticals, *Wissenschaftliche Verlagsgesellschaft*, Stuttgart, CRC Press, Boca Raton, Florida.

Bócsa, I., Karus, M. and Lohmeier, D. (2000). Der Hanfanbau – Botanik, Sorten, Anbau und Ernte, Märkte und Produktlinien, vollständig überarbeitete und ergänzte 2. Auflage; Landwirtschaftsverlag GmbH Münster-Hiltrup.

Council Decision 2000/766/EC (2000). *Official J.*, **306**, 32.

Dachler, M. and Pelzmann, H. (1999). Arznei- und Gewürzpflanzen, Österreichischer Agrarverlag, Wien.

de Vrije., T., Buitelaar, R.M., Bruckner, S., Dissevelt, M., Durad, A., Gerlagh, M., Jones., E.E., Luth, P., Costra, J., Ravensberg, W.J., Renaud, R., Rinsema, A., Weber, F.J. and Whipps, J.M. (2001). The fungal biocontrol agent *coniothyrium minitans*: production by solid-state fermentation, application and marketing. *Appl. Microbiol. Biotechnol.*, **56**(1–2), 58–68.

FAO (2002). FAO – Statistics – Primary production, fibre plants, cotton, sisal, jute, hemp, flax, http://apps.fao.org/form?collection = Production. Crops. Primary & Domain.

Gawande, P.V. and Kamat, M.Y. (1999). Production of *Aspergillus xylanase* by lignocellulosic waste fermentation and its application. *J. Appl. Microbiol.*, **87**, 511–519.

Goldstrand, R.E. (1988). In: *Edible Meat By-Products, Advances in Meat Research* (eds A.M. Pearson and T.R. Dutson), **5**, pp. 1–13. Elsevier Applied Science, London and New York.

Hang, Y.D. and Woodams, E.E. (1998). Production of citric acid from corncobs by *Asp. niger*. *Bioresource Technol.*, **65**, 251–253.

Hardin, M.T., Mitchell, D.A. and Howes, T. (2000). Approach to designing rotating drum bioreactors for solid-state fermentation on the basis of dimensionless design factors. *Biotechnol. Bioeng.*, **67**, 274–282.

Hesseltine, C.W. (1972). Biotechnology report: solid-state fermentations. *Biotechnol. Bioeng.*, **14**, 517–532.

Knipe, C.L. (1988). In: *Edible Meat By-Products, Advances in Meat Research*, Vol. 5, 5 (eds A.M. Pearson and T.R. Dutson), **5**(5), pp. 147–165. Elsevier Applied Science, London and New York.

Matsuhashi, T. (1998). Agar. In: *Polysaccharides, Structural Diversity and Functional Versatility* (ed. S. Dumitriu), pp. 335–375. Marcel Dekker Inc., New York.

Marquard, R. and Siebenborn, S. (1999). The red which came from the root, German research 1/2000, 18–20.

Marquard, R. and Kroth, E. (2001). Anbau und Qualitätsanforderungen ausgewählter Arzneipflanzen, *AgriMedia*, Bergen Dümme.

Mudgett, R.E. (1986). Solid state fermentations, In: *Manual of Industrial Microbiology and Biotechnology* (eds A.L. Demain and N.A. Solomon), pp. 66–81. Am. Soc. Microbiol., Washington.

Ockerman, H.W. and Hansen, C.L. (2000). In: *Animal By-Products Processing and Utilization*, pp. 1–354, CRC Press, New York.

Pandey, A., Soccol, C.R. and Mitchell, D. (2000). New developments in solid state fermentation: I-bioprocesses and products. *Process Biochem.*, **35**, 1153–1169.

Proposal for a regulation of the European Parliament and the Council laying down the health rules concerning animal by-products not intended for human consumption. 2000/0259(COD).

Robinson, T., Singh, D. and Nigam, P. (2001). Solid-state fermentation: a promising microbial technology for secondary metabolite production. *Appl. Microbiol. Biotechnol.*, **55**(3), 284–289.

Scheer-Triebel, M. and Léon, J. (2000). Industriefaser – Qualitätsbeschreibung und pflanzenbauliche Beeinflussungsmöglichkeiten bei Faserpflanzen: ein Literaturreview. *Pflanzenbauwissenschaften*, **4**, 26–41. Verlag Eugen Ulmer GmbH & Co., Stuttgart.

Siebenborn, S. (2001). *Untersuchungen zur Inkulturnahme und Qualitätsverbesserung von Rubia tinctorum L.*; Shaker, Aachen.

Soares, M., Christen, P., Pandey, A., Raimbault, M. and Soccol, C.R. (2000). A novel approach for the production of natural aroma compounds using agro-industrial residue. *Bioprocess Engin.*, **23**, 695–699.

Viswanathan, P. and Surlikar, M.R. (2001). Production of α-amylase with *Aspergillus flavus* on Amaranthus grains by solid-state fermentation. *J. Basic Microbiol.*, **41**(1), 57–64.

Wagner, H. (1999). *Pharmazeutische Biologie 2 – Drogen und ihre Inhaltsstoffe*, Gustav Fischer Verlag, Stuttgart, New York.

Wilimowska-Pelc, A., Olczak, M., Olichwier, Z., Gładysz, M. and Wilusz, T. (1999). *Comp. Biochem. Physiol.*, **B 124**, 281–288.

5

Energy from Renewable Resources (Bio-Energy)

Mehrdad Arshadi

In co-authorship with *Elias T. Nerantzis, Paavo Pelkonen, Dominik Röser and Liisa Tahvanainen*

5.1 Introduction

Bio-energy is probably the most well-known aspect of the use of renewable resources. It has received a lot of attention in the political arena and in the international media because of its relation with environmental problems (global warming, CO_2 emission, green energy labels, etc). Further, the discussion of the use of nuclear energy is connected to the alternative energy sources, of which renewable resources are promising.

In this chapter an overview of the broad area of bio-energy will be given, including:

- The different bio-energy sources
- Solid fuels and upgraded fuels
- Agro-forestry techniques and implications
- The development of new energy crops.

However, the political and technical discussion of the different kinds of crops that may potentially meet some of the world's energy demands cannot be covered completely in this chapter.

Renewable Bioresources: Scope and Modification for Non-food Applications. Edited by C.V. Stevens and R. Verhé
© 2004 John Wiley & Sons, Ltd   ISBNs: 0-470-85446-4 (HB); 0-470-85447-2 (PB)

5.2   Renewability in Energy Production and Use

Renewable energy sources (RES) have a long tradition. Long before the discovery of coal and oil, renewable biomass resources were used for construction, clothing and energy. Wind and hydropower as energy sources have been used in agriculture, and wood and peat were common resources for thermal energy. With the advancement of industrialization the demand for energy rose leading to an increased use of oil and coal. The introduction of nuclear power in the 1950s further developed the energy market. By the year 1973, the share of nuclear power in total primary energy supply in the world was 0.9% and had risen to 6.8% by the year 2000.

The need for more research into alternative energy sources became obvious during the energy crisis in the 1970s, when the shortage of energy showed how vulnerable the economy was. In addition, predictions of oil prices reaching US$35–50 per barrel in the 1980s led to further efforts. An excess of raw materials, a higher energy demand, political considerations and pressures from international environmental non-governmental organizations eventually opened the way for national policies towards biomass energy production, with economic and strategic issues as driving forces (Hall and Coombs, 1987).

At present, the discussion is dominated by increasing CO_2 and other greenhouse gas emissions and the resulting climate change. This development has given renewable energy supporters strong arguments since renewable energy systems are considered neutral with regard to CO_2 emissions.

The contribution of RES today represents 10.8% of the total primary energy production in the European Union (EU). The more intensive use of energy from biomass resulted from the growth of combined heat and power (CHP) generation mainly in the northern countries, as well as the direct use in the domestic sector. It showed an accelerating progression since 1990 to reach 7% of total primary production in 1997. Hydroelectricity and wind energy have remained relatively stable since 1995 and represent 2.8% of the primary production.

The 1997 White Paper of the European Communities on Renewables aims for an ambitious and optimistic target. The Paper sets a target of 12% of the EU's gross inland energy consumption by 2010, coming from RES. It is planned to increase the use of RES to 12% by 2010 mainly by encouraging a large increase in the use of wind energy, photovoltaics and energy from biomass. For example, the use of wind is expected to expand by nearly 20% by the year 2010 and by 7% after 2010. It was proposed that biomass energy in total in the EU could contribute an additional 3.8 EJ annually by 2010, compared to the current contribution of about 2.2 EJ p.a. This challenging target can be met only with a significant amount of financial and political support from member states and other regions of the EU (European Commission, 1999).

Additional benefits offered are that RES are typically decentralized and thus promote employment opportunities in rural areas and, therefore, support the EU policy on rural development. Furthermore, there is potential to reduce a country's dependency on imported energy carriers (and the related improvement of the balance of trade). In the long run, Renewable Energy Systems contribute to diversification as well as to the independence and security of EU energy supply, better waste control, and potentially benign effects with regard to biodiversity, desertification and an increase of the recreational value of natural areas. If issues related to the practical exploitation are considered wisely, then there is the potential to significantly improve sustainable development.

5.3 Role of Renewable Energy Systems in Climate Change

In the past few decades, there has been an increase in concern for nature and for the environmental effects of energy use, improving the attitude towards Renewable Energy Systems. A practical result was the Kyoto Protocol in 1997, which has already had an immense influence on present bio-energy policy and will influence policies in the future. RES offer the possibility to reduce greenhouse gases by switching from the use of fossil fuels.

Today, 80% of the greenhouse emissions come from energy use. Under the Kyoto Protocol, the EU is required to cut its combined emissions of the six greenhouse gases to 8% below their 1990 level by the 2008–2012 period. Renewable Energy Systems play an important role in achieving that commitment, since the conversion processes in Renewable Energy Systems result in smaller amounts of pollutant emissions in the form of CO_2, SO_2 and NO_x.

Another example of a cost-effective method to significantly reduce CO_2 emissions in the future is to replace power generation based on coal with co-generation of electricity with heat from biomass. Solid, liquid and gaseous bio-fuels have the potential to replace fossil fuels in almost every application if sustainable production and efficient conversion of bio-fuels is assured. Today's major sources of bio-fuels come from forest and agricultural products. In future, energy plantations have the potential to provide additional sources, which would also open new opportunities for agriculture and forestry on the energy market.

If biological sources and sinks would be included in the accounting of national greenhouse gas emissions, biomass would offer great potential through the long-term storage of carbon in forests or in wooden products. Biomass can be considered as part of a closed carbon cycle. The mass of biospheric carbon involved in the global carbon cycle provides a scale for the potential of biomass mitigation options. Through the combustion of fossil fuels 6 Gigatonnes of Carbon (GtC) are released into the atmosphere each year. The net amount of carbon that is taken up from and released into the atmosphere by terrestrial plants amounts to about 60 GtC annually (corresponding to a gross energy content of about 2100 EJ p.a., including biomass), whereas an estimated 600 GtC is stored in terrestrial living biomass.

Without the additional land use for biomass there are a number of options for improved use of existing biomass resources for energy. Examples are the use of residues from forestry and agriculture, the food processing industry and the biomass fraction of municipal solid waste (paper, landfill gas, disposed wood products). It is estimated that in the EU over 250×10^6 t of municipal waste are produced each year and more than 850×10^6 t of industrial waste. A large fraction of the globally available biomass residues (representing a potential for about 40% of the present energy use of 406 EJ p.a.) could be available for bio-energy. The resource size of recoverable crop, forest and dung residues has been estimated to offer a yet untapped supply potential in the range of 40 EJ p.a., which could meet about 10% of the present global primary energy demand.

Providing residues for energy use is not the only advantage of forest products, wood is also widely used for long-life products, with a CO_2 mitigation benefit. First, it can be used as a substitute for more energy-intensive products (e.g. concrete and steel), which leads to an indirect replacement of fossil fuels. Furthermore, the stock of carbon in wood products can be increased, and finally wood products can be used as energy sources at the end of their life cycle – contributing to the displacement of fossil fuel (IEA, 1998).

5.4   Renewable Energy Sources

5.4.1   Biomass and Waste

All organic matter is known as biomass, and the energy released from biomass when it is eaten, burnt or converted into fuels is called biomass energy. Biomass energy technologies produce an array of products including electricity, liquid, solid and gaseous fuels, heat and chemicals. Biomass and waste are the only renewable forms of energy that compete directly with fossil fuels because they are both solid fuels and they share similar conversion processes.

Biogas

Biogas is the product of organic material (solid and/or liquid including garden waste, manure, etc.) decomposition (anaerobic digestion), composed mainly of methane and carbon dioxide. Biogas is produced in large quantities in landfill sites, and it has potential as a fuel to generate electricity or to provide process heat.

At the end of the 1990s biogas energy production amounted to approximately 65 000 TJ annually in the EU, predominantly from sewage sludge gas and landfill gas. Electricity generation from biogas was 3200 GWh, primarily from landfill gas, whereas heat production amounted to about 18 000 TJ.

Solid Bio-Fuels

Solid bio-fuels are solid RES from living organisms. They are to be distinguished from solid fossil fuels which are also of biological origin but non-renewable. Solid bio-fuels include: wood, straw, energy crops and organic wastes.

Liquid Bio-Fuels

Liquid bio-fuels are liquid RES converted from organic material. Liquid bio-fuels are transport fuels, primarily bio-diesel and bio-ethanol/ETBE (ethyl tertiary butyl ether), which are processed from agricultural crops and other renewable feedstocks. Other liquid bio-fuels are bio-methanol/MTBE (methyl tertiary butyl ether) and bio-oil or pyrolysis oil. Primary energy production from liquid bio-fuels has increased and attained 17 000 TJ in the late 1990s in the EU. France, Germany and Austria are the most important contributors.

Ethanol and methanol are two different liquid bio-fuels which can be synthesized by biomass. These alcohols can be prepared from agricultural residues, wood and many other industrial wastes. Although the price of bio-ethanol is still not competitive with fossil fuels, a breakthrough in the use of cellulose and hemicellulose to produce fermentable sugars may lead to a major price decrease so that it does become competitive. This aspect is one of the real challenges for the research in renewable resources.

Ethanol can replace gasoline and diesel oil in both personal cars and cargo transports. There are even some advantages of using ethanol instead of fossil fuels, for example (Månsson, 1998):

- A relatively lower toxicity compared to gasoline.
- Carbon dioxide released from the combustion of ethanol has no net effect on the environment.
- It can be added to gasoline (5–10%) and be used in cars without any modification of the engine.

Today, it is also possible to replace at least 5% of the fossil fuels by ethanol in motor vehicles without any extra costs, thereby reducing the emission of CO_2 into the atmosphere significantly.

Energy Crops

Energy crops are selected plants cultivated to provide biomass that can be used as a fuel or be converted into other fuels or energy products. In the 1990s, biomass/wastes were the most important renewable energy source in the EU. They contributed about 2 000 000 TJ of primary energy production, amounting to 64% of the total RES energy production. Heat production reached 1 800 000 TJ and electricity production, 27 200 TWh. France is the largest producer of energy from biomass/wastes, accounting for 22% of total energy production. The country producing the most electricity from biomass/wastes is Finland, with approximately 7000 GWh.

5.4.2 Wind

Wind, which is a form of solar energy, is a low-density source of power. The quantity of energy developed is the product of the density of the air and the wind speed. Energy can be extracted from the wind by transferring the momentum to rotor blades. The energy is then concentrated into a single rotating shaft. The power of the shaft is used in many different ways – for example, large modern turbines convert it into electricity. Wind can also be used for other purposes like pumping water or grinding grain. Wind turbines operate together in wind farms to produce electricity for utilities. Small turbines can be used by homeowners and remote villages for meeting local energy needs.

The installed capacity of wind energy in the EU amounted to 4500 MW in the late 1990s. That represented 7300 GWh of electricity. Since 1989, installed capacity has increased more than ten times. The biggest share, 43%, of this capacity is installed in Germany. In Denmark, the share of wind energy amounts to 25%.

5.4.3 Geothermal

Geothermal energy originates from the natural heat of the earth. This heat is stored in rock and water within the earth and can be extracted by drilling wells to tap anomalous concentrations of heat at depths shallow enough to be economically feasible. Low enthalpy resources (50–150 °C) can be used for heating purposes: large base load demands such as district heating, horticulture and recreational uses such as swimming pools. Medium and high enthalpy resources (>150 °C) are used for electricity production.

Electricity production and installed capacity of geothermal power plants at the end of the 1990s were around 4000 GWh, which is an increase of 24.7% in generation and 8.5% in capacity, since 1989. Primary production from geothermal electricity was 2000 ktoe (tonnes of oil equivalent), which represents 3% of the total RES primary energy. Due to high enthalpy geothermal resources, electricity generation from geothermal energy is almost exclusively found in Italy (3905 GWh).

As mentioned above, geothermal energy has a different use as well. The direct end-use of low enthalpy geothermal heat is widely spread across Europe and serves mainly in district heating and agriculture.

5.4.4 Hydropower

Hydropower in general is produced from the movement of a mass of water: streams; rising and falling of tides through lunar and solar gravitation; wave energy; and energy of sea currents.

Different types of hydropower are:

Impoundment Typically, a large hydropower system uses a dam to store river water in a reservoir. The water may be released either to meet changing electricity needs or to maintain a constant reservoir level.

Diversion (sometimes called run-off river) A facility that channels a portion of a river through a canal or a penstock. It may not require the use of a dam.

Pumped Storage When the demand for electricity is low, a pumped storage facility stores energy by pumping water from a lower reservoir to an upper reservoir. During periods of high electrical demand, the water is released back to the lower reservoir to generate electricity.

Hydropower is the second largest renewable resource in the EU in terms of primary energy production, accounting for almost 32% of the total RES energy production at the end of the 1990s. By the end of 1997, installed capacity reached 93 395 MW. The growth potential in the EU has already been exploited, so that only limited growth can be realized in the future.

5.4.5 Solar Energy

Solar technologies use the sun's energy and light to provide heat, light, hot water, electricity and even cooling, for homes, businesses and industry. This energy comes from processes called *solar heating*, *solar water heating*, *photovoltaic energy* (converting sunlight directly into electricity) and *solar thermal electric power* (the sun's energy is concentrated to heat water and produce steam, which is used to produce electricity). Conversion of light to heat can be achieved through passive systems or active systems (mechanically transferring heat by means of a working fluid such as oil, water or air). Total installed surface of solar collectors was close to $8\,000\,000\,m^2$ in the late 1990s. Primary energy production amounted to 320 ktoe, which was 0.4% of the total RES primary energy in the EU.

5.4.6 Photovoltaic Energy

The sun's energy can also be made directly into electricity using *photovoltaic (PV) cells*, sometimes called *solar cells*. PV cells make electricity without moving, making noise or polluting. They are also used in calculators and watches. They provide power for satellites, electric lights and small electrical appliances such as radios. PV cells are increasingly used to provide electricity for homes, villages and businesses. Some electric utility companies are building PV systems into their power supply networks. One disadvantage of PV cells is that the installation costs are comparatively high. Installed capacity of PV panels was around 50 300 kWp. Electricity generation has risen by factor ten from 1989 to 1997. Primary energy from PV panels was 3.5 ktoe.

5.4.7 Peat

The definition of peat is 'a partially decomposed organic matter that has accumulated in a moist environment'.

5.4.8 Hydrogen

Being the third most abundant element on the earth's surface, hydrogen is primarily found in water and organic compounds. In general, it is produced from hydrocarbons or water. In the case of hydrogen being burned as a fuel or transformed into electricity, it reacts with oxygen to form water again.

Through the application of heat, hydrogen can be produced from natural gas, coal, gasoline, methanol or biomass. Another option is the production of hydrogen through photosynthesis by bacteria or algae. Water can also be split into hydrogen and oxygen utilizing electricity.

A great number of applications of hydrogen in fuel and energy usage are known. Some common applications are the empowerment of vehicles, turbines or fuel cells for electricity production or heat generation in small-scale heating systems (Eurostat, 2000).

How Fuel Cells Work?

Fuel cells can be utilized to produce clean energy from hydrogen. A fuel cell has two electrodes, a negative electrode (anode) and a positive electrode (cathode), that are encased by an electrolyte. During the process, hydrogen is fed to the anode and at the same time oxygen is fed to the cathode. The hydrogen atoms, activated by a catalyst, are separated into protons and electrons, which take different paths to the cathode. The movement of electrons through an external circuit creates a flow of electricity. The protons travel through the electrolyte to the cathode where they again combine with oxygen and the electrons to generate water and heat.

A number of advantages are associated with hydrogen utilization. The most important one is that there is no pollution associated with the use of hydrogen. Even biomass hydrogen originates from water and oxidizes back to water. The fuel cell converts the chemical energy directly into electricity with greater efficiency than any other current power system. Furthermore, the hydrogen systems operate nearly silently and can range from very small to large fuel cells according to the needs of the user.

Europe has been a driving force in promoting the wider use of hydrogen. There have been efforts, starting in 1991, to make resources available to research into the use of hydrogen in various end-use technologies. Nowadays the momentum from those early efforts has been replaced by private initiatives, especially in Germany.

Production of Hydrogen from Microalgae – A Case Study

The term 'microalgae' is used in biotechnology and includes the group of cyanobacteria, which are prokaryotic microorganisms, and the unicellular green algae, which are eukaryotic. Both categories of these organisms share the same property: they are microscopic and unicellular. The growth of microalgae is much faster than that of plants, they also possess higher photosynthetic capability per volume and their cultivation is possible in seas and deserts. Besides their capability to produce pure hydrogen, the microalgae can also produce other bioactive compounds, as well as food or food ingredients and a plethora of useful metabolites.

Hydrogen on the other hand is an environmentally clean energy source due to its properties in its production and utilization. In fact, hydrogen is renewable and does not involve the production of CO_2 during its combustion. It liberates large amounts of energy per unit weight and is converted to electricity by fuel cells. There are two ways to produce hydrogen:

1. by using a photoelectrochemical synthesis; and
2. biologically.

The production of hydrogen biologically presents a lot of advantages over the photoelectrochemical process. The biological process requires the use of a simple photobioreactor, which is a bioreactor illuminated by a light source. This light source can either be artificial or solar. The photobioreactor can be easily constructed and has low energy input requirements. In contrast, the photoelectrochemical process requires high electron inputs.

The biological processes are preferred because although they exhibit low conversion efficiencies, they present low initial investment costs and low energy inputs.

The other advantage is that the biological processes consist entirely of clean technologies:

- The biological process of hydrogen production operates without causing any environmental pollution problem.
- It contributes to the reduction of CO_2, a cause of the greenhouse effect.

To conclude this overview, although large amounts of solar energy are irradiated to the earth's surface – approximately 5.7×10^{24} J/yr, which is about 10 000 times more than the total energy consumed by human beings – solar energy is difficult to be captured and utilized.

The effective energy concentration (energy/unit area) of solar energy at any one point on the earth's surface is only 1 kW/m^2. This low efficiency to captivate the available energy and consequently the low conversion rates by the existing systems are the main reasons why the use of solar energy is costly to the point where it is not economically attractive. Any system that can contribute to the increased utilization of this vast amount of solar energy is considered as a candidate for further investigation.

Therefore, the first step to utilize the solar energy is to accumulate it efficiently and then use it. But collecting the energy with the available technology, to PV cells or to solar panels for thermal energy, collectors require a large capture area. Since the energy concentration is small, we need very large surface areas to collect a sufficient amount of energy for efficient systems. On the other hand, the photosynthetic systems require very simple means to produce highly sophisticated products, which humans could imitate only if decidedly complicated equipment and methods were used.

5.5 Solid Fuels from Renewable Resources

5.5.1 Wood-based Fuels

Wood-based fuels represent a large share of fuels from renewable energy due to the long history and tradition of utilizing wood sources. The wide variety and availability of wood fuels as well as environmental considerations are also in favour of the fuel. Wood fuels support the local economy, representing a decentralized energy system, and create a market for small diameter roundwood and logging residues that otherwise would remain commercially unutilized. At the same time, jobs are created for rural inhabitants. There are also a number of silvicultural advantages in first thinnings and afforestations. On the other hand, there is an ongoing debate on how much the nutrient out-take affects future increment and growth.

Forests and Trees

Forests are the main source of wood-based fuels. Forest biomass is defined by Young (1980) as 'the accumulated mass, above and below ground, of the wood, bark, and leaves of living and dead wood shrub and tree species'. In Europe, the tree types can be divided into hardwood and softwood. The main components of forest biomass are carbohydrates and lignin, produced by the trees during the process of photosynthesis using carbon dioxide and water and at the same time storing solar energy.

The forest industry, as the largest consumer of wood, mainly utilizes stems that meet certain standardized dimensions and quality requirements. The rest is a potential 'leftover' for the wood fuel energy sector. There are three types of residues. First, the silvicultural residues that arise during pre-commercial thinning of young stands. Logging residues are the second type and they arise during commercial harvesting operations. The tops and branches also referred to as residual forest biomass are very commonly used for wood fuels. The last source of residues is the industrial process residues such as bark, sawdust, slabs, cores and lignin-based black liquor from the primary forest industries. Usually they are called primary process residues. Secondary process residues come from secondary industries, e.g. furniture manufacturing (Hakkila and Parikka, 2002).

Biomass components of a tree

Biomass distribution naturally varies from tree to tree. Every tree is a product of the interaction between the genetic structure of the tree and the environment. To get an accurate estimation of the potential of biomass in the future, it is essential to have knowledge of how the biomass is distributed in tree components. The amount of branches on a tree and therefore the ratio of crown mass to stem mass are an important measure. The main factor affecting this ratio is stand density. The composition and quantity of the crown mass differs from species to species and depends largely on, e.g. anthropogenic and geographic factors. One of the most significant differences between hardwoods and conifers is the foliage component. In boreal regions, firs (*Abies* spp.) and spruce (*Picea* spp.), being more shade tolerant, have a much greater share of foliage and crown mass compared to pines (*Pinus* spp.).

Short Rotation Forestry

Since the 1960s, short rotation forestry (SRF) has become an interesting topic due to the high fossil fuel prices and an increasing shortage of pulpwood. This global problem of creating new sources of energy from non-fossil fuels was especially taken on by Swedish scientists, and they have succeeded in moving this approach from the status of mere speculation to being a part of the national energy plans (Dawson *et al.*, 1996). Today, Sweden has a leading role in the SRF sector in Europe. In 2002, there were around 18 000 ha of Salix plantations and annually approximately 2000 ha were harvested to provide fuel for 38 heating plants.

In SRF, a closely spaced stand of fast-growing tree species is established. The rotation period varies between 1-year cutting cycles, in some extreme cases such as poplar or willow grown under the so-called 'wood grass' regimes, and 20-year rotations of southern pines. The differences are due to factors such as climate, soil type and management objectives. Other characteristics of SRF are intensive cultivation practices, repeated harvesting, using short cutting cycles as well as making use of sprouts and suckers to regenerate for the following crops. Especially, appropriate spacing and rotation length are crucial when optimizing the biomass over time. Typical species used for SRF in Europe are found among the genera *Salix, Populus, Castanea, Quercus, Alnus, Betula* and *Eucalyptus* and *Robinia* in southern Europe.

There are demands and characteristics that species have to meet in order to make SRF plantations profitable. They include rapid growth, high re-sprouting capacity after harvesting and good shoot-morphology. SRF species have to compete with fast-growing weeds, a reason for the need to be fast growing especially in the early years of the rotation. Another aspect is the resistance to fungi and insects as well as frost resistance, at least in the northern parts of Europe.

Site factors are of great importance to make the plantation profitable. Deep, fertile and well-drained soils are necessary to maximize outputs. There has to be an adequate supply of water during the growing season and the groundwater table should be relatively high. Poorly drained and waterlogged sites have a lower productivity and are more susceptible to soil compaction. The pH of the soil should be at least 5 in the case of willows. Poplars, in comparison, require at least a pH of 6–7 for optimal growth. The site should also be on flat land and have a history of farming (Mitchell, Stevens and Watters, 1999).

Also, the physical soil properties are an important factor. For SRF plantations in general, silt and loam soils provide a good supply of water and nutrients. Sandy soils are not suitable due to relatively poor supply of water and nutrients. Clay lies in between the two, usually having ample supply of water and nutrients, but creating the so-called 'hard pan' during drought. The aeration is also poor in clay soils. Nutrient cycling and degradation in an SRF stand is fast.

There are a number of problems associated with the coppicing system. The main concern today is the optimization of coppicing, meaning a possibility of successive harvests. Successive harvests seem to be connected with stump mortality and the survival of the root system. One of the more serious threats to durable production of biomass willows is the development of leaf rust, *Melampsora* spp., a disease not treatable with a fungicide. Frost injury is also a serious obstacle to expanding the use of fast-growing willows in the countries of the northern Temperate Zone.

Salix species have a high capacity to accumulate heavy metals such as cadmium (Cd) and zinc (Zn) in high concentrations. SRF plantations can thus act as a vegetation filter for wastewater treatment or sewage sludge by filtering out unwanted components of the water. At the same time, the wastewater is working as a fertilizer. This characteristic of the Salix species can also be utilized for soil purification from heavy metals and cadmium, or to stop the metals from reaching the groundwater. It is not recommended to use ashes as a fertilizer since Salix stems have high Cd concentrations.

5.5.2 Agro Fuels

Reed Canary-Grass

Reed Canary-grass (*Phalaris arundinacea* L.) is a perennial, about 2-m tall grass with sturdy, upright straw, broad leaves and long panicle. The stem is partially surrounded by a sheath and is divided by nodes into shorter segments called internodes. This grass naturally grows in Europe, Asia and North America, especially in wet and humus rich soil. Reed Canary-grass is a new promising bio-energy crop (Hadders and Olsson, 1997).

Energy efficiency (Energy output/input ratio) for Reed Canary-grass is around 14 in studies conducted in Sweden, which means that 14 times more energy is produced from Reed Canary-grass than the energy consumed to produce the grass (Venendaal, Jørgensen and Foster, 1996). The mean harvest level of Reed Canary-grass is 6–8 t dry material per ha and the mean caloric heat value is 4.9 MWh/t of dry material (for salix this is 4.5 MWh/t of dry material) (Larsson, 2003).

Miscanthus

Miscanthus is a perennial grass, which is adapted to warmer climates. The most northern regions to grow this crop are the southern part of Sweden, Denmark, southern UK and Ireland. Growing Miscanthus for pulp and energy purposes was performed for the first time in the late 1960s. Miscanthus is most often harvested in spring when it is dry. Two main barriers have prohibited the development of Miscanthus production. The first is that Miscanthus has a low first winter survival, particularly in the northern part of EU. The second is the costs of establishment, which are much higher than that for other perennial crops like Salix or Reed Canary-grass. The energy output/input ratio for Miscanthus in Germany was 15–20 and in Denmark, 18 (Venendaal, Jørgensen and Foster, 1996).

Switch-Grass

Switch-grass (*Panicum virgatum*) is a perennial C_4 species, which can be used as an energy crop. This grass is grown in Central USA as fodder crop or for soil conservation. As Switch-grass forms fertile seeds, it has the advantage over, for example, Miscanthus that it can be established at low costs.

Rapeseed Oil

Oil-seed rape can be used to produce rape oil. The oil is obtained by pressing of the rapeseeds or by extraction. After reaction (transesterification) of the oil with methanol and a base, the product, which is called Rape Methyl Ester (RME), can be used as an alternative for fossil diesel oil in conventional diesel engines. The by-products are oil cake and glycerine, which are respectively used as cattle feed (protein rich) and as feedstock for sectors of the chemical industry such as cosmetics (Saha and Woodward, 1997). Rapeseed oil production is a commercial activity in many countries such as Denmark, Ireland, Germany, Belgium, France, Finland, Austria and the UK. The utilization of RME, e.g. in Germany (and also in France and Austria) is favoured by tax exemption. In 1996 the production costs for RME (1.17–1.4 €/l) were higher than that for fossil diesel oil (0.57 €/l). The Energy output/input ratio for rape bio-diesel is between 1 and 6 (results from different countries in Europe) (Venendaal, Jørgensen and Foster, 1996). However, the EU has some restrictions about how many hectares each country is allowed to use for the cultivation of oil crops, e.g. in Sweden it is 120 000 ha, which gives around 80 000–100 000 m^3 rape oil or RME (Månsson, 1998).

5.5.3 Peat Fuels

Biological Composition

The decay of organic material into peat is called the humification process. Microorganisms are partly involved in this process and occur in places where there is an abundance of water. The resulting lack of oxygen causes the partial decomposition of the organic matter (Lappalainen and Zurek, 1996).

Environmental Aspects

Peat and climate

The debate as to whether or not the burning of peat as a fuel has contributed to the so-called greenhouse effect is still a subject of discussion. Complex relationships between carbon binding in peat-lands and emissions of methane and carbon dioxide, as well as the impact of, e.g. ditching are being analyzed around the world. A more detailed discussion of the development and carbon balance of peat-lands can be found, e.g. in Korhola and Tolonen (1996). Peat is clearly a slowly renewable resource. Therefore to keep peat utilization and peat accumulation in balance, with respect to carbon dioxide emissions and mitigation, the area utilized must be rather small compared to total peat-land area. On the other hand, virgin peat-lands are emitting methane, whose global warming potential is far higher than that of carbon dioxide. In Finland, for example, energy taxes on peat fuel are between those on wood (no taxes) and fossil fuels. It should be noted that without anthropogenic interference, there is a continuous paludification process on mineral soils adjacent to existing peat-lands. Countries other than Sweden use peat for combustion, and many others use it for other purposes, for example as a cultivation substrate or a soil improvement agent.

5.5.4 Waste Fuels

Household Waste

Extracting energy from waste is of great importance for the handling of waste. Household waste can be used for energy production. Co-combustion of bio-fuel and the dry fraction of household waste might be an alternative in the future in many rural areas (Olsson *et al.*, 2001).

Industrial Waste

Many industrial processes result in large amounts of waste. Many industries using timber, i.e. sawmills, packaging materials in the form of paper and cardboard, produce tonnes of trash. These, together with huge amounts of municipal waste and logging residues are potential bio-fuels from which only small amounts are being utilized as bio-fuels today.

5.6   Properties

5.6.1   Wood-based Fuels

Chemical Properties of Wood

Photosynthesis needs carbon dioxide (CO_2), which the plant takes from the atmosphere, and water (H_2O) which comes from the soil. During the process, carbon dioxide and water are transformed into simple sugars and monosaccharides. Carbon (48–52%), oxygen (38–42%) and hydrogen (6–6.5%) represent the principal elements of biomass. Nitrogen, mineral elements and ash are also present in small quantities, making up about 0.5–5% of the total. Cellulose, hemicellulose and lignin are products of a further conversion from monosaccharides. The fuel value of biomass products depends on the proportional distribution of these elements.

Carbohydrates and lignin are made up of different combinations of carbon, oxygen and hydrogen, and they are the primary cell wall constituents of wood. The carbohydrate portion of biomass is made up of cellulose and hemicellulose and is also known as holocellulose. The main component of cellulose is glucose ($C_6H_{12}O_6$), which is produced by the tree during the process of photosynthesis. The difference between cellulose and hemicellulose is that cellulose is composed exclusively of glucose units, whereas a great number of monosaccharides are part of hemicellulose. In the case of softwood, galacto-glucomannans are the main part of hemicelluloses, and xylans have the greatest share in hardwood hemicelluloses (Hakkila, 1989).

Lignin adds rigidity and increased stiffness to cell walls in the stem, roots and branches. Lignin is the one cell wall component that differentiates wood from other cellulosic materials found in plants. In normal wood, lignin accounts for about 20–30%, varying from species to species. In many countries lignin is used for energy production, as a main component of black liquor (produced during paper production, p. 166).

Apart from the three main components, there are also a number of other extraneous components that are found in the woody cell walls and cell lumina. These are terpenes, resins, fats, waxes, gums, starches, sugars, oils and tannins, just to name a few. In general, these so-called extractives amount to about 5–10% of the dry mass of wood. In some extreme cases, for example in some tropical species, extractives may amount to much higher percentages of the dry mass.

In all tree species, there are a number of extraneous inorganic components. Foliage and roots extract minerals mainly from the soil. The highest concentration of minerals is found in parts of the tree where life processes take place. In the Temperate Zone, the amount of ash in stems is rarely ever higher than 0.5–0.6%. The ash content in bark and foliage is generally much higher. Calcium (Ca), Potassium (K) and Magnesium (Mg) represent the majority of mineral elements in tree biomass. Inorganic components can cause problems in chemical recovery systems, harvesting and the energy use of wood.

Energy Content of Wood

There is an important relationship between the chemical composition of wood and the thermal energy content. Wood releases energy either when it is decomposing or during combustion. In both cases, the energy is released by breaking the high-energy bonds between carbon and hydrogen. Thus, the higher the carbon and hydrogen content the higher the heating value. Other elements such as oxygen, nitrogen and inorganic elements do not have an effect on the heating value, whereas high contents of lignin and extractives increase the heating value.

Ash

Ash itself is an undesirable product of the burning process. When biomass is burned, most of the inorganic components form ash. If the temperature is right, nitrogen and sulphur escape in the form of NO_x and SO_x during combustion. The content of ash in different tree components varies greatly. Bark has a much higher ash content than stem wood. This is important regarding the combustion of wood chips. The higher the bark content of the chips the more ash will be produced. The amount of ash of uncontamin-ated whole-tree chips from small-sized trees is about 1%, producing about 4–6 kg of ash per m^3 of solid fuel (Hakkila and Parikka, 2002). The amount actually produced is greater due to impurities. There are also great differences in the amount of ash that actually accumulates after combustion. The value depends on factors like composition of the chips, impurities, furnace controls, the ash separation apparatus and storage and handling of the chips. Through the use of ash as a fertilizer the depletion of plant nutrients is radically reduced.

Heating Value of Wood

During combustion, oxygen combines with carbon to produce carbon dioxide (CO_2), water (H_2O) and thermal energy. The actual heating value of a specific fuel is a measure of the maximum amount of energy that can be released on burning a given quantity of the fuel under given conditions. The metric system is most commonly used across Europe to express heating value. The units used are MJ/kg or kWh/kg. Other units to express the heating value are the British Thermal Units per pound Btu/lb and toe in which the fuel is compared to the heating value of crude oil.

When dealing with the heating value there must be differentiation between the calori-metric or higher heating value, and the effective or lower heating value. During the combustion process, some of the heat is lost to the vaporization of water. The water comes from moisture that is in the fuel and from the combination process of hydrogen and oxygen. The total heat is referred to as the calorimetric or higher heating value, and is independent from the moisture content of the biomass. The energy spent on vaporiza-tion is subtracted from the calorimetric heating value and is the so-called effective or lower heating value. It is very much affected by the amount of moisture and hydrogen in the fuel.

In general, wood has a moisture content varying between 45 and 58% after felling, which results in a 15–20% lower effective heating value per kg of dry mass than the calorimetric value. The calorimetric heating value is typically about 20.5 MJ/kg and the effective heating value is about 19.2 MJ/kg of dry wood.

Stages in the Combustion Process

During the first stage of the combustion process moisture is evaporated to dry the wood, and then the volatile components are driven off and burned. Evaporation of 1 kg of water at 100 °C needs approximately 2.3 MJ of energy. Upon heating, solid biomass transforms into volatiles and solid char. The volatiles are made up of organic tar, light permanent gases, water vapour, light hydrocarbon gases and organic gases. At the levels of heat flux, tar is the predominant product of devolatilization. Typically, 80–90% of the solid biomass burns as volatile matter, which has a high tar fraction. The primary volatiles formed during devolatilization undergo secondary reaction to produce a wide range of highly oxygenated, predominantly single ring compounds, which become the fuel gas.

The last step in the combustion process is char oxidation. Following devolatilization, oxygen diffuses to the surface of a residual char particle, in its pores, and reacts with the char forming CO and possibly CO_2, which diffuse into the free stream. Typically, the highest temperatures found in biomass combustors are those of the oxidizing char particles.

Energy Efficiency of Combustion

The moisture content is the main issue affecting the efficiency of combustion. Boiler efficiency in the combustion of biomass is generally less, since there is more water in biomass than in fossil fuel. The moisture content and other above-mentioned fuel properties tend to vary erratically. Therefore, combustion efficiency does not only depend on the properties of the fuel, it also greatly depends on the furnace and its controls. Energy is lost because stack gases carry heat with them and escape unburned. Incompletely burnt carbon is then left as charcoal among the ashes. Heat also escapes through radiation and convection.

Conversion processes

There are two processes for the conversion of biomass to energy. They are either thermochemical or biological. Thermochemical conversion includes combustion, gasification, pyrolysis and liquefaction. Gasification is the process of heating wood in a chamber until all volatile gases such as CO, H_2 and O_2 are released from the wood and combusted. The emitted wood gases are then superheated and mixed with air or pure oxygen for complete combustion. Gasification has the great advantage of having extremely high combustion efficiency and thereby generating minimal emissions.

In the process of pyrolysis, heat is used to chemically convert biomass into fuel oil. It occurs when biomass is heated in the absence of oxygen. After pyrolysis, biomass turns into a liquid, which is called pyrolysis oil that can be burned like petroleum. Liquefaction is a thermochemical conversion process resulting in liquid products (methanol) from biomass by direct or indirect processes. Indirect, through gas phases; direct, without gas phase, for example by rapid pyrolysis.

Density

The density of the biomass is an important aspect of the combustion process. It determines how good a fuel from a certain species is. Conifers usually contain more lignin and resin than hardwoods and, therefore, their heating values per unit mass of stem wood are higher. However, hardwoods are the better fuel since the density is much higher and the heating value is directly proportional to the wood density.

5.6.2   Agro Fuels

Energy crops are agricultural and woody crops, which can be used for energy production. There are some advantages to using energy crops:

- they are renewable; and
- they can be grown on set-aside land.

Studies of different grasses for energy production in Sweden started in 1981. These studies exposed that Reed Canary-grass was the most interesting species because of its high yield, good quality and sustainability. Miscanthus could grow only in the south of Sweden because of the warmer climate compared to the northern sector.
   Some advantages of Reed Canary-grass are:

- easy to cultivate;
- low establishment costs and the possibility of using conventional machinery for production; and
- an extra income for farmers by upgrading the fuel on a small-scale industry level (e.g. briquettes).

Reed Canary-grass consists not only of cellulose, hemicellulose and lignin; but also of proteins, lipids and a relatively high content of inorganic materials (Paulrud and Nilsson, 2001). The first large-scale productions of pulp (8 t of pulp) from Reed Canary-grass were performed in 1999, the critical properties of short-fibre pulp were better for Reed Canary-grass than for birch pulp. The lignin can be burned as bio-fuel for heat production in pulp production system (Paavilainen *et al.*, 1999). Reed Canary-grass briquettes with various chemical compositions have been burned and examined (Paulrud and Nilsson, 2001).

5.6.3   Peat Fuels

The chemical composition of peat varies and depends on the formation process, the quality of the bedding, the groundwater flow, the amount of precipitation and, to some degree, the impact of airborne particles. Decomposition of the peat, i.e. the degree of humification, determines the peat's chemical composition and end-use possibilities. Low-humified peat contains mostly cellulose, while highly humified peat contains various humus substances.
   Well-decomposed peat is used for energy, balneological therapy and soil improvement. Slightly decomposed peat is used as a growing media in horticulture, in manure treatment

in livestock farms, in plant disease control, in industrial filters and oil absorption, as well as in municipal wastewater and sludge treatment. Peat can also be chemically converted to a variety of products, including vehicle fuel.

The heating value of peat is similar to that of wood, but lower than that of fossil fuels. The largest peat-land areas in the EU are found in Sweden (10.4 Mha) and Finland (8.9 Mha) (Lappalainen, 1996). In the EU, the largest peat-harvesting areas are found in Ireland and Finland (Lappalainen, 1996).

5.6.4 Waste Fuels/Solid Recovered Fuels

Our household waste is an important energy source. The combustible dry fraction of household waste, together with other bio-fuel, might be an alternative to fossils fuels in the future. The results from the co-combustion of briquettes made by energy grass, *Reed Canary-grass*, mixed with a very well separated dry fraction of household waste, shows that newly developed commercial boiler technology, 100–1000 kW, has been well designed with high efficiency and low emissions (Olsson *et al.*, 2001).

5.7   Procurement, Harvesting and Handling

5.7.1 Wood-based Fuels

When talking about biomass for green energy production, a lot of aspects have to be considered. In this section, a more detailed overview will be given of issues that need to be considered when forest residues, as an example, are applied for energy production.

The production of wood fuels can be separated into three different production steps (Andersson *et al.*, 2002). The primary production of raw materials takes place in the forest and is managed by silvicultural practices. Forest and transport operations mark the secondary production including the harvest, collection of forest biomass, the preparation for transport and transport of forest biomass itself. Tertiary production is then performed by industry, which includes the processes within the wood consuming industries such as drying or storage.

Silviculture

There are a number of factors that have a great influence on the silvicultural practices that take place in our forests. One of the most important is the objective of the forest owner, taking into consideration economics, risk and markets. The actual forest type sets limitations to forestry, since different site characteristics and species require different management techniques to optimize harvesting systems and generate high yields. Environmental and social issues have gained in importance over the last 20–30 years as well. Clearcuts, for example, have been a very controversial topic as society demands close to nature forestry and multiple use of forests.

Natural, mature and immature forests with high timber value have a very low potential for bio-energy production except for logging residues. In the case of naturally regenerated dense, immature and mature stands with relatively low timber value, the potential for bio-energy production is much higher. Growth of biomass for energy as the major objective is achieved in the case of plantations with short rotations as described in section 3.1.

Silvicultural systems designed for bio-energy production have a long history. Especially when transport of resources was limited before industrialization, communities had to rely on local resources for fuel and construction purposes. Coppicing and coppice-with-standards are silvicultural systems that satisfy the need for both fuelwood and construction wood. The selection forest, historically practised by farmers, is also a silvicultural system that produced a steady supply of fuelwood as well as timber for construction.

Harvesting

Harvesting includes the cutting and collection of the biomass at the actual forest site. The harvesting operation can be classified according to its silvicultural goal and treatment. In thinnings, low-quality trees are removed from the site in order to minimize competition with the remaining high-quality trees. During pre-commercial thinnings, unmerchantable trees are cut down to reduce the number of stems per hectare and there is no commercial value for the cut trees. In commercial thinnings, merchantable trees are harvested and then used primarily in the pulp and paper industry. In selection cuts, single trees or small groups of trees are removed from uneven-aged stands with the aim to mix age and size classes for reproduction. In the clear-cut operations, trees are removed from a mature stand in one procedure and the site is then left for either artificial or natural regeneration. A modification of this method to promote natural regeneration is to leave some trees standing for seeding and the protection of the established seedlings later on. That operation is either called Seed Tree Cut or Shelter Tree Cut.

Industries

The forest industries are converting the forest biomass into merchantable products that can be divided into three key groups based on the end product. The first group is made of the solid wood industries that includes sawmills and veneer plants. This industry has very high demands on the raw material, such as structural and aesthetic properties.

The pulpwood industries consisting of pulp mills, fibreboard and chipboard, among others, represent the second group. The quality requirements of the raw material are typically lower since the resource is dissolved during the process and unwanted material can be filtered out. The last group is made up of the energy wood industries; they include bio-fuel heating and CHP plants. Their set of demands differs from the two industries mentioned above and focuses mainly on caloric value and storing, processing and handling abilities of the raw material. All of these industries are usually located close to the resource in order to minimize transport costs. Most of these industries are interrelated, for example a large number of pulp mills are producing their electricity by burning residues of the pulping process or black liquor. Heating and power plants utilize the residues and sawdust produced by sawmills.

Procurement and Economy of Forest Processed Chips

Forest chips represent a large share of forest fuels. Although the chipping of whole trees is easier, logging residues from clearcuts represent the largest share of forest fuels. There are quite a few techniques and methods to harvest forest chips that will be discussed in further detail. Sources of forest chips include residues from forest fellings, small-diameter trees from thinnings of young forests or seedling stands and chips made from stemwood, branches and needles.

Methods

There are different places where forest chips are generally produced. The production can happen at the stand level, at an intermediate storage place, at a chipping terminal or at the power plant where the material is then utilized.

Chipping at the stand level

The advantage of chipping at stand level is that residues or whole trees do not have to be transported out of the stand. The machine moves along the logging roads and thereby chips the residue piles directly into its own container or to an interchangeable container. The container is then transported to the roadside where it is tipped into a bigger container, which is then used for the long-distance transport. The productivity of a stand chipper lies between 15 and 20 bulk m^3 per operation-hour. Practical advantages of chipping at stand level are: the moving costs of machinery are considerably lower; the biomass can be collected more thoroughly than in an intermediate storage method; there is small demand for storage space; and the chips and the roadsides are kept cleaner. On the other hand, the machine unit is relatively expensive and transportation is not profitable if the transport distance is over 300 m.

Chipping at roadside or intermediate storage

In this method, the forest residuals are transported to the roadside storage area, generally by a forwarder equipped with a special load space and grapple. The chippers used in the operations are normally truck-adapted and heavy-duty vehicles because they can operate on site and on flat ground. The chips are blown through a tube to the containers of a waiting truck or to free containers. The productivity ranges from 15 to 30 loose m^3/hour. The advantages are that the chippers are more powerful compared to the stand-chipping method and that the larger feed opening makes processing much easier, especially in the case of chipping logging residues. There is a need for a large storage area that in practice is located next to the road and interferes with traffic.

Chipping at terminal or power plant

At the terminal or power plant the chipper operates continuously at the same place and heavy-duty chippers can be used. In general, the storage areas are rather large and the terminals are often located at peat swamps, which allows the mixing of peat with the chips, a method used commonly in northern European countries. The productivity is relatively high with easy maintainability of the chipper. Quality control is made a lot easier and the biomass can be preserved better. The system faces a major challenge: long distances make the transport of forest residues unprofitable since loads are generally of low density and the moving of residues at a terminal also increases costs.

All-in-one systems

There are also systems available especially for small worksites where the forest transport, chipping and long-distance transport are performed by the same unit. These systems work only on flat and bearing terrain, and the long-distance tranport has to be less than 50 km in order to be profitable.

Slash log method

The last method discussed is the slash log method where forest residues are bundled into 'logs' that are approximately 3-m long and have a diameter of 60–80 cm. The energy content of one 'log' is about 1 MWh, which represents approximately 1.5 bulk m³ of chips. On one hectare of clear cutting, about 60–150 slash logs can be harvested. There are a number of advantages associated with the slash log method, especially regarding transport since costs are significantly decreased because of bigger load sizes.

Economic Considerations

Transport

Long-distance transport plays an important role when considering the profitability of using forest residues. The following graph (Figure 5.1) illustrates how costs increase steadily with longer transport distances.

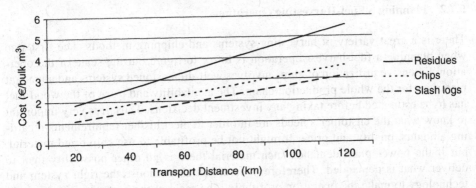

Figure 5.1 *Long-distance transport costs for residues, chips and slash logs (Reproduced from Brunberg, Uppdragsprojekt Skogsbränsle: slutrapport – Forest Bio-energy Fuel – Final report of Commissioned Project, p. 60 (SkogForsk. Redogörelse Nr. 6); published by The Forestry Research Institute of Sweden, 1998)*

The forest transport distance also affects the productivity of the operations. When the transport distance increases from 200 to 400 m the productivity decreases by approximately $\frac{1}{4}$ (MWh m³/h), at the same time the unit cost increases by about $\frac{1}{3}$ (Euro/m³).

Productivity can also be significantly boosted if the same machinery transports residues and merchantable wood. In general, procurement of timber and energy wood should be integrated closely in order to maximize profits. A preliminary bunching of residues is needed in order to optimize profits.

The amount of residues per hectare is a key value. Profits increase if more residues are available per hectare. This is one of the reasons why spruce stands are preferred over pine stands.

Moisture content

The moisture content has a significant effect on energy content. Fresh logging residues have moisture content of approximately 50–60% (mass percent). By storing the chips in the summer time, moisture contents of less than 30% can be achieved, whereas in winter time snow and ice increase the moisture content. The calorimetric value also decreases with the so-called green logging residues (residues with needles).

Storage

One option to decrease the moisture content and increase the heating value is to store the material. The storage can take place either at the site, at the landing, at the terminal or at the plant. Green chips are very sensitive to exothermic microbiological, physical and chemical degradation, which can cause health hazards and a substantial loss of raw material and fire by self-ignition. In spruce, for example, storage can cause 20–30% loss due to defoliation.

5.7.2   Planning a Fuel-Harvesting Operation

There is a great variety of harvesting systems and chipping methods. The first step when planning a fuel-harvest operation is to try to figure out the scale of the operation. Extensive utilization of chips involves well-dimensioned systems and abundant resources for the whole production chain. The availability and price of the wood fuel has to be estimated before taking any investment decisions. Thus, it is very important to know what the customer's needs are in order to meet his/her requirements regarding amounts, quality and price. It might not be profitable to use chip dried material, but if the power plant demands such material there is no other possibility than to deliver what is demanded. Therefore, it is essential to choose the right system and technology to make the operation profitable. The scale of the operation significantly affects the cost of the forest fuel and filters out choices for technology and method selection.

5.7.3 Agro Fuels

Several crops can be evaluated as energy crops, considering the climate, the type of soil, etc. For every crop, a complete description can be made. In this section, one crop will be commented on as an example that is suitable for a colder climate – Reed Canary-grass. Also some aspects of Miscanthus will be discussed.

Background

Research and development of new methods for cultivation of spring-harvest Reed Canary-grass (*Phaláris arundinácea* L.) have been proceeding during the last decade. Most of the experiments were performed on mineral soils in the north of Sweden. Therefore, the following instruction is particularly valid for mineral soils in the northern part of Sweden. For the rest of Sweden and other types of soil the knowledge is limited (Olsson, 2003).

The Origin of Reed Canary-Grass

The grass occurs in the wild in the Scandinavian countries, especially on wet and humus-rich soil. Reed Canary-grass is a 2-m tall grass with high yield. Until today, Reed Canary-grass has been used mostly as animal feed. These animal feed varieties of Reed Canary-grass are used for energy purposes in the USA and Canada.

The Type Suitable for Energy Production

Research has been carried out into the improvement of new varieties of Reed Canary-grass for energy and industry production since 1990. A new variety, which is called 'Bamse', has been developed. This variety of Reed Canary-grass has a 20% increased yield than Palathon, which is the best-feed variety of Reed Canary-grass. But the most available data and calculations are from Palathon, which is used as feed.

Establishment of Reed Canary-Grass Leys

In the construction of Reed Canary-grass leys, the same demands are required as for normal grass leys. It is best to cultivate Reed Canary-grass alone. The proper amount of seed to cultivate is around 15 kg/ha. Cultivation should be done in spring time or in early summer time without a companion crop (protective crop). Cultivation in the summer time is not recommended because it will cause a weaker stock in the next year. Germination is dependent on daylight and that is the reason for the seed to be around 0.5–1 cm under ground. It is necessary to roll the soil over before and after cultivation. The seed will grow up slowly and it is quite sensitive to dryness.

Weed Killing

Reed Canary-grass will establish very gently. The seed is therefore sensitive to competition with weed. It is necessary to kill weeds using convenient weed-killers for the establishment of grass leys. To avoid the use of weed-killers, a mechanical method may also be used.

Fertilization of Reed Canary-Grass Plantations

Until today, experience of the cultivation of Reed Canary-grass shows that Reed Canary-grass is able to keep the necessary nutriment for growing if we apply the spring-harvest method. It is necessary to have a well-fertilized land/crop system in order to reach a high yield of Reed Canary-grass. If the land has a phosphorus–potassium content equal to a class II–III (2–8 mg phosphorus and 4–16 mg potassium/100 g dry soil), it should be supplied with 200 kg of nitrogen, 40 kg of phosphor and 150 kg of potassium per hectare during the first 2 years. Afterwards, it is necessary only to compensate for the amounts removed by the grass. It is important to keep the level of fertilization above 50% of the requirements in the first few years, as a poorly established Reed Canary-grass is not easy to improve. Table 5.1 shows the average values of nutrients that should be added to soil for yielding a well-established Reed Canary-grass.

Harvest

Harvesting time

Reed Canary-grass for the production of bio-fuel or fibre raw material can be harvested for the first time in the spring, 2 years after cultivation. The harvesting conditions are normally very good. The water content of this wither grass is often below 15%. Since the air is dry and there are a lot of sunny days, it is possible to harvest and handle a field-dried crop. Generally, it is possible to harvest as soon as the frost has ended and the land has dried. If the harvest is started too early, it will cause damage to the soil packaging and in the worst-case it will kill the grass. In the case that it is started too late the new crop will start to grow, which will increase the water and ash content in the harvested material. The harvesting time is varied from year to year, depending on the part of the country and also

Table 5.1   The amount of fertilization for an optimal cultivation of Reed Canary-grass

Element	Fertilization[1] year 1 + 2 (Kg/ha)	Average values of nutrients removed by harvest (%)	Fertilization[2] year 3–10 (Kg/ha)
Nitrogen	200	0.80	60
Phosphor	40	0.11	8
Potassium	150	0.27	20

[1] Maximum 50% of nitrogen should originate from farmyard manure or sewage.
[2] The amount of nutrients which should be added on each hectare with a harvesting yield of 7500 kg/ha.

on the condition of the land. In many cases, one cannot consider having more than two weeks to harvest.

Harvesting technique

The dried Reed Canary-grass is very fragile in the spring. An ungentle treatment can cause a big loss. Therefore, the grass must be harvested as gently as possible. The aim should be that the mover puts the grass in a compact and loose string, which can be pressed without the use of a windrower. Reed Canary-grass can be picked up by the normal grassland harvesters. Even combine harvesters have been tested. Primary tests performed on the spring-harvest method have shown that waste products are around 20% or more. With a usual mover and conditioner modified for a mild treatment, the amount of waste can be reduced to 50%.

Baling of the Reed Canary-Grass

Spring-harvested Reed Canary-grass in string can then be pressed by both round-bale press and big-bale press techniques – the so-called HD-press (High Density). The HD-technique has been used with good results. The big square bales, with better forms and higher energy density, require lower handling costs than the round bales and also provide the possibility of outside storage under tarpaulin. To press by HD-press requires a good land carrying capacity and the land, therefore, needs good drainage.

Storage of Reed Canary-Grass in Big Bales

The best way to store round bales or HD-bales is in a storage hall if available. Full-scale tests have shown that it is possible to store HD-bales outdoors under cover, if it is done in the right way. This method will result in low storage price and acceptable water content inside the bales.

Miscanthus

The dryness of Miscanthus depends on the harvesting season. In most countries, Miscanthus is harvested in spring to prevent risk of mould or fungal attack. The moisture content is usually 15–25%. In Austria and Greece it is harvested in autumn, since this has environmental benefits, the moisture content is much higher (up to 66%). There are several different harvesting methods for Miscanthus such as forage harvesting, mowing, baling, bonding and also pelletizing. The harvesting and processing of Miscanthus can be done in two ways. The first method is cutting, transporting and then baling. The second method is cutting and baling in one step. In the first method, some loss of material may occur and there are also high transportation costs. In the second method, some dirty or wet material may be baled as well (Venendaal, Jørgensen and Foster, 1996).

5.7.4 Peat Fuels

Worldwide about 50% of the peat harvested is used for fuel and 50% is used as growth medium, soil conditioner or for environmental protection purposes. Peat production is very similar to agricultural operations, i.e. peat is harvested in summer and the machines used resemble those used in soil preparation and crop harvesting. Peat production methods are different but in general it entails the peat being removed from the deposit (milled peat) and ends with the stockpiling of extracted peat. In one method, raw peat is excavated from a depth of 0.5–3.0 m using a digging machine. After cutting, the peat mass is macerated and pressed out through nozzles into sods of different size and shape. After about two to four weeks of drying, the sods, at about 30–40% moisture content, are collected into stockpiles (Lappalainen, 1996).

5.8   Upgraded Fuels

5.8.1   Raw Materials

Wood powder is a bio-fuel produced using sawdust, shavings and bark. The raw material is crushed, dried and milled until it becomes fine in order to obtain the best fuel properties. There are many wood powder qualities depending on different physical properties such as particle size distribution, particle shape and also moisture content. Raw materials and type of mill will affect these properties (Paulrud, Mattsson and Nilsson, 2002).

The use of bio-energy is steadily increasing in the world and by-products from sawmills have so far served as a raw material source for up-graded bio-fuels due to a low raw material price. An expanding market will however result in a shortage of by-products from sawmills. Not only cutting reminders from the forestry are considered as an important raw material but also spring-harvested Reed Canary-grass, which gives a natural dry raw material, has the potential for being an important part of the future raw material base.

Many industrial processes result in large amounts of waste. Many industries using timber, i.e. sawmills, packaging materials in the form of paper and cardboard, produce tonnes of trash. These together with a huge amount of municipal waste and logging residues are potential bio-fuels, and today a small amount of these are being utilized as bio-fuels. Briquettes of these industrial wastes can be used as fuel for energy production.

5.8.2   Techniques

Classification of Densification

Depending on the type of equipment used, densification can be divided into four main categories: (1) piston press densification; (2) screw press densification; (3) roll press densification; and (4) pelletizing. The definition of a briquette is a cylinder of organic matter, which has been compressed, with a minimum diameter of 25 mm. If the diameter is smaller they are called pellets. Pellets normally have a diameter of 6–12 mm (Lehtikangas, 2001).

Briquetting Plant

Briquetting improves some properties of the material; it makes it denser and, therefore, more easy to handle and it increases the heating value. Heating value means caloric heating value, which is a measurement of the energy content (Lehtikangas, 2001).

The piston in the briquette machine compresses the material and the temperature increases in the briquette due to friction between particles. The increased temperature makes the fibres softer and the lignin more fluid. Lignin and/or hemicellulose behave as glue at 80–200 °C and bind the material within the briquette. The water content of the material entering the densification process greatly affects the properties of the briquettes. A moisture content that is too high results in steam formation in the briquette machine and steam explosion may occur when the pressure is released, which breaks the briquettes. There may also be shell formation (the formation of a hard surface of briquettes), which can result in problems if the hard surface is stuck in the nozzle of a briquette press. Water makes lignin more fluid and improves the thermal conduction, increasing the rate of the binding process. Water content that is too low makes the briquette fragile (Petterson, 1999).

Pellet Production

Biomass fuels can be processed not only as briquettes as mentioned above but also as pellets. By-products from the wood industry like sawdust or cutter dust (cutter shaving) are the raw materials. Sawdust usually contains up to 50% of water. In the first step, other products like stone or coarse sand and metals are removed from the raw material. In the second step, the raw materials disintegrate in a different mill to suitable size. In the third step, the raw materials are dried to more than 90% dryness. In the last step, the dry materials will be pressed. When the wood materials are pressed, the temperature will rise and lignin will be released which will work as a glue to bind the material together (Nilsson, 2000; Olsson, 2002).

5.8.3 Properties and Handling

Briquetting of Biomass

The EU today imports half of its energy demand. This share is calculated to reach 70% within the next two decades if no measures are taken. The European Communities Commission (November 1997) considers this increase unacceptable and suggests a strategy and action programme for the energy sector. This white book includes a doubling of RES from 6 to 12% by 2010. To be able to reach this goal, the commission claims that the use of bio-fuels must triple.

Agricultural and forestry residues and other waste materials are often difficult to use as bio-fuel because of their uneven, bulky and troublesome characteristics. This drawback can be overcome by means of densification of the residues into compact regular shapes. There are many advantages to this process:

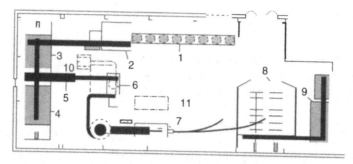

Figure 5.2   *Plan of a fuel plant (Bio-fuel Technology Centre (BTC) at SLU, Umeå, Sweden): 1 = conveyor, 2 = shredder, 3 = silo nr 1, 4 = silo nr 2, 5 = mixing equipment, 6 = separator, 7 = briquette press, 8 = fuel storage, 9 = boiler storage, 10 = mil (not installed) and 11 = pellet press*

- The process increases the net caloric content of the material per unit volume, producing a product uniform in size and with a well-defined quality.
- The product is much easier and cheaper to handle, to transport and to store.
- This makes it easier to optimize the combustion, resulting in a higher efficiency, a lower emission and a smaller amount of ashes.
- Considerably lower investment in furnaces and purification equipment.

The bales of raw material are put on a conveyor and transported to a shredder and stored in a silo (Figure 5.2). The shredder cuts the material and this will give rise to particles ranging from microscopic to 15 mm. When the production of briquettes starts, material from the silos are transported through mixing equipment and a separator to a buffer silo above the briquette press. At the bottom of the buffer silo there is an auger, which transports the material to the press. The briquette press has a capacity of 450–500 kg/h. A haymatic H_2O tester is used to get an estimation of the humidity of the raw materials. This instrument measures the conductivity in the raw materials and gives a rough estimation of the moisture content at the surface of a sample. It is also possible to measure the moisture content of the raw material continually by online Near Infrared Spectroscopy (NIR-Spectroscopy) in combination with multivariate data analyses (Siesler *et al.*, 2002; Osborne, Fearn and Hindle, 1993).

Pellet Process

There are some advantages to using pellets instead of briquettes (Lehtikangas, 2000):

- lower moisture content;
- homogeneity causes fewer variations in moisture content;
- low transport costs;
- easy to feed into burners;
- low storage costs;

- less uneven combustion with unnecessary emissions;
- higher density and better combustion properties;
- renewable fuel that does not provide any net contribution to the greenhouse effect;
- the raw materials for manufacturing fuel pellets are shavings, sawdust and also bark, wood chips, agricultural stalk crops (straw and grass), sorted household waste, paper and wood waste.

One drawback is the higher production costs compared to briquettes.

Combustion of pellets

It is possible to convert small oil-burning boilers to pellet burners. The burner costs in Sweden are between €1700 and 2000, in addition another storage system will be needed which costs an extra €2000. This means that it is possible to convert around 600 000 oil boilers to pellet burners in Sweden (Nilsson, 2000).

5.9   Conclusion

This chapter has highlighted that energy from RES is already an important part of energy systems today. Its importance will further increase in the future due to its nature of being a local source of energy, which therefore promotes regional development. Furthermore, it has a significant advantage over fossil fuel energy sources with regard to greenhouse gas emissions. There is a further need to develop more kinds of RES, since together they can cover the various demands of a society. Depending on the country, the combination of RES can differ greatly. The challenge of the future is to develop RES as a whole, depending on the conditions in each country. RES can be divided into two major parts – wood-based energy sources and agro-based energy sources.

Wood as a source of biomass is an important aspect of the planning of the EU's Energy policy for the future and particularly in the northern countries where wood is an essential factor in the energy systems. In comparison with other fossil fuels, the non-uniform nature of wood fuels is a considerable disadvantage. Technological development has been rapid and today it is possible to burn a wide variety of fuels at different moisture levels in the same boiler. Nevertheless, to make the most of them in the future, more research is necessary to improve utilization and efficiency levels. Upgraded fuels such as pellets and briquettes have proven to be a good solution and there is great potential to further optimize their production, for example by enhanced process control and use.

Development has been rapid over the last decade and efforts need to be continued to ensure further growth in the wood energy sector. However the wood-based energy sources alone will not be able to cover the whole bio-energy demand in the future due to the limitation on raw materials such as sawdust, shavings and bark for production of wood-based fuels, e.g. pellets. Today, there is a need to transfer and exchange existing know-how to avoid the repetition of mistakes that have been made before.

Other sources of solid fuels such as waste and energy crops are less developed, even though there exists a large amount of available resources. Substantial efforts are still required to further increase their utilization. Today, the pellet industry is looking for new raw materials besides wood-based materials, therefore agro-based fuels as well as waste fuels will play an essential role in the future.

With regard to procurement of wood fuels, harvesting and transport costs are major obstacles to a further increase in the share of wood fuels. Consequently, the research community together with forest industries is constantly working on the fine-tuning of existing technology and on the development of new harvesting and transport systems. There is a clear need to implement research results that are already present in the every-day practices of forest management.

Among RES, biogas and ethanol (which can be produced from both wood-based and agro-based materials) are two main bio-fuels since these fuels can be used in motor vehicles instead of fossil fuels and thereby reduce the amount of CO_2 in the atmosphere.

The overall advantage of a further development of RES is the creation of jobs in numerous related fields such as technology development, improvement of fuel quality and sales and marketing of RES. However, it remains to be seen how well RES will be supported by political decisions in different countries, since nowadays most of the RES are not competitive with the fossil fuels. So far, the development has been positive and at present the future looks bright. If the necessary actions are taken, RES can contribute to improvements in many fields, but most of all in support of regional development, the creation of new jobs and the protection of the environment.

References

Andersson, G., Asikainen, A., Björheden, R., Hall, P.W., Hudson, J.B., Jirjis, R., Mead, D.J., Nurmi, J. and Weetman, G.F. (2002). Production of forest energy. In: *Bio-Energy from Sustainable Forestry: Guiding Principles and Practice* (eds J. Richardson, R. Björheden, P. Hakkila, A.T. Lowe and C.T. Smith), p. 344. Kluwer Academic Publishers, The Netherlands.

Brunberg, B. (1998). *Uppdragsprojekt Skogsbränsle: slutrapport – Forest Bio-energy Fuel – Final Report of Commissioned Project*, p. 60 (SkogForsk. Redogörelse Nr. 6, 1998). Uppsala, Stiftelsen Skogsbrukets Forskningsinstitut.

Dawson, M., Isebrands, J., Namkoong, G. and Tahvanainen, J. (1996). International evaluation of Swedish research projects in the field of short rotation forestry. National Board for Industrial and Technical Development. Stockholm, Sweden.

European Commission (1999). Renewable energy systems – new solutions in energy supply. Directorate General for Energy. Brussels, Belgium.

Eurostat (2000). Renewable energy sources statistics in the European Union 1989–1997, p. 62. Luxembourg Office for Official Publications of the European Communities, 2000.

Hadders, G. and Olsson, R. (1997). Harvest of grass for combustion in late summer and in spring. *Biomass and Bio-Energy*, **12**, 171–175.

Hakkila, P. (1989). *Utilization of Residual Forest Biomass*, p. 568. Springer Series in Wood Science. Springer. Heidelberg, New York.

Hakkila, P. and Parikka, M. (2002). Fuel resources from the forest. In: *Bio-Energy from Sustainable Forestry: Guiding Principles and Practice* (eds J. Richardson, R. Björheden, P. Hakkila, A.T. Lowe and C.T. Smith), p. 344. Kluwer Academic Publishers, The Netherlands.

Hall, D.O. and Coombs, J. (1987). Biomass energy in Europe. In: *Biomassan uusia jalostusmahdollisuuksia. (New possibilities to process biomass.)*, pp. 27–49. VTT Symposium 75, Espoo.

IEA (1998). The role of bio-energy in greenhouse gas mitigation. A position paper prepared by IEA Bio-energy Task 25 "Greenhouse Gas Balances of Bio-Energy Systems". Austria, 4pp.

Korhola, A. and Tolonen, K. (1996). The natural history of mires in Finland and the rate of peat accumulation. In: *Peatlands in Finland* (ed. Vasander), pp. 20–26. Finnish Peatland Society. 168pp.

Lappalainen, E. (1996). In: *Global Peat Resources*. International Peat Society, Jyskä, Finland, 1–354.

Lappalainen, E. and Zurek, S. (1996). Peat in other European countries. In: *Global Peat Resources* (ed. E. Lappalainen), pp. 153–162. International Peat Society. 358pp.

Larsson, S. (2003). Modelling of the potential for energy crop utilization in Northern Sweden. Licentiate thesis. Unit of Biomass Technology and Chemistry, SLU, Sweden, 1–18.

Lehtikangas, P. (2000). Storage effects on pelletized sawdust, logging residues and bark. *Biomass and Bio-Energy*, **19**, 287–293.

Lehtikangas, P. (2001). Quality property of pelletized sawdust, logging residues and bark. *Biomass and Bio-Energy*, **20**, 351–360.

Mitchell, C.P., Stevens, E.A. and Watters, M.P. (1999). Short-rotation forestry – operations, productivity and costs based on experience gained in the UK. *Forest Ecology and Management*, **121**, 123–136.

Månsson, T. (1998). *Rena fordon med biodrivmedel*, Series in KFB-Rapport 1998:1, AB C O Ekblad & Co, Västervik.

Nilsson, B. (2000). *Woodpellets in Europe*, Industrial network on wood pellets, pp. 1–9, UMBERA GmbH, A-3100 St. Pölten, SchieBstattring 25.

Olsson, M. (2002). *Wood Pellets as Low-emitting Residential Biofuel*, Thesis for the degree of licentiate of engineering, Department of Chemical Environmental Science, Chalmers University of Technology, Sweden.

Olsson, R. (2003). Production manual for reed canary grass (Phalaris Arundinacea), *Internal Report*. Unit of Biomass Technology and Chemistry (BTK), Swedish University of Agricultural Sciences, Umeå, Sweden.

Olsson, R., Marklund, S., Nilsson, C., Burvall, J. and Hedman, B. (2001). Energy production from burning of mixture of source-sorted waste fractions and biofuels, *BTK-rapport* no. 1.

Osborne, B.G., Fearn, T. and Hindle, P.H. (1993). *Practical NIR Spectroscopy with Applications in Food and Beverage Analysis*, Longman Group UK Limited. 1–220.

Paavilainen, L., Tulppala, J., Finell, M. and Rehnberg, O. (1999). Reed canary-grass pulp produced on mill scale. *Tappi Pulping Conference*, Orlando Fl., USA, Proceedings Tappi Press, 31 Oct.–4 Nov. 1999, Vol. 1, pp. 335–341 [Atlanta, GA, USA: TAPPI Press, 1999, 1, 252 pp, 3 vol].

Paulrud, S. and Nilsson, C. (2001). Briquetting and combustion of spring-harvested reed canary-grass: effect of fuel composition. *Biomass and Bio-Energy*, **20**, 25–35.

Paulrud, S., Nilsson, C. and Öhman, M. (2001). Reed canary-grass ash composition and its melting behaviour during combustion. *Fuel*, **80**, 1391–1398.

Paulrud, S., Mattsson, J.E. and Nilsson, C. (2002). Particle and handling Characteristics of wood fuel powder: effect of different mills. *Fuel Processing Technology*, **76**, 23–39.

Petterson, M. (1999). Briquetting of biomass: A compilation of techniques and machinery, *Degree project in forest technology, The Swedish University of Agricultural Sciences*, no. 22, 1–17.

Saha, B.C. and Woodward, J. (1997). *Fuel and Chemicals from Biomass*, American Chemical Society, 172–208.

Siesler, H.W., Ozaki, Y., Kawata, s. and Heise, H.M. (2002). *Near-Infrared Spectroscopy, Principal, Instruments, Applications*, Wiley-VCH Verlag GmbH, D-69469 Weinheim (Germany), 1–328.

Venendaal, R., Jørgensen, U. and Foster, C. (1996). Synthesis report of the European energy Crops, *Overview Project, EU FAIR* Contract no: FAIR1-CT95-0512.

Young, H.E. (1980). Biomass utilization and management implications. In: *Weyerhaeuser Science Symposium 3*, pp. 65–80. Forest-to-Mill Challenges of the Future.

Definitions of Renewable Terms

Ash content Refers to the non-combustible materials in a fuel, which reduces the heating value per unit of weight.

Bio-energy Energy sources derived from organic matter. These include wood, agricultural waste and other living cell material that can be burned to produce heat energy. They also include algae, sewage and other organic substances that may be used to make energy through chemical processes.

Biomass   All forms of organic matter including wood, agricultural crops and residues, animal dung, human waste and biomass products such as ethanol.

Briquetting   Densification of loose organic material such as rice husk, sawdust, coffee husks, to improve fuel characteristics including handling and combustion properties.

Carbonization   The destructive distillation of organic substances in the absence of air resulting in the removal of volatile constituents and leaving a residue high in carbon, e.g. coke, charcoal.

Charcoal   Solid residue consisting mainly of carbon obtained by the destructive distillation of wood in the absence of air.

Co-generation   Sequential production of both heat and power using the same fuel. An example is the use of expanded steam left over after generating electricity for heating purposes. The concept is increasingly being applied in many wood and agro-processing industries.

Combined heat and power generation (CHP)   Simultaneous production of heat and power. The difference between CHP generation and co-generation is that the generation of heat and power may be done as parallel processes, which results in a lower overall efficiency.

Combustion   Chemical reaction between a fuel and oxygen which usually takes place in air. More commonly known as burning. The products are carbon dioxide and water with the release of heat.

Crop residues (or agricultural residues)   By-products of the agricultural production system, such as straws, husks, shells stalks and animal dung, with a large number of uses including energy production. Residues can be divided into two groups: field residues, which remain on the field after harvest, e.g. cotton stalks; and process residues, generated off-field at a central production site, e.g. rice husk.

ECU   The ECU was the precursor of the new single European currency, the Euro, which was introduced on 1 January 1999.

End-use   The user application for which energy is required, e.g. lighting, cooking, space heating, drying, milling and sawing.

Energy balance   Table accounting for all the energy produced and consumed for a certain time period in a system, e.g. a country, region, factory or process. The energy balance is used to represent the production, conversion and consumption of all fuels in the system in one table and in one unit.

Energy content (or heating value)   The amount of energy per physical unit, e.g. joule per kilogram (J/kg) or tonnes of oil equivalent per litre. For combustibles, this is the amount of energy in the form of heat that is released when the fuel is totally burned. For wood fuels, the heating value can vary widely due to the density and moisture and ash content of the wood. The density of the wood varies for different wood species, e.g. hardwood has a higher density than softwood, and so hardwood has a higher heating value per unit of volume.

Energy conversion   In physical terms, a process which transforms energy from one form into another, for example, the conversion of wood into heat by combustion. In energy terms, it refers to production processes that transform one fuel into another fuel that can be transported or transmitted, for example electricity generation from coal, or charcoal production from wood.

Energy efficiency   Conversion ratio of output to input energy of energy production technologies and end-use appliances. The lower the efficiency, the more the energy lost.

Energy unit   Unit used to express the quantity of heat, energy or work. Common units are the joule (J), ton of oil equivalent (toe), calorie (cal), kilowatt-hour (kwh).

Fossil fuels   The non-renewable energy resources of coal, petroleum or natural gas or any fuel derived from them.

Fuelwood (or firewood)   All wood in the rough used for fuel purposes, including trunks, branches and twigs of trees as well as residues from wood processing industries used as energy source.

Gasification   Conversion of solid fuels (biomass and coal) to combustible gases at high temperatures in the absence of air.

Moisture content   Quantity of water in a material, expressed as a percentage of the material's weight. Moisture content is the most critical factor determining the amount of useful heat from biomass combustion, because the water is evaporated before heat is available for the application.

Peat fuel   Peat is a bio-fuel whose properties lie between those of wood and lignite, but closer to wood, according to its physical and chemical properties.

Primary energy   Energy form as it is available in nature.

Residue   Any organic matter left as residue, such as including agricultural and forestry residue, but not limited to conifer thinnings, dead and dying trees, commercial hardwood, non-commercial hardwoods and softwoods, chaparral, burn, mill, agricultural field, industrial residues and manure.

Renewable energy   Any form of primary energy, for which the source is not depleted by use. Wind and solar energy are always renewable, biomass can be renewable if its consumption is based on sustainable management.

Renewable Resources   Renewable energy resources are naturally replenishable, but flow-limited. They are virtually inexhaustible in duration but limited in the amount of energy that is available per unit of time. Some, for example geothermal and biomass, may be stock-limited in that stocks are depleted by use, but on a timescale of decades, or perhaps centuries, they can probably be replenished. Renewable energy resources include: biomass, hydro, geothermal, solar and wind. In future they could also include the use of ocean thermal, wave and tidal action technologies.

Secondary energy   Energy after conversion in a form ready for transport or transmission.

Wood energy (or wood fuel)   Refers to all forms of energy derived from wood.

Woodfuel flows   The distribution and marketing of wood fuels from the production site to end-users.

Woody biomass   Stems, branches, shrubs, hedges, twigs and residues of wood processing. Non-woody biomass refers to stalks, leaves, grass, animal and human waste.

6

Identification and Quantification of Renewable Crop Materials

Anton Huber and Werner Praznik

6.1 Introduction

In order to introduce and use renewable resources in several applications, there is a constant need to analyze and study the renewable raw materials. In this chapter, attention is paid to the technical aspects of how these compounds can be analyzed. Of course, this is a very broad area and there are thousands of methods available to study all the different types of materials.

In the first section, the more fundamental aspects are discussed, such as why nature is organized the way it is and why nature has developed such a variety of biomolecules for all its purposes.

In the second section, an overview of the different technical techniques that can be used to gain information on the structure and conformation of the biomolecules is given. Due to the diversity and the complex structure of polysaccharides the chapter focuses on the methods for the analysis of polysaccharides.

Next to the fundamental concept of the organization of nature, the following aspects are discussed:

- Identification and classification of cell compositions.
- Analytical strategies for crop materials (non-starch crops, fructan crops, starch crops).
- Experimental approaches to analyzing the molecular characteristics.

Renewable Bioresources: Scope and Modification for Non-food Applications. Edited by C.V. Stevens and R. Verhé
© 2004 John Wiley & Sons, Ltd   ISBNs: 0-470-85446-4 (HB); 0-470-85447-2 (PB)

6.1.1 Ecosystem Earth: A Complex System

Nowadays natural sciences and system theory consider the Ecosystem Earth as a Complex System (CS) – however, not just intuitively 'complex', but complex as a well-defined technical term (Figure 6.1) with a wide range of consequences for system characteristics.

Complex Systems consist of a huge number (10^{20} and more) of material components and may contain complex subsystems with various qualities and quantities of interaction between constituting components. Additionally, such systems form and modify specifically permeable and structurized system borders for communication purposes with the system environment, which operates as an integral feature of CSs.

The system environment provides CSs with energy input in a very particular way. The energy transfer takes place permanently, far away from thermodynamic equilibrium and supercritical in terms of quantity. CSs transform imported energy to maintain or develop their structure, functionality and genuine properties, and by doing so produce large amounts of low-quality energy (entropy) which needs to be exported.

The system border resembles the interface between CSs and the system environment. It is formed, maintained and permanently adjusted by the system (not by the system environment), representing system-specific communication capacities; typically small communication windows for structural coupling of CSs to the system environment.

Complex Systems transform high-level energy input by appropriate internal structures – made up, tested, selected and fixed by self-organization mechanisms – to lower-level energy and finally to entropy. Such appropriate structures are maintained and adjusted by:

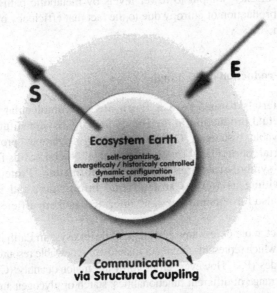

Figure 6.1 *Ecosystem Earth – a Complex System or Dynamic Process constituted by: (1) a huge number of material components; (2) long-range interaction between constituting components; (3) selectively permeable system borders for communication with system environment via Structural Coupling; (4) system environment which permanently provides energy (E) in supercritical quantities; (5) permanent production of entropy (S) which constitutes any aspects of time (history, age, turnover, time horizon, irreversibility → vector of time) and (6) entropy (S) export from system to environment*

- Appropriate energy input from the system environment (input level which allows maintenance of established organization) 'supplies' the actual system organization and results in optimized internal organization by spontaneous evolution in the direction of stabilizing (full dimensional) attractors.
- Established by 'challenges': supercritical or subcritical energy input to CSs which results in the necessity to explore new energy management options. This results either in collapse and subsequent liberation of constituting components or in intermediate fragmentation and reconstruction of more efficient structures under dissipative chaos conditions – a spontaneous evolution in the direction of de-stabilizing (fractal) attractors → dissipative chaos.

Organization of CSs is non-hierarchical but controlled by quality and quantity of contained energy and by system history: CSs are self-organizing processes – dynamic configurations of energetically/historically controlled material compounds (Prigogine, 1979).

Complex Systems are 'machines' which manage permanent high-level energy input from the system environment. In the case of the Ecosystem Earth, selected contributions of radiation energy from the sun (~680 nm) will permanently be transformed into matter (glucose) by photosynthesis and, in this way, biomass or renewable material is formed. Energy management of CS Earth results in:

- Fixation of radiation energy as matter, dominantly by photosynthesis.
- Fixation of energy by establishing long-range order interactions.
- Transformation of energy input to lower levels by metabolic pathways – at anytime correlated with production of entropy due to the fact that efficiency of transformation is always less than 100%.

6.1.2 Biomass Production of CS Earth

Many of the materials formed by CS Earth are macromolecular compounds and, due to their perpetual production in the life cycles of CSs, constitute the majority of biomass or renewable resources. Carbohydrates/polysaccharides provide structurizing voluminous material, particularly for energy storage. Nucleic acids form genetic code libraries and are symbol-structures containing information. Proteins are important biocatalysts, providing as well as structurizing information and acting as storage materials. Lipids also form membranes and interfaces and are efficient energy storage materials.

As a matter of fact, more or less any subsystem of the Ecosystem Earth is a power plant for a class of materials which represents the major quantity of renewable resources: carbohydrates (CH)/ polysaccharides (PS). However, crops form the major quantity (CH/PS) (Table 6.1) which have a wide range of different functionalities: starch or glycogen are energy resources of crops and mammals; cellulose and hemicellulose are structurizing fibre-forming components in crops; chitin/chitosan form the protective structures of insects and maritime organisms against mechanical damage; hyaluronic acid is a lubricant of joints; chontroidin and dermatansulphate are the basic materials of the bones and cartilages of mammals; glyco-conjugates form specific surface patterns on biological interfaces, which are a major requirement for

Table 6.1 Crops: annually assimilated biomass with details on carbohydrate/polysaccharides

Annually assimilated biomass	10^9 t (Gt) dry matter annually	
Lignin	~20–80	~20%
Lipids	~2–8	~2%
Proteins	~2–8	~2%
Others	~2–8	~1%
Carbohydrates/ polysaccharides	~75–300	~75%
Non-branched $\beta(1 \rightarrow 4)$ linked glucan	Cellulose: 50–200	~45% of CH/PS
	Hemicellulose 20–100	~20–25% of CH/PS
$\alpha(1 \rightarrow 4)$ linked $+ \alpha(1 \rightarrow 6)$ branched glucan	Starch: 1–5 ~0.02 $\rightarrow$ industrial utilization	~2–5% of CH/PS
Others	mannan, galactan, fructans, . . .	
Compared to annual yield of petroleum	~2	
Annually petroleum-based prod. synthc polym	~0.1–0.2	~5–10%

immunological activity; many bacteria produce exopolysaccharides (EPS) as metabolic products (Metzler, 1977; Atkins, 1985; Burchard, 1985).

However, due to this widespread range of functionalities, utilization of these materials is rather tricky as many heterogeneities exist and are superimposed, which obstructs any simple processing. The only way to overcome this problem is comprehensive analysis in terms of identification and quantification at any stage of processing these materials.

6.2 Analytical Approaches for Raw Materials from Crops

Many analytical approaches for identification and quantification of crop components have been published within the last few decades. Standardized and evaluated techniques have been developed and published by public and commercial organizations (AOAC, 2003), (ASTM, 2003), (DIN, 2003), (TAPPI, 2003). Accredited laboratories have been installed all over the world to run standardized analytical techniques to develop significant results by improving individual approaches and by round-robin tests to guarantee secure high-quality agronomy products (ISO International, 2003).

Most of these standardized analytical techniques are rather sumptuous with respect to applied instrumental equipment and time, a fact which in particular causes problems if modifications of applied techniques (depending on the crop variety actually screened) are necessary.

In this chapter, strategies for a simplistic yet comprehensive analysis of different crop materials will be provided in schemes with selected details. Particular focus is given to water-soluble polysaccharides for which a concept will be presented which provides comprehensive information about molecular background of macroscopic material qualities. The data obtained are supposed to considerably facilitate the profiling of native raw materials from crops for appropriate utilization.

6.2.1   Identification/Classification of Cell Composition

Staining of Cell Walls for Bright-field Microscopy

Basic classification of crop raw materials may be achieved by different stainings of cell walls and identification by bright-field microscopy (Flint, 1988) (Table 6.2a).

Staining of Cell Walls for Fluorescence Microscopy

Discrimination of fibre forming polysaccharides and structurizing proteins in the cell walls of crops may be achieved by means of staining and fluorescence microscopy (Fulcher, Irving and deFrancisco, 1989) (Table 6.2b).

Scanning Electron Microscopy and Atomic Force Microscopy

Crystallinity, symmetry, order/disorder and packing characteristics of cell components may be identified by scanning electron microscopy (SEM) and atomic force microscopy (AFM) (Gallant *et al.*, 1992; Jenkins and Donald, 1995; Baldwin *et al.*, 1996; Gallant, Bouchet and Baldwin, 1997). Packing of starch granules, for instance, may be discussed with respect to the packing of their long-chain-branched (lcb) (amylose-type) and short-chain-branched (scb) (amylopectin-type) glucans. All kinds of crystallinity within a starch granule represent more or less ordered structures on a more or less dominant amorphous background with amorphous single-chain and ordered double-helix glucans (Gidley and Bociek, 1985). scb-Glucans are assumed to form crystalline lamellae by parallel double helices with branching positions in amorphous regions. lcb-Glucans are preferably

Table 6.2a   Cellwall staining for bright-field microscopy analysis

Component	Stain	Colour
Cellulose	Thionin	Violet
Protein	Light green	Green
Lignin	Phloroglucinol	Red
Pectin	Ruthenium red	Rose

Table 6.2b   Cellwall staining for fluorescence microscopy analysis

Component	Stain	Colour
Mixed linked β-glucans	Calcofluor	White/blue
	Congo red	Red
	Immuno-staining	Green
Arabino xylans	Immuno-staining	Green
Proteins	Acid fuchsin	Red

located in the amorphous layers (Jenkins *et al.*, 1994) and are subject to complex formation with lipids. Additionally, there exists limited co-crystallization of scb- and lcb-glucans forming small (~25 nm) and large (80–120 nm) blocks.

Element Analysis

Subsequent to isolation, purification, homogenization and drying of crop raw materials, elementary analysis provides information about content of nitrogen (N), carbon (C) and sulphur (S). The content of macroelements in ash may be obtained according to standard methods (AOAC: Association of Analytical Communities, TAPPI: Technical Association of the Pulp and Paper Industry), by means of atomic absorption spectroscopy (AAS), which provides information about the concentration of sodium (Na), calcium (Ca), magnesium (Mg), iron (Fe) and aluminium (Al) within investigated samples. The content of phosphorus (P) is determined from hydrochloric acid-soluble components of ash; silica (Si), from the insoluble residue of the hydrochloric acid extraction. Finally, the content of trace elements may be determined according to established extraction procedures combined with AAS and/or inductively coupled plasma-mass spectrometry (ICP-MS).

Protein-containing Crop Materials

The protein content in crop raw materials may be determined by several standardized methods, which quantify nitrogen (N) as an equivalent of protein. The most commonly applied method is the determination of the total nitrogen according to Kjeldahl and nitrogen-specific elementary analyses. However, using these techniques, discrimination between low-molecular nitrogen-containing compounds such as nitrate, nitrite, amino acids and peptides, and proteins is impossible. For the correlation of nitrogen with specific classes of compounds, additional preparation steps, sample-specific pre-treatment and particular analytical techniques need to be applied. Soluble proteins, for instance, may be extracted by means of particular buffer systems and can be quantified by photo-metric methods or by biochemical assays. Electrophoresis and chromatography combined with particular staining and labelling provides detailed information about protein composition and sample-specific enzyme patterns.

6.2.2 Analytical Strategies for Crop Materials

Non-Starch-containing Crop Materials

For the processing of non-starch-containing crop materials (Table 6.3a) such as hemp, flax, millet and grasses, careful selection of crop components is probably the most important step. However, in addition, representative sampling from selected components is equally important for analysis. To obtain reproducible results, materials of interest immediately need to be cleaned, pre-selected, weighed and stabilized at the location of harvesting. The sample components need to become homogenized and dried for subsequent analysis (Linskens and Jackson, 1996).

Table 6.3a *Analytical strategy for non-starch-containing crop material*

	Non-starch-containing crop material	
	Sampling	*Selection of plant components, determination of dry matter content and determination of moisture*
Freezing/drying/cooling milling/ sieving deactivation of enzymatic activities storage conditions	(pre) treatment	
	Homogenized dry material	
	Apolar extraction	*In hexane/acetone: identification/quantification/ analysis of dissolved lipids, phenolic compounds and chlorophyll; elimination of rest of the moisture*
In alcohol/aqueous media: identification/quantification/ analysis of dissolved mono-/di-/ oligosaccharides, peptides/proteins and salts	Polar extraction	
	Insoluble residues	Identification/analysis: *solid body/surface spectroscopy*
ALKALINE TREATMENT		ACIDIC TREATMENT
Soluble components	**Insoluble components**	**Soluble components**
Analysis/quantification of **dissolved hemicelluloses**	**Ligno-celluloses**	Degradative dissolved cellulose and hemicellulose Acid-soluble lignin
	Acidic treatment Analysis/quantification of **degradative dissolved cellulose**	
	Residue: lignin	Residue: lignin (Klason-lignin)

The initial step in the analysis of these dry and powdery raw materials is an apolar extraction to eliminate low-molecular phenolic compounds and, in particular, chlorophyll. Subsequent polar extraction with different ratios of alcohol and water dissolves (in the case of carbohydrates/polysaccharides) the mono-, di-, oligo- and polysaccharides in a step-by-step process (Linskens and Jackson, 1988).

The remaining residue represents hemicellulose, cellulose, lignin and other aqueous, insoluble cell wall components. Identification and quantification of these materials may be achieved by TAPPI methods with respect to investigated crop variety and available sample quantity.

Hemicelluloses, which dominantly consist of the pentoses xylose and arabinose, are extracted by alkaline treatment and are identified/quantified by chromatography after total acidic hydrolysis: fast identification of monomers may be achieved by thin layer chromatography (TLC); quantification by reversed-phase high performance liquid chromatography (HPLC), anionic exchange chromatography combined with pulsed amperometric detection (HPAEC-PAD; Dionex) or gas–liquid chromatography (GLC) after previous derivatization (BeMiller, Manners and Sturgeon, 1994; Kalra, 1997).

Characterization of lignin requires total hydrolysis of cellulose in lignocellulose (holocellulose), which is achieved by 72% w/w of sulphuric acid (H_2SO_4). In this process, cellulose is degraded to glucose and lignin gets sulphonated. This Klason-lignin is cleaned with hot water in order to remove the sulphuric acid, then dried and quantified gravimetrically. An optional approach to characterizing the cellulose matrix is the oxidative elimination of phenolic compounds (treatment with sodium hyperchlorite or peracetic acid) and gravimetric quantification of the remaining residue.

Fructan-containing Crop Materials

Approximately 25% of the known crop varieties store fructans as reserve polysaccharides. Actually, the most interesting from the industrial point of view are composite varieties for the production of $\beta(2{\to}1)$-fructan inulin. The annual processed quantity of chicory already exceeds 150 000 t, with applications in both food and non-food areas. However, the cultivation of Jerusalem artichoke is increasing as the yield/ha is extremely promising. As the tuber components can be fermented easily, these tubers are perfect raw materials for alcohol production. Different grasses and agaves are additional fructan-plants which provide fructans with varying branching characteristics and, thus, maybe new perspectives for industrial applications (Suzuki and Chatterton, 1993).

An analytical approach to fructan-containing crop materials is provided in Table 6.3b. Apolar components, in particular in samples from the leaves and stem, are eliminated from the dried and homogenized crop material by apolar extraction with hexane/acetone. Subsequently, hot water extraction at 80 °C dissolves the water-soluble components, in particular the carbohydrates. These mono-, di-, oligo- and polysaccharides may be identified by enzymatic test kits after total hydrolysis. The degree of polymerization of the high-molecular polysaccharides is determined by means of calibrated analytical size-exclusion chromatography (SEC). Structural analysis with respect to polymer-constituting glycosidic linkages can be achieved by reductive methylation analysis combined with GLC, coupled with flame ionization detection (FID) or mass spectrometry (MS).

Starch-containing Crop Materials

As starch occurs in many quite different crops, there are a number of approaches established to obtain this material. All of these approaches have aqueous processing of tubers, roots or seeds as a basic treatment in common; however, individual processes are established for industrial handling of potato, maize and wheat, which provide starch in good yield and high purity from these raw materials. Additionally, there are worldwide efforts to improve the quality of starch-containing crops for industrial processing by breeding as well as by genetic modifications (Aspinall, 1982, 1983, 1985).

These efforts need to be qualified and analyzed to evaluate the results obtained either by continuous screening or by analysis of individual samples. Both approaches may be applied at two ends – either on samples that have already undergone industrial processing or on samples that have been taken directly from crop source materials at the location of breeding and/or harvesting. From wherever the samples have been acquired, careful preparation for analysis is the first step in their laboratory characterization. For raw

Table 6.3b *Analytical strategy for fructan-containing crop material*

Fructan-containing crop material		
Tubers/roots/leaves/stem: Jerusalem Artichoke, chicory		
Smashing/suspending in water separation from fibres by sedimentation, freeze-drying and storage conditions	(pre) treatment	
	Homogenized dry material	identification/analysis: *solid body/ surface spectroscopy*
	Apolar extraction (defatting)	dissolution in hexane/acetone: *lipids, phenolic compounds and chlorophyll; elimination of rest of moisture (in particular for leaves and stem)*
Identification/quantification/ analysis dissolution in aqueous media @ 80 °C: *mono-/di-/ oligo-/polysaccharides, peptides/ proteins and salts*	Polar extraction	
	Insoluble residues	*Hemicellulose, cellulose and lign*
	Aqueous dissolved components	
Acidic/enzymatic hydrolysis	Enzymatic analysis of Glu + Fru + Suc	Analytical fractionation: size-exclusion chromatography
Enzymatic + chemical Glc + Fru-analysis quantification of Glc + Fru + fructans	**Fructan**	Molecular weight distribution and degree of polymerization distribution Non-destructive molecular analysis:
	Quantification of Glc + Fru + fructans	dimension, conformation and interactive properties

materials with a high lipid content, such as starch granules from wheat, rye, oat or barley, defatting is a pre-requirement before carbohydrate and polysaccharide analysis can be started. Extracted co-materials such as lipids, proteins and salts should be identified and quantified according to standardized methods.

Optimum stability of starch granules depends on the source and the water content; cereal starch granules show optimum stability with 10–12% water content, whereas potato starch granules need 14–18%. The size and shape of starch granules is characteristic of different crop varieties and, thus, classification of starch according to the geometry of the granules by microscopy is a simple but an effective first approach.

Air-dried starch granules swell rather fast in water with an increase in their diameter by 30–40%. Irreversible modifications of the starch granules may be achieved by raising the temperature of such suspensions so that re-organization of supermolecular structures (gelatinization) is induced.

The gelatinization temperature is an individual characteristic of different crop varieties. Further elevation of the temperature yields optical transparent starch suspensions and subsequent formation of opalescent solutions. Under these conditions, granules disintegrate and take up extreme quantities of water (20–40 g/g) forming a colloidal solution. Continued thermal stress finally provides more and more homogeneous suspensions, a fact which may be monitored by a decreasing viscosity. Gelatinization characteristics of starch

Table 6.3c *Analytical strategy for starch-containing crop material*

Starch-containing crop material		
Tubers/roots →potato; seeds: cereals;		
Smashing/suspending in water separation from fibres by sedimentaion, drying and storage conditions	(pre) treatment	
	Homogenized dry material	Identification/analysis: *solid body/surface spectroscopy*
	Apolar extraction (defatting)	Dissolution in propanol: *lipids; elimination of rest of the moisture*
	Dissolution in polar media	**insoluble residues**
Technological treatment: *suspending in pure water; steam processing*		
Analysis/quantification: *short-chain branched (scb), long-chain branched (lcb)* and *non-branched (nb)*	Dissolved/suspended in polar medium	Dissolution for analysis: aqueous solution @ elevated temperature alkaline (NaOH) dimethyl sulphoxide (DMSO)
Acidic/enzymatic hydrolysis	Physico-chemical analysis of technological characteristics	(semi-) preparative fractionation — Analytical fractionation: SEC
Enzymatic+chemical Glc-analysis	Visco-elasticity: rheology disintegration/ re-organization: DSC thermal resistance: swelling, gelatinization order/ crystallinity: X-ray	Fractionation of aqueous soluble glucans according to differences in excluded volume
Quantification: Glc + glucans		Destructive/ non-destructive branching analysis: chain-lengths distribution — Non-destructive molecular analysis: dimension, conformation and interactive properties

suspensions are typically monitored by applying standardized temperature and time programmes by means of a Brabender Viscoamylograph.

Since the basics of the technological starch properties are supposed to be located in molecular characteristics, a number of 'molecular' techniques such as physico-chemical approaches, pure chemical and enzymatically supported techniques may supply valuable information about starch polysaccharides. Table 6.3c provides a scheme for the steps of an analytical approach for starch-containing crop materials. Table 6.4 lists a number of experimental techniques which provide appropriate information about the molecular and supermolecular level of the investigated starch polysaccharides. A summary of the characteristics on different levels (molecular, supermolecular and macroscopic-technological) will constitute a datasheet. It also provides information on the appropriate utilization of individual fractions – for instance, about their sensitivity to oxidation, esterification, etherification or specific enzymatic modification.

Table 6.4 *Experimental approaches to analyze molecular and supermolecular characteristics of starch polysaccharides*

Experimental approach	Obtained information
Thin layer chromatography (TLC); Enzymatic assays reversed phase high performance liquid chromatography (rpHPLC); High performance anionic-exchange chromatography – pulsed amperometric detection (HPAEC-PAD)	Identification/quantification of soluble components; Mono-saccharides, di-saccharides, oligosaccharides; Obtained from direct extraction or enzymatic and acidic hydrolysis
Enzymatically controlled step-by-step fragmentation + fragment analysis by means of rpHPLC or size-exclusion chromatography (SEC)	Branching characteristics: constituting chain lengths distribution, branching percentage
(semi-) preparative SEC staining/ complexing + spectroscopy	Mass fractions separated according to excluded volume for subsequent processing or analysis; Fraction specific branching characteristics
SEC – mass + standards	Calibrated molecular weight distribution; Degree of polymerization distribution; Molecular weight averages
SEC – mass/light scattering/viscosity	Absolute molecular weight distribution; Absolute degree of polymerization distribution; Excluded volume distribution; Polysaccharide-coil packing density distribution
Quantitative terminal labelling of polysaccharides + SEC – mass/molar	Absolute molecular weight distribution of molar fractions
Intrinsic viscosity	Excluded volume, overlapping concentration
Viscosity at varying shear stress	Viscous + elastic contributions (visco-elasticity); gel-properties
Viscosity at increasing thermal stress	Gelatinization temperature (range); disintegration temperature (range)
Viscosity at varying thermal + const mech. stress (Brabender Viscograph)	Stability/resistance towards applied energy (temperature, shearing, periods); disintegration/ re-organization capacities
Differential scanning calorimetry (DSC)	Modifications in conformation; phase transitions
Photon correlation spectroscopy (PCS)	Diffusion mobility of molecular and supermolecular polysaccharide structures

6.2.3   Analysis of Carbohydrates/Polysaccharides: Experimental Approaches for Analyzing Molecular Characteristics

Polysaccharides from natural sources such as cellulose, hemicellulose, starch or fructans are heterogeneous materials in more or less any characteristic at the molecular level. Polysaccharides represent a mix of superimposed distributions controlled by two major classes of influences: systematics of biology in biosynthesis (Metzler, 1977) and heterogeneities caused by polymer chemistry. Characterization of polysaccharides therefore needs to consider heterogeneities as generic qualities of these materials, and appropriate strategies must include component analysis by means of separation techniques such as chromatography and bulk techniques to monitor components during interaction. Application of this strategy will be demonstrated in more detail for starch polysaccharides. Table 6.4

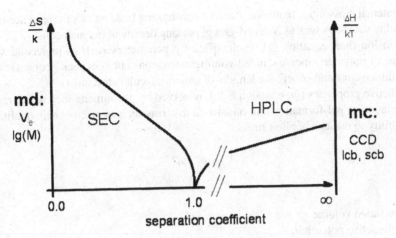

Figure 6.2 *Separation criteria in liquid chromatography (LC). Entropy-controlled separation (ΔS/k) according to differences between individual components in excluded volume (Vₑ), size-exclusion chromatography (SEC), Enthalpy-controlled high performance liquid chromatography-separation (ΔH/kT) according to differences between individual components in interaction potential with liquid chromatography matrix, S: entropy, H: enthalpy, T: Temperature, k: Boltzmann constant, mc: molecular conformation, CCD: chemical composition distribution, lcb: long-chain branched, scb: short-chain branched, md: molecular dimension, Vₑ: excluded volume, lg(M): logarithm of molecular weight*

contains an overview of experimental approaches for starch polysaccharides to obtain comprehensive information about fraction properties (chromatographically supported techniques), as well as about building blocks and branching characteristics (destructive techniques), and molecular plus supermolecular organization characteristics (supported by rheology, thermal and mobility scanning).

Separation techniques, in particular liquid chromatography (LC) techniques, are major tools in the analysis/characterization of carbohydrates and polysaccharides. Separation criteria of LC techniques may be distinguished into two groups (Figure 6.2): whereas enthalpy-controlled HPLC techniques provide powerful separation mechanisms for mono-, di- and oligosaccharides due to different interaction potentials of individual components with LC-matrix-ligands, application of entropy-controlled SEC enables determination of molecular dimensions of oligo- and polysaccharides in terms of molecular weight/degree of polymerization distribution and excluded volume distribution.

Molecular Weight Distribution by Means of SEC

The SEC-separation criterion (separation of components according to differences in excluded volume (V_e) and molecular dimension (md)) matches perfectly with the major property of polysaccharides, that is filling volume in a more or less regular way. The excluded volume for each component is controlled by a fine-tuning mechanism at the molecular level (equation 6.1a):

- Conformation (*mc*; equation 6.1a): molecular symmetries such as helices; beta-sheets; branching pattern (short-chain, long-chain branches, number of branching points); external and internal cross-links; oxidation status of constituting building blocks; presence of compatibility structures such as *N*-acetyl-group; packing density of polymer coils.

- Dimension (*md*; equation 6.1a): classified by parameters such as molecular weight/ degree of polymerization/excluded volume; transition states between geometric molecular dimensions and coherence lengths of supermolecular structures.
- Interactive properties (*ip*; equation 6.1a): observed by phenomena such as aggregation/ association or gel-formation and quantified in terms such as visco-elastic qualities and capability to manage applied stress.

$$V_e = ip \cdot md^{mc} \tag{6.1a}$$

where

V_e excluded volume
ip   interactive potential
md   molecular dimension
mc   molecular conformation.

However, although variations of *md*, *mc* and *ip* already provide a countless variety of polysaccharides, diversity is increased even more by distributions of each of these features. Each distribution additionally needs to be considered either as distribution of mass fractions (m_V_eD; equation 6.1b) or distribution of molar fractions (n_V_eD; equation 6.1c).

$$m\_V_eD = ipD \cdot m\_mdD^{mcD} \tag{6.1b}$$

where

m_V_eD mass fractions of excluded volume distribution
ipD   distribution of interactive potentials
m_mdD   mass fraction of molecular dimension distribution
mcD   distribution of molecular conformation.

$$n\_V_eD = ipD \cdot n\_mdD^{mcD} \tag{6.1c}$$

where

n_V_eD molar fractions of excluded volume distribution
ipD   distribution of interactive potentials
n_mdD   molar fraction of molecular dimension distribution
mcD   distribution of molecular conformation.

Starch polysaccharides are a superimposed heterogeneous mix of regular and irregular modules with highly symmetrical helices (multiple helices): primarily formed by lcb-glucans, irregular 'fractal' structures; primarily established by scb-glucans, compact and internally H-bond stabilized structures; and dominantly formed by scb-glucans (crystallinity). Additionally, there exist minor compact 'amorphous' domains with pronounced re-organization capability, dominantly formed by lcb-glucans.

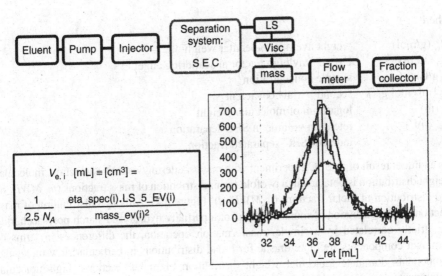

Figure 6.3 *Determination of excluded volume distribution (m_V_eD) of starch polysaccharides by means of SEC – mass/light scattering/viscosity – triple detection: mass by refractive index DRI (–△–), scattering intensity as Rayleigh Factor @ $\Theta = 5°$ (–○–) and specific viscosity via differential pressure (dp)/inlet pressure (ip) (–□–)*

Due to this background, application of SEC for analysis/characterization of polysaccharides is supposed to provide information about actually excluded volume (V_e) of each fraction with details on *ip*-, *md*- and *mc*-contributions if appropriate detection systems are coupled to SEC. Excluded volume (V_e) and sphere equivalent radius of excluded volume (R_e) of polysaccharide fractions may be obtained by means of triple detection coupled to previous SEC: mass concentration ($conc_{mass}$ [g/L]); scattering intensity for scattering angle $\Theta \to 0$ [$R_\Theta \to$ g/mol]; and specific viscosity η_{spec} [mPa $\to$ mL/g].

Figure 6.3 shows a scheme for such an experimental setup which consists of an SEC section with 2–5 columns of different pore sizes in series, followed by a detection section with one or more detectors in series and a numerical data processing section where parameters of interest ($V_{e,i}$) are computed from initially detected elution profiles from different detectors. SEC separation of polysaccharides is typically achieved by aqueous eluents with defined salt content (e.g. 0.005–0.1 M NaCl) at isocratic conditions and with constant flow rates between 0.3 and 1.0 mL/min. Manually, or via autosampler, applied volumes of sample solutions range between 50 and 300 μL with typical concentrations between 0.1 and 0.3% w/w. If selected fractions are transferred to subsequent offline analyses, such as staining/complexation for instance, a fraction collector is the terminal component of such an SEC-multiple-detection system (Barth, 1984; Cooper, 1989; Provder, 1993).

Combined processing of elution profiles from mass and scattering detectors (equation 6.2) provides absolute molecular weight for each SEC-separated polysaccharide-fraction ($M_{w,i}$) and enables the establishment of an absolute molecular weight calibration function (lg(M) vs V_{ret}) for the utilized SEC system.

$$M_{w,i} = \left[\left[\frac{K \cdot c_i}{R_{\Theta,i}} \right]_{\substack{c \to 0 \\ \Theta \to 0}} - 2A_2 c_i \right]^{-1} \to \text{lg}(M) \text{ vs } V_{ret} \tag{6.2}$$

where

M_w (g/mol)	weight average molecular weight
R_Θ (1/cm)	Excess Rayleigh Factor at scattering angel Θ
c (g/L)	sample concentration
A_2 (mL mol g^{-2})	second Virial coefficient
lg (m)	logarithm of molecular weight
V_{ret} (mL)	retention volume in SEC separation
i	index of SEC-separated fraction.

As an initial result of an SEC experiment with mass/scattering/viscosity detection, molecular weight distribution is obtained in a twofold form: distribution of mass fractions (m_MWD_d) and distribution of molar fractions (n_MWD_d) (Figure 6.4). Although distribution of mass fractions and distribution of molar fractions of a partially hydrolyzed starch polysaccharide are strictly correlated by a simple transformation operation, the differences in terms of dominant components are quite drastic for broad distributions in particular: low-molecular components dominate molecular weight distribution of molar fractions; high-molecular components dominate molecular weight distribution of mass fractions.

As indicated in Figure 6.4, molecular weight averages may be computed as representatives for obtained molecular weight distributions. Number average molecular weight M_n and weight average molecular weight M_w are computed according to equations 6.3a and 6.3b respectively. The ratio of M_w to M_n is assigned as polydispersity, a mean indicator of heterogeneity.

$$\overline{M_n} = \frac{\sum n_i M_i}{\sum n_i} \tag{6.3a}$$

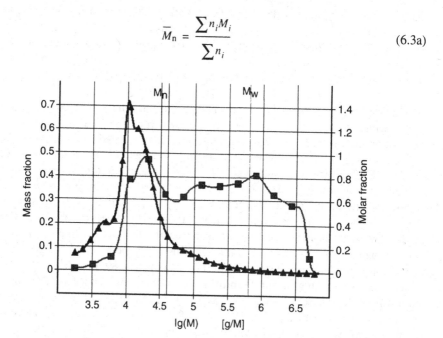

Figure 6.4 *Molecular weight distribution of a partially hydrolyzed starch polysaccharide. (–■–) differential distribution of mass fractions (m_MWD_d); (–▲–) differential distribution of molar fractions (n_MWD_d); weight average molecular weight (M$_w$ = 625 000 g/mol); number average molecular weight (M$_n$ = 39 700 g/mol); polydispersity M$_w$/M$_n$ = 15.7*

$$\overline{M}_w = \frac{\sum n_i M_i^2}{\sum n_i M_i} \tag{6.3b}$$

where

M (g/mol)   molecular weight
n (mol/L)   mol number
i index of SEC-separated fraction.

Processing of molecular weight distribution and intrinsic viscosity elution profile from SEC ($[\eta] = K \cdot M^a$) according to equation 6.4a provides distributions of excluded volumes for the investigated starch-hydrolyzed components: once again, and analogous to molecular weight distribution, distribution of mass fractions ($m\_V_eD\_d$) as well as distribution of molar fractions ($n\_V_eD\_d$) of excluded volumes may be obtained. A simple transformation according to equation 6.4b provides sphere equivalent radii (R_e) of investigated polysaccharides from their excluded volume (V_e).

$$V_{e,i} = \frac{KM_i^{a+1}}{2.5N_A} \rightarrow m\_V_eD\_d \wedge n\_V_eD\_d \tag{6.4a}$$

$$R_{e,i} = \left[\frac{3V_{e,i}}{4\Pi}\right]^{\frac{1}{3}} \rightarrow m\_R_eD\_d \wedge n\_R_eD\_d \tag{6.4b}$$

Finally, for the case of known molar mass (M_i) and excluded volume ($V_{e,i}$) for each polysaccharide fraction, packing density may simply be obtained as the ratio ($M_i/V_{e,i}$). Even slight changes in these distributions due to technological processing or applied modifications strongly influence material properties and, thus, observation of these distributions provides good indicators for appropriate/inappropriate utilization of the fractions.

Branching Characteristics of Starch Polysaccharides

The stability of molecular and supermolecular polysaccharide structures depends on a huge number of influences, however the branching pattern is one of the most important parameters. Branching analysis may be achieved by means of (semi-)preparative separation techniques (combined with staining and classification of obtained chromophores by spectroscopy) or by controlled fragmentation and subsequent fragment analysis. Both approaches are focused on parameters such as the kind of branching (scb, lcb), number and/or percentage of branching positions and the homogeneity of the branching positions within individual molecules. For comprehensive information, both approaches typically provide complementary data.

Starch polysaccharide-fractions may be stained with polyiodide anions, as the hydrophobic helical caves of $\alpha(1 \rightarrow 4)$-linked glucans incorporate these polyanions and shift the extinction maximum from 525 nm for dissolved iodide to 640 nm for the complexed form. An extinction ratio E_{640}/E_{525} profile for SEC-separated fractions provides a correlation of branching characteristics with the molecular dimensions (Figure 6.5): the smaller the E_{640}/E_{525} ratio the more pronounced the scb-characteristics. In a qualitative classification, values in the vicinity of 0.5 are significantly scb-glucans, typically for waxy-type cereal

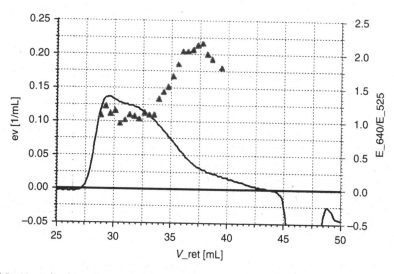

Figure 6.5　Normalized SEC-elution profile (ev: area = 1.0) of wheat starch polysaccharides: heterogeneous mix of lcb + scb-glucans (—); offline iodine-stained fractions, vis-monitored extinctions at λ_{640} (complexed polyiodide by lcb-glucan) and λ_{525} (free iodine) as ratio of E_{640}/E_{525} ($\triangle$)

starches. However, values up to 1.2 indicate scb-type polysaccharides. The E_{640}/E_{525} values between 1.2 and 1.5 represent transition-state-glucans between scb and lcb, values exceeding 1.5 are lcb-type, typically found for amylose-type starches from pea or maize. Results of such investigations for an lcb/scb mixed-type wheat starch are illustrated in Figure 6.5 showing scb-characteristics for polysaccharides with large excluded volume, via transition-state-polysaccharides in the midrange section, and lcb-characteristics for the fractions with small excluded volume.

A parallel destructive approach by controlled fragmentation combined with subsequent fragment analysis directly provides information about the chain-length composition (A-, B-, C-chains), the mean number of molecule-constituting segments (branches) and the number of branching points (typically expressed in terms of branching percentage (Table 6.5)). Therefore, two fractions of the starch polysaccharide are obtained by complexation with n-butanol, which yields lcb-glucans as precipitate with subsequent precipitation of scb-glucans from the supernatant by addition of an excess of methanol. For analysis of the branching characteristics, the obtained fractions are redissolved and stabilized in aqueous buffer systems at pH 6. For the wheat starch, for example, mass fractions of 22% lcb-glucans and 78% scb glucans could be separated.

Composition of starch polysaccharides in terms of lengths (degree of polymerization) of the backbone C-chain branches, primary B-chain branches and the terminal secondary A-chain branches are obtained by step-by-step specific hydrolysis and spectroscopy of specifically labelled fragments. Molar concentration (number) of C-chains (n_C) is obtained from calibrated spectroscopy of quantitatively labelled terminal semiacetals of fully intact molecules. Hydrolysis of terminal A-chains by β-amylase (EC 3.2.1.2) provides mass percentage of A-chains (m_A). Subsequent debranching of $\alpha(1 \rightarrow 6)$ linkages by iso-amylase (EC 3.2.1.68) and pullulanase (EC 3.2.1.41) provides chain-length distribution of B+C-chains. Second-step labelling of newly formed terminal semiacetals from a

Table 6.5 Branching characteristics of a wheat starch polysaccharide obtained from destructive step-by-step fragmentation and subsequent fragment analysis

	%	dp$_n$	n Glc	seg	Full turns	br_%
n-BuOH → 22%						
Molecule		254				3.6
C-chains	46	112	112	basic 1	1×2–15	
B-chains	14	39	39	1	1×6	
A-chains	40	13	104	8	8×2	
A + B-chains	54					
MeOH → 78%						
Molecule		968				6.3
C-chains	5	50	50	basic 1	1×2–7	
B-chains	25	40	241	6	40×7	
A-chains	70	13	677	52	52×2	
A + B-chains	95					

% – percentage; dp$_n$ – number average degree of polymerization; n Glc – number of anhydro-glucose units; seg – number of segments (branches); full turns – number of helical full turns for complexation; br_% – branching percentage as percentage of glucose units with branching at 6-position.

debranching action provides the number of B+C-chains $(n_B + n_C)$. Simple subtraction $(n_B + n_C) - n_C$ yields the molar concentration (number) of primary B-branches n_B. Parallel debranching of an equivalent of fully intact polysaccharides, labelling of terminal semia-cetals and spectroscopy provide the molar concentration of $n_A + n_B + n_C$-chains. The number of A-chains n_A may be computed simply as the difference of $(n_A + n_B + n_C) - (n_B + n_C)$. Thus, by combinatory fragmentation percentages of A-, B-, C-chains, the chain-length distribution of these chain categories and the number of branching-positions within mean molecules can be achieved. Branching percentage (br_%) may be calculated according to $(100 \cdot \Sigma \mathrm{seg})/dp_n$.

As an example, the results for a mixed-type wheat starch polysaccharide are listed in Table 6.5. Initial precipitation fractionation yields a 22% fraction of lcb-starch glucans and 78% fraction of scb-glucans. Detailed destructive analysis supports the initial classification: dominant mass contributions (46%) in basic C-chains for the n-butanol-precipitated fraction with a mean molecule consisting of 1 basic segment and 9 branches (1C + 9B) results in a branching percentage of br_% = 3.6%, whereas the methanol-precipitated fraction dominantly consists of A-chains with a mean molecule consisting of 59 segments (1C + 6B + 52A) and a branching percentage of br_% = 6.3%.

Stability and Re-Organization Capacity of Starch Polysaccharides

Typically, stability investigations are performed by means of different rheological/viscosimetric approaches. In general, stress is applied on a polysaccharide solution or suspension and the correlated changes in viscosity are interpreted in terms of less pronounced stability and re-organization capacity.

Application of increasing thermal stress on starch polysaccharides provides information about the so-called gelatinization temperature: a critical temperature at which disintegration

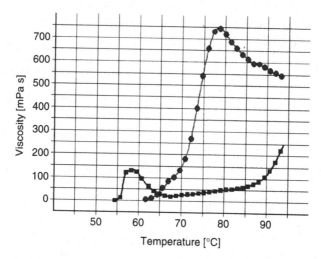

Figure 6.6a *Stability of starch glucans on applied thermal stress – Viscosity at increasing temperature. wheat (–■–): minor initial stability; disintegration peak: 56–62 °C; T$_{max}$ = 58 °C, waxy maize (–●–): high initial stability; disintegration peak: 65–85 °C; T$_{max}$ = 78 °C*

of the supermolecular glucan structures starts. Differences in gelatinization and disintegration behaviour are illustrated in Figure 6.6a for two starch polysaccharides with different branching patterns: lcb/scb mixed-type wheat starch is disintegrated at temperatures (56–62 °C; T$_{max}$ = 58 °C) significantly lower than scb-type waxy maize glucans (65–85 °C; T$_{max}$ = 78 °C). A much more pronounced increase in viscosity was observed upon disintegration of scb-type waxy maize starch compared to lcb-type wheat starch, indicating a comparably better glucan/glucan stabilization of the investigated scb-type compared to the lcb-type starch.

Monitoring the so-called Brabender viscosity at simultaneously applied varying thermal and constant mechanical stresses provides stability and re-organization capacities as time/ temperature programme profiles. A typically 3-state temperature programme (heating, holding, cooling) is illustrated in Figure 6.6b for the same two samples as investigated before for their gelatinization behaviour. As an important result of such investigations, a significant difference between the two starches becomes apparent: they have a quite different re-organization capacity once they have been disintegrated. Although more energy is needed to disintegrate scb-type waxy maize glucans in the initial heating period, once liberated from supermolecular formations, they do not tend to re-establish super-molecular structures, neither in the high temperature holding nor in the subsequent cooling period. Quite a different behaviour is observed for lcb/scb mixed-type wheat starch glucans: disintegration is achieved much easier than for scb-glucans, however (indicated by enormously increasing viscosity in the final cooling period) tendency of liberated lcb-glucans to re-establish supermolecular structures is much more pronounced.

Finally, information obtained about molecular, supermolecular and technological characteristics may be compiled in a comprehensive datasheet as suggested for a wheat starch sample (Table 6.6). An easily accessible (e.g. WWW) database of such information

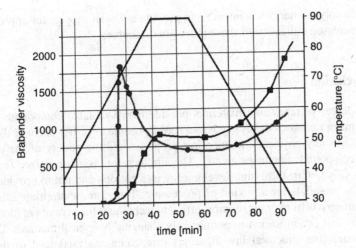

Figure 6.6b *Brabender viscosity: constant shear deformation; applied temperature programme: heating –*
30 → 90 °C within 40 min, holding – 90° for 15 min, cooling – 90 → 30 °C within 40 min; wheat glucans
(–■–): broad disintegration → re-organization of supermolecular structures; re-organization on cooling;
waxy maize glucans (–●–): more sharp disintegration → minor re-organization or supermolecular structures

Table 6.6 *Selected molecular and supermolecular characteristics in a datasheet for the*
investigated wheat starch polysaccharide sample

Source	Wheat starch
Origin	Chamtor, France
Glucan content (%)	98.3
moisture (%)	9.8
Type of X-ray diffraction pattern	A
Branching characteristics: scb-fraction (mass %)	78
Branching characteristics: lcb-fraction (mass %)	22
Percentage of branching – lcb-fraction (%)	6.3
Percentage of branching – scb-fraction (%)	3.6
Percentage of branching – mean (%)	5.7
ß-amylolysis → terminal A-chains (mass %)	70.1
Excluded sphere equivalent radii $V_e \rightarrow R_e$ from SEC range (nm)	2–55
Maximum of molar fractions distribution (nm)	5
Maximum of mass fractions distribution (nm)	38
Range of apparent absolute molecular weight from SEC-mass/LS (g/mol)	10–120×10^7
Range of apparent absolute molecular weight from SEC-mass/molar (g/mol)	32 000–380 000
Corresponding weight average molecular weight M_w (g/mol)	204 000
Corresponding number average molecular weight M_n (g/mol)	147 000
Disintegration upon thermal stress: temperature range/T_{max} (°C)	56–62/58
Re-organization capacity after disintegration	+++

for different polysaccharides is intended to provide essential support for appropriate and therefore increased utilization of these renewable resources.

6.3 Conclusion

Crops as sources of renewable materials provide many valuable compounds, however, with some major drawbacks: they occur as a hard-to-handle mix; they occur as distributions to more or less any physical and chemical property; and the quality of the matrix and individual compounds fluctuates greatly. The aim therefore is an optimized functionality of crops as CSs to transform high-level energy into entropy and not to provide pure and identical raw materials for any kind of processing. Therefore, appropriate utilization of renewable materials from crops requires efficient techniques for permanent identification, classification and quantification in permanently changing biological matrices. Destructive and non-destructive structural investigations (branching analysis) lead to information about molecular dimensions. Determination of excluded volume and coherence lengths provides data about molecular conformation, and investigation of response capability upon applied stress (stability, re-organization capability and lifetime of supermolecular structures) leads to data about the interactive properties. As polysaccharides are rather broadly distributed, particular attention is paid to the application of separation techniques (SEC) and to the distribution characteristics (mass and molar fractions), as even minor shifts in fraction percentages influence technological properties significantly.

A major argument for the utilization of renewable resources is the fact that they are not limited to a few geographical locations, but rather are ubiquitous and annually available. However, cost-covering and sustainable processing of renewable resources requires total utilization, that is producing as little waste as possible. There is no doubt that this goal can be achieved only using biotechnology, which needs to become implemented much more than it is already. Analytical methods must also be adapted for this challenge: to handle distributions; to handle complex matrices; and to provide sensitive characteristics for specific material qualities.

References

AOAC (2003). International, Association of Official Analytical Chemists, Official Methods of Analysis, 481 North Frederick Avenue Suite 500, Gaithersburg, Maryland 20877, USA, www.aoac.org.

Aspinall, G.O. (ed.) (1982, 1983, 1985). *The Polysaccharides*, Vol. 1–3, Academic Press, New York.

ASTM (2003). International, American Society for Testing Materials, Standard methods, Headquarters, 100 Barr Harbor Dr, PO Box C700, W. Conshohocken, PA 19428–2959, USA, www.astm.org.

Atkins, E.D.T. (ed.) (1985). Polysaccharides – topics in structure and morphology, Vol. 8. In: *Topics in Molecular and Structural Biology*. VCH, Weinheim.

Baldwin, P.M., Frazier, R.A., Adler, J., Glasbey, T.O., Keane, M.P., Roberts, C.J., Tendler, S.J.B., Davies, M.C. and Melia, C.D. (1996). Surface imaging of thermally sensitive particulate and fibrous materials with the atomic force microscope: a novel sample preparation method. *J. Microsc-Oxfor.*, **184**(2), 75–80.

Barth, H.G. (ed.) (1984). *Modern Methods of Particle Size Analysis*, John Wiley & Sons, New York.

BeMiller, J.N., Manners, D.J. and Sturgeon, R.J. (eds) (1994). *Methods in Carbohydrate Chemistry*, Vol. 10, Enzymic Methods, John Wiley & Sons.

Burchard, W. (ed.) (1985). *Polysaccharide*, Springer-Verlag, Berlin, Heidelberg.

Cooper, A.R. (ed.) (1989). Determination of Molecular Weight, Vol. 103. In: *Chemical Analysis*. John Wiley & Sons, New York.

DIN (2003). Methods, Deutsches Institut für Normung, www.din.de.

ISO International, The International Organization for Standardization, www.iso.org.

Flint, F.O. (1988). The evaluation of food structure by light microscopy. In: *Food Structure – Its Creation and Evaluation* (eds J.M.V. Blanshard and J.R. Michtell), pp. 351–365. Butterworth, London.

Fulcher, R.G., Irving, D.W. and deFrancisco, A. (1989). Fluorescence microscopy: applications in food analysis. In: *Fluorescence Analysis in Foods* (ed. L. Munck), pp. 59–109. Logman Scientific & Technical, Singapore.

Gallant, D.J., Bouchet, B. and Baldwin, P.M. (1997). Microscopy of starch: evidence of a new level of granule organization. *Carbohydr. Polym.*, **32**, 177–191.

Gallant, D.J., Bouchet, B., Buleon, A. and Perez, S. (1992). Physical characteristics of starch granules and susceptibility to enzymatic degradation. *Eur. J. Clin. Nutr.*, **46**, S3–S16.

Gidley, M.J. and Bociek, S.M. (1985). Molecular organization in starches: a ^{13}C CP/MAS NMR study. *J. Am. Chem. Soc.*, **107**, 7040–7044.

Jenkins, P.J. and Donald, A.M. (1995). The influence of amylose on starch granule structure. *Int. J. Biol. Macromol.*, **17**, 315–321.

Jenkins, P.J., Cameron, R.E., Donald, A.M., Bras, W., Derbyshire, G.E., Mant, G.R. and Ryan, A.J. (1994). In situ simultaneous small and wide-angle X-ray scattering: a new technique to study starch gelatinization. *J. Polym. Sci. Polym. Phys. Ed.*, **32**, 1579–1583.

Kalra, Y.P. (eds) (1997). *Handbook of Reference Methods for Plant Analysis*, CRC Press, Boca Raton, FL.

Linskens, H.F. and Jackson, J.F. (1988). *Modern Methods of Plant Analysis*, Vol. 8, Springer-Verlag, New York.

Linskens, H.F. and Jackson, J.F. (1996). *Plant Cell Wall Analysis*, Springer-Verlag, New York, Incorporated.

Metzler, D.E. (1977). *Biochemistry – The Chemical Reaction of Living Cells*, Academic Press, New York, San Francisco, London.

Prigogine, I. (1979). Vom Sein zum Werden: Zeit und Komplexität in den Naturwissenschaften. Piper Verlag, München, Zürich. Originally published as 'From Being to Becoming – Time and Complexity in Physical Sciences.'

Provder, T. (ed.) (1993). *Chromatography of Polymers – Characterization by SEC and FFF*, ACS Symposium Series 521, ACS, Washington DC.

Suzuki, M. and Chatterton, N.J. (1993). *Science & Technology of Fructans*, CRC Press, Boca Raton, FL.

TAPPI (2003). Technical Association of the Pulp and Paper Industry, Test Methods & Technical Information Papers, www.tappi.org.

7

Industrial Products from Carbohydrates, Wood and Fibres

Christian V. Stevens

In co-authorship with *Piotr Janas, Monica Kordowska-Wiater, Paul Kosma, Zdzislaw Targonski and Carlos Vaca Garcia*

7.1 Introduction

Carbohydrates are the most widespread class of compounds in nature and are produced in enormous amounts by all plants. Seventy-five percent of the annual biomass production by photosynthesis, i.e. 170×10^9 t, are carbohydrates. However, only 6% of this annual production is being used by human beings (62% for food, 33% for energy and housing and only 5% for non-food applications) (Röper, 2002). Not only are carbohydrates the primary products formed by photosynthesis and the energy storage compounds of plants and bacteria, they also play a very important role in the functions of cells and are starting materials for many vital compounds in biochemical pathways, in cell–cell recognition strategies and in cellular transport. Further, carbohydrates form an immense resource pool for organic synthetic chemists in the synthesis of many complex natural compounds, such as antibiotics (Nicolau and Mitchell, 2001).

In this chapter, next to some fundamental characteristics of the compounds discussed, most attention will be focused on the large-scale use of carbohydrates and modified carbohydrates as industrial products and the technology involved in these transformations. The following classes of carbohydrates and their applications will be discussed:

- The occurrence and use of polysaccharides and fibres, including wood, cotton, paper, starches and chitin.
- The applications of oligosaccharides such as maltodextrins, fructans and cyclodextrins.

Renewable Bioresources: Scope and Modification for Non-food Applications. Edited by C.V. Stevens and R. Verhé
© 2004 John Wiley & Sons, Ltd   ISBNs: 0-470-85446-4 (HB); 0-470-85447-2 (PB)

- The occurrence and use of disaccharides such as sucrose.
- The modification of monosaccharides (glucose, fructose, xylose and arabinose) into useful chemicals.

The use of carbohydrates for non-food applications in industry can have several advantages with respect to the sustainable development of industry in the Western World, and also in developing countries. The net zero balance towards carbon dioxide emission during the utilization of renewable resources, the reduction of overproduction of food in the more developed countries and the chance for a higher degree of biodegradation of naturally derived products are only some advantages of the use of carbohydrates in industry. The production of renewable resources for industry by agriculture may also become socially important for the agricultural community, in order to prevent the gradual exodus of young people from agricultural activity (as discussed in Chapter 2).

The topic of this chapter covers a vast area of chemistry and the aim of this work is not to give a complete overview, but rather to provide a flavour of the area covering several examples of the use of carbohydrates as industrial products.

7.2   Use of Polysaccharides

7.2.1   Wood (Cellulose, Hemicellulose, Lignin) and Fibres

Talking about industrially important natural polysaccharides immediately refers us to the use of wood, fibres and their derived products, since wood is a complex network of the three most important natural polysaccharides, i.e. cellulose, hemicellulose and lignin. First, some information will be provided on the structure of the wood and fibres of important industrial crops, then the technology used to transform wood into paper will be described and finally the industrially derived products from wood will be discussed.

Morphology Characteristics of Natural Fibres and its Components

Softwood and hardwood

Wood is a naturally grown composite. Both hardwoods (*dicotyledonous* or *angiosperms*) and softwoods (coniferous wood or *gymnosperms*) are composed of strengthening, conducting and storing cells, which are necessary for all functions of the tree. Important differences are not only observed between softwoods and hardwoods, but also between various species and even within one sample, depending on growth rings, earlywood and latewood, pores, etc. These phenomena result from the growth of the wood tissue.

Softwood exhibits quite a simple structure; it contains fewer cell types than hardwood, with less variation in the wood structure, and consists of longitudinal tracheids which constitute 90–95% of the volume of most types of softwoods. The tracheids, which are long and slender cells with flattened or tapered closed edges, are arranged in radial files and their longitudinal extension is oriented in the direction of the stem axles. During passage from earlywood to latewood the diameters of the cell become smaller and the cell wall thicker. At the beginning of the next growth period large-diameter tracheids are developed by the tree. These changes are visible as an annual or growth ring. The thick-walled tracheids provide strength, while earlywood tracheids conduct water with minerals within the tree. In the small and thin-walled parenchyma cells storage and food transport take place. These cells are arranged in softwoods in radially running rays. Softwood can be categorized by the presence or absence of resin canals (Fengel and Wegener, 1984).

Basic hardwood tissue contains fibres and fibre tracheids that provide strength. Within this strengthening tissue conducting vessels are distributed. These vessels, typical for hardwood, are long pipes ranging from a few centimetres up to some metres in length and consist of single cells with open or perforated ends. These elements may be distributed in cross-section in ring porous, semi-ring porous or diffuse porous patterns. The basic elements of hardwood tissue-fibres are smaller and have thicker cell walls than those of the softwood tracheids. The differences in wall thickness between earlywood and latewood are also not as extreme as in softwood. The number of parenchyma cells in hardwoods is higher than in softwoods. The density of wood is determined by the thickness of the fibre walls or tracheids, and the number and diameter of the vessels as well as the number of parenchyma cells.

Cellulose is the key polymeric component in the structure of the cell wall. It measures up to 1 μm for a native plant, with a degree of polymerization which can range from 7000 to 15 000. The stabilization of the long molecular chains originates from the three hydroxyl groups on each monomeric unit, which are able to interact with each other resulting in strong inter- and intramolecular hydrogen bonds. These linkages are responsible for the formation of supramolecular structures. The primary structures formed by hydrogen bonds are fibrils, which make up the wall layers and finally the whole cell wall. Fibres are also able to form hydrogen bonds with each other. These bonds between fibre surfaces determine mechanical properties such as the strength of a pulp or paper sheet. The smallest units are elementary fibrils with an average diameter of 3.5 nm. These basic units are associated in higher systems with diameters of 10–30 nm (microfibrils). The fibrillar units can be split into subunits and single molecular chains by chemical or mechanical treatment.

Three basic models of organization within fibrillar units are known (Figure 7.1). Ordered regions formed by longitudinally arranged chains are common to all models, but differences are present in less ordered regions (Fengel and Wegener, 1984). In the first

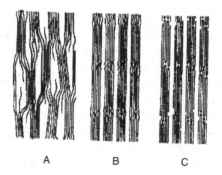

A   B   C

Figure 7.1 *The basic models for the arrangement of cellulose molecules within the fibrillar units (Reproduced from Fengel and Wegener, Wood: Chemistry, ultrastructure, Reactions; published by Walter de Gruyter GmbH & Co. Kg, 1984)*

model, there are ordered regions with transitional molecular chains changing from one ordered region to another (A). In the second, the structure consists of individual fibrillar units with a sequence of ordered and less ordered regions (B). In the third model, fibrillar units consist of folded cellulose chains (C).

In addition to cellulose, hemicelluloses are important polysaccharides present in wood as well as in other plant tissues. They are composed of various sugar units forming shorter chains than cellulose and are very often branched. The main chain of hemicellulose can consist of only one unit (homopolymers), e.g. xylans, or of two or more units (heteropolymers), e.g. glucomannans. Some of the units like 4-*O*-methylglucuronic acid and galactose are side groups of the main chain. Softwood and hardwood differ in the percentage of total hemicellulose and in the percentages and composition of individual hemicellulose. Softwoods have a high proportion of mannose units and more galactose units than hardwoods. Hardwoods have a high proportion of xylose units and more acetyl groups than softwoods (Table 7.1).

After cellulose, lignin is the most abundant and important polymer in the plant world. The amount of lignin present in wood ranges from 20 to 40%. The primary precursors and building units of all lignins are coumaryl, coniferyl and sinapyl alcohol. Lignin permeates cell walls and intercellular regions, giving wood its relatively high hardness and rigidity and acts as a glue for all wood cells. The complicated structure has a pronounced effect on the resistance of lignin to various kinds of biotransformations (Targonski, Rogalski and Leonowicz, 1992).

The structure formed by cellulose, hemicellulose and lignin is relatively similar in different hardwoods and softwoods. It exhibits a layered structure with a thin primary (first-formed) wall, and a thicker secondary wall composed of three or four sublayers: S1, S2 and S3 layers and a last fibrillar layer named the tertiary wall (T). The basic framework of the secondary wall layer (S1–S3) consists of cellulose with varying amounts of hemicellulose, pectin and lignin.

Individual cells are joined by a highly lignified thin layer, the middle lamella, which glues the cells together to form the tissue. Although single fibrils may cross the middle lamella, this layer is in principle free of cellulose.

Annual fibre plants

Annual plants are plants which grow, reproduce and die in 1 year. This group consists of cereals such as wheat, barley, oats, rye, rice, maize, etc. and other fibre plants such as hemp (*Cannabis sativa*), flax (*Linum usitatissimum*) and sunflower (*Helianthus annuus*). Cereal straw, which is used for fibrous mass production, consists of stem, leaves and ear. Depending on the straw species, it consists of 45–70% of the total straw weight. The biggest amount of straw is produced in rye, followed by wheat, barley, oats and rice. The cell walls of straw cells consist of a primary wall and a secondary wall with three layers: an external, middle and internal layer, which are similar to those in wood cells. In the

Table 7.1 *Carbohydrate contents in hardwoods and softwoods (%)*

Lignocellulose	Glucan	Mannan	Galactan	Xylan	Araban	Uronic acid	Acetyl groups
Hardwoods	48 ± 6	2.5 ± 0.7	0.8 ± 0.3	18 ± 4.0	0.5 ± 0.1	4.1 ± 0.6	3.6 ± 0.5
Softwoods	46 ± 1	11 ± 1.0	2.0 ± 1.2	5.8 ± 1.7	1.3 ± 0.7	3.6 ± 0.5	1.3 ± 0.3

fibre cells of the cereals, the secondary wall constitutes over 90% of the fibre cell and the cellulose content is about 50%, hemicelluloses about 34% and lignin in the range of 16–19%. The rice straw is over 40% cellulose, 32% hemicelluloses and 20% lignin.

Some *Dicotyledons*, for example hemp, flax and jute have stems consisting of bark, wood and core. The bark that is the outer cover of the stem is divided into original bark and phloem with fibre cells. The fibres of the above-mentioned plants are on average 70–80% cellulose, 3–9% lignin, 10% hemicelluloses, 1–3% waxes and lipids, and 1–2% mineral substances. However, there is little difference between the ultrastructure of these fibres and the cell walls of wood cells, for example there is no change of the order of fibrils in the same layer. Elementary fibres are grouped and connected with each other by pectic substances and they form a technical fibre. Technical fibre of flax is up to 1-m long, but the average is about 0.5 m. It is built from elementary fibres, 20–40-mm long and 16–32-μm thick. They are similar to cotton fibres but are less flexible and are more difficult to stain. On the other hand, they are more resistant to rupture and are better conductors of heat. In terms of chemical composition they are similar to softwood. Technical hemp fibre is 1–3-m long and the elementary fibre is 15–25-mm long and has a diameter of 20–28 μm. In terms of chemical composition, they are characterized by a sizable proportion of cellulose (about 80%). Corn or sunflower fibre is less important for industry.

Cotton

Cotton (*Gossypium* sp.) fibre, which is used extensively in the textile industry, looks like spiral twisting band with smooth walls without pits. The length of the fibres is about 30–31 mm (middle fibre cotton) or 40–42 mm (long fibre cotton). The width of the fibre ranges from 12 to 42 μm. Mature cotton fibres consist of cellulose that forms 92–94% of the dry biomass. Cotton cellulose fibres contain about 10% of different non-cellulosic compounds such as hemicellulose, pectic substances, proteins, waxes, pigments, lignin-containing impurities and mineral salts. Almost all non-cellulosic compounds are located in the outer layers of the fibre, in the cuticle and in the primary wall, on the fibre surface.

Commercial cottons produce marketable lint fibres (varying between species and cultivars, but approximately 25 mm in length) and shorter-thickened fuzz fibres of limited value. Both types of fibre originate as outgrowths (fibre initials) of epidermal cells of the ovule. The wall of the cotton fibre consists of primary and secondary layers. Changes in the composition of cotton fibre cell walls are observed during development, with a maximum amount of fructose, galactose, mannose, rhamnose, arabinose, uronic acid and non-cellulosic glucose residues at the end of the primary wall formation or at the beginning of the secondary wall formation. Only the absolute amounts of xylose and cellulosic glucose residues increase until the end of the fibre development.

Industrial Processes for Dissolving Pulps and Cellulosic Fibres for Paper Production

Introduction

Since 1900, the human population has increased from 1.6 to more than 6 billion, imposing a continuing demand on textile fibres. Traditionally, cellulose provides the basis for the production of textile fibres, originating from either cotton or wood. Generation of fibres from natural resources, however, has to compete with food production with respect to soil and water usage, and with synthetic fibres whose market share is constantly increasing (synthetic fibres 54%, cotton 38%, other cellulosic fibres 5% and wool 3%) (Johnson, 2001). The total production estimates for viscose rayon, cellulose acetates, carboxymethylcellulose, cellulose nitrates and cellulose ethers are in the range of 4.7×10^6 t/yr (Klemm *et al.*, 1998).

The production of pulp, paper and cellulosic fibres from wood is based on a variety of complex and costly chemical processes. Cellulosic fibre manufacture (the annual production of paper and chemical pulps amounts to $\sim 350 \times 10^6$ t/yr) comprises multiple steps covering from harvesting, storage and chopping of wood, to pulping and bleaching steps, dissolution of cellulose and spinning, and finally finishing of the resulting textile fibres. Thus, it is mandatory to improve the economic basis of fibre production from renewable resources by enhanced utilization of by-products from biomass and process effluents.

Pulping processes

The disintegration of wood into a fibrous mass is achieved by various processes termed "pulping". The input of thermal and mechanical energy and the effects of added chemicals lead to the breakage of bonds within the wood structure and its main constituents, liberating cellulosic fibres which may still contain varying amounts of hemicelluloses and lignin (Gellerstedt, 2001). The major effects of pulping are aimed at dissolving the lignin at the middle lamella without inducing too much damage to the fibres.

Mechanical pulping Mechanical pulping is used for the production of newspaper and print materials. The mechanical energy is provided by grinding the wood with a rotating stone or refining of wood between metal discs, leading to mechanical pulps (MP) or thermo-mechanical pulps (TMP). The yield of the pulping process is usually >90% since most of the lignin is retained, which however implies a reduced light stability of the paper. In order to reduce the high energy needed for the disintegration of the wood structure, chemicals such as Na_2SO_3/Na_2CO_3 or NaOH may be added (chemothermomechanical pulping, CTMP, or chemomechanical pulping, CMP).

Kraft pulping Kraft (German: strength) pulping is the dominant technology employed in the production of paper- and board-grade pulps. A variant of the process, including a pre-hydrolysis stage (pre-hydrolysis Kraft), is also being used for the preparation of dissolving pulps to be further converted into viscose rayon, cellulose acetates (cigarette filters, films, fibres), cellulose nitrates (explosives) and carboxymethyl celluloses (detergents, food additives, cosmetics). The process tolerates various wood sources (soft- and hardwoods) and comprises an efficient recovery cycle of chemicals. The yield of the Kraft process is usually in the range of 44–49% based on the wood raw material, and the resulting pulps contain approximately 80% cellulose, 10% hemicellulose and 10% lignin, which have to be removed in the following bleaching steps. The remaining lignin content of the pulp, which is expressed as the so-called Kappa number, is determined by titration with $KMnO_4$.

In the cooking step, performed in batch or continuous mode, the wood chips are treated at 150–170 °C for 1.5–2 h under pressure (7–11 bar) with "white liquor", containing Na_2S and NaOH. The alkaline conditions induce elimination of α-arylether bonds in lignin, generating quinone methide intermediates which may further react with nucleophilic species such as HS^- and OH^-. However, the strong alkaline conditions also affect the integrity of the polysaccharide constituents by a stepwise "peeling reaction" starting from the reducing end of the sugar chains.

After the cooking the pulp is separated and washed, and the effluent ("black liquor") is concentrated and transferred into the recovery system. In the recovery boiler the organic material is combusted, whereas the inorganic salts are obtained as Na_2S (from reduction of Na_2SO_4) and Na_2CO_3. Finally, in a causticization step, Na_2CO_3 is reacted with $Ca(OH)_2$ to yield $CaCO_3$ and NaOH, which together with the recovered Na_2S are fed into the "white cooking liquor" for the next cooking cycle or preferably as a split charge during the cooking process to compensate for a decrease in the concentration of sulphite ions (Figure 7.2).

In the variant pre-hydrolysis Kraft process employed for the preparation of dissolving pulps, wood chips are pre-treated with steam at 140–170 °C or alternatively with dilute acid at 110–120 °C. The hydrolysis is needed to remove the fibres from the primary wall, whereby hemicelluloses are cleaved. This leads to a reduction of wood mass by ~20%, which is due mainly to degradation of glucomannans in the case of softwood pulping, whereas hardwood pre-hydrolysis mainly affects the lignin components. Prolonged hydrolysis times, however, are responsible for fibre depolymerization as well as undesirable condensation reactions of lignin, leading to *de novo* formation of chromophores which are notoriously difficult to be removed in the bleaching steps.

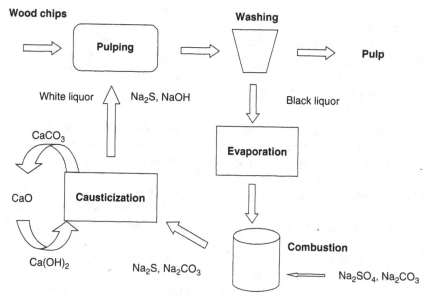

Figure 7.2 *Simplified flow diagram of the Kraft process*

Sulphite-based pulping processes   The requirements for the production of pure dissolving pulps not only comprise the efficient removal of lignin but also the extraction of hemicelluloses and the adjustment to appropriate viscosity values for the cellulose. Sulphur dioxide-based pulping technologies are well suited to provide chemical pulps of high purity but put more demand on the quality of the wood sources (beech, spruce). The outcome of the processes depends mainly on the pH of the cooking liquor, bound and total SO_2 content (equation 7.1) and the temperature profile (135–170 °C, 5–7 bar) during the cooking step.

$$SO_2 \cdot H_2O + H_2O \rightarrow HSO_3^- + H_3O^+ \tag{7.1}$$

The counter-ion of the sulphite salt determines the overall pH of the liquor, with Mg^{2+} and Na^+ being the most commonly employed cations. Previously used cooking liquors containing calcium are less common nowadays, since they may not be subjected to recovery of pulping chemicals.

Typically, the pH used in the magnesium bisulphite process is in the range of 4–5. Delignification under acidic conditions occurs by hydrolysis of α-hydroxyl and α-ether bonds at the α-carbon of the phenylpropane units of lignin, followed by sulphonation of the intermediate benzylium cations by hydrated sulphur dioxide or bisulphite ions (Sjöström, 1981). With increasing acidity and decreasing bisulphite concentration undesirable competing side reactions set in, which are due to self-condensation reactions or reactions of the carbocations with reactive phenolic extractives. This limits the use of wood raw materials such as pine heartwood or eucalyptus, which are rich in extractives such as pinosylvine (3,5-dihydroxy-*trans*-stilbene).

Furthermore, the acidic conditions are responsible for cleavage of glycosidic bonds, leading to a decrease of the molecular weight of the cellulose chains and also to the formation of acetic acid, which is liberated from acetyl groups of galactoglucomannans in softwood or glucuronoxylans in hardwoods. In addition, furfural is formed from pentoses such as xylose and may be exploited as a substitute for petrochemicals. Xylose itself may be directly recovered from the spent liquor after the cooking step, and is further treated by catalytic reduction with hydrogen to give xylitol (a sweetening food additive) on a commercial scale.

Sulphite pulping is also performed at neutral or alkaline pH conditions to furnish high-yield pulps, which still contain a high proportion of residual lignin. The recovery of pulping chemicals depends on the type of base used in the sulphite process. The spent liquor is concentrated to ~50–65% of solids and combusted. In the magnesium bisulphite process, MgO and SO_2 are regenerated from the combustion gases, while sodium-based spent liquors are burned in a Kraft-type furnace to give a smelt containing Na_2S and Na_2CO_3.

Miscellaneous pulping processes   The pre-hydrolysis Kraft process may also be performed without sulphur containing cooking liquors – as in soda-anthraquinone variant, where 0.02–0.1% anthraquinone is added. Anthraquinone serves as a stabilizing agent for the alkaline-sensitive carbohydrates, which are oxidized to give aldonic acid end groups. The reduced form, anthrahydroquinone, in turn is re-oxizided by lignin components. Furthermore, anthraquinone oxidizes terminal hydroxymethyl groups of lignin to give aldehyde derivatives, which then induce bond cleavage of α-aryl ether linkages via β-elimination and Retro–Aldol reactions. Polysulphide may also be added to improve delignification and prevent peeling reactions. In another modification of the process (ASAM-alkaline sulphite), the cooking liquor contains methanol which is recovered from the effluent of the cooking step. For pulping of eucalyptus Organosolv Pulping (employing acetic acid [Acetosolv] or a mixture of formic acid and acetic acid [Formacell]) has also been used. Biopulping refers to the action of enzymes from fungi as a pre-treatment of wood and wood chips (Scheme 7.1).

Bleaching

Bleaching is performed in several stages in order to remove residual lignin and to improve the brightness of cellulosic materials. Alternatively, only the chromophoric groups in lignin may be selectively destroyed, thus preserving a larger amount of lignocellulose (high-yield pulps). For chemical pulps removal of lignin is required, resulting in fibres with high and stable brightness properties. Bleaching also comprises extraction steps to remove hemicelluloses as well.

Whereas bleaching protocols in European mills frequently employ totally chlorine-free chemicals (TCF-bleaching), in North America elemental chlorine-free bleaching (ECF-bleaching) is in use, which has replaced the former bleaching step with elementary chlorine. Common bleaching chemicals (with their abbreviations) are chlorine (C), hypochlorite (H), chlorine dioxide (D), oxygen (O), hydrogen peroxide (P), ozone (Z), xylanase (X), sodium hydrosulphite (Y) and formamidinesulphinic acid (FAS). In ECF-bleaching chlorine dioxide is applied followed by an alkaline extraction in the presence of oxygen or hydrogen peroxide.

The use of chlorine-containing chemicals, however, may lead to the formation of various chlorinated organic compounds such as chlorinated phenols or highly toxic chlorinated benzodioxines (TCBD). This is certainly true in particular for elemental chlorine, where both chlorination and oxidation reactions occur during bleaching, whereas chlorine dioxide treatment results in the formation of chlorite and hypochlorous acid after reaction with phenolic hydroxyl groups. TCF bleaching, on the other hand, avoids the contamination of effluents with chlorinated hydrocarbons by using ozone, oxygen and hydrogen peroxide. Under alkaline conditions the latter two agents generate hydroperoxide ions which may attack lignin structures, forming organic hydroperoxides, which then decompose via side chain elimination or ring opening of aromatic units with formation of carboxylic groups.

These agents are less selective towards lignin and may also affect the cellulosic fibres, particularly at the end of bleaching sequences. In addition, the presence of metal ions such as

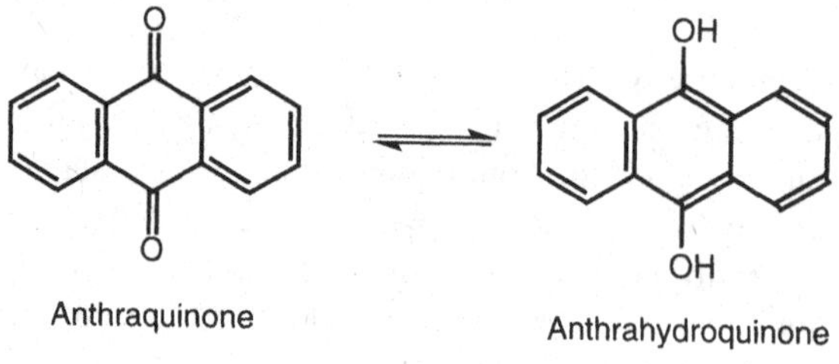

Anthraquinone   Anthrahydroquinone

Scheme 7.1   *The anthraquinone–anthrahydroquinone equilibrium*

Fe^{2+}, Co^{2+}, Cu^{2+} or Mn^{2+} should be avoided (by adding chelate complex-forming additives such as EDTA) to prevent the formation of the highly reactive hydroxy radical $HO^{\bullet}$, which may also degrade the cellulosic fibres. Ozone (Z) bleaching steps are carried out at low pH to limit the decomposition reaction with formation of reactive radical species (equation 7.2).

$$2O_3 + HO^- \rightarrow O^{\bullet-}_2 + HO^{\bullet} + 2O_2 \qquad (7.2)$$

Dissolution of cellulose

Dissolving cellulosic fibres requires the breaking of the strong inter- and intramolecular hydrogen-bond network by powerful solvating agents. For analytical purposes treatment of celluloses (of reduced molecular weight) with *N,N*-dimethylacetamide/LiCl results in the formation of clear solutions, amenable to the determination of molecular mass distribution by gel permeation chromatography with appropriate detection systems (refractive index, MALLS = Multiple Angle Laser Light Scattering).

For the industrial preparation of artificial silk and medical fabrics the cuprammonium process is still being used, but does not match the large-scale output of the viscose rayon process. The complex formation is believed to occur by coordination of the copper ions of $[Cu(NH_3)_4](OH)_2$ with the 2- and 3-OH groups of cellulose.

Viscose process

In the viscose process the pulp prepared from beech or spruce by sulphite processes, or from eucalyptus by Kraft pulping, is first steeped in aqueous sodium hydroxide (17–19%), which leads to a conversion of the cellulose I modification into alkali cellulose and removes residual hemicelluloses. The lye is then removed by pressing and the material is subjected to shredding in order to provide a uniform surface. During the pre-ripening step, performed at 40–50 °C, partial oxidative depolymerization and "peeling" reactions set in which lead to a decrease of the degree of polymerization of the pulp from DP (degree of polymerization) 750–850 to ~270–350. The swelling and ageing process also renders the fibres much more accessible to the subsequent chemical derivatization step with gaseous carbon disulphide to give the highly viscous yellow-orange cellulose xanthate. The sodium xanthate is then dissolved in dilute aqueous NaOH at a lower temperature (8–12°). During the ensuing ripening steps, the distribution of xanthate groups is changed from the kinetically formed C-2 and C-3 substituents to the more stable C-6-substituted derivatives. After filtration and degassing, the spinning dope is spun through spinnerets (30 000–50 000 holes of 40–60 μm diameter) into a spinning bath containing H_2SO_4 and additives such as Zinc-sulphate and modifiers. The coagulated rayon filaments are stretched to improve the tensile strength, washed, desulphurized and finished.

Lyocell process

Whereas the old viscose process relies on a chemical derivatization which requires a considerable input of chemicals such as NaOH and carbon disulphide, the lyocell process

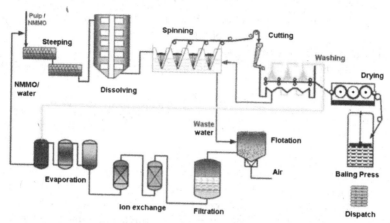

Figure 7.3   *Flow chart of the lyocell process*

(Figure 7.3) utilizes a direct dissolution of the pulp in *N*-methylmorpholine-*N*-oxide (NMMO) at higher temperature (90–120 °C).

Regarding the environmental aspects, the lyocell technology has distinct beneficial features. The solvent may be recycled to a near quantitative yield, it is non-toxic and the effluents of the process are non-hazardous and fully biodegradable. Starting from small pilot plants during the early 1980s, the industrial production has meanwhile reached a capacity of ~100 000 t/yr. The process tolerates a variety of pulp sources and pulp qualities and the regenerated fibres show increased crystallinity and a high degree of orientation, resulting in a relatively high tensile strength in the wet and also in the dry state. Due to the good fibre and performance characteristics, the fibres not only find applications in fabric and garment development (also blended with other fibres), but are also used in non-wovens and films.

In principle the pulp is first shredded and then fed into the mixer to give a slurry in an aqueous solution of NMMO. Evaporation and heating lead to the removal of water, until the solution range of the ternary system NMMO/water/cellulose is reached, to give a dark-coloured viscous solution of roughly 10–18% cellulose content. Over-heating and the presence of degradation-inducing agents have to be avoided to prevent exothermic events. Stabilizing agents are also added to keep the process under control (Rosenau *et al.*, 2001). After filtration the solution is spun through an air gap into an aqueous spinning bath. The regenerated fibres are then cut, washed, dried and finished, which may also include treatment with cross-linking agents in order to reduce fibrillation.

Use of Fibres for the Production of Composites

The use of natural resources and fibres for human clothing has been known as long as humans have been living (i.e. use of animal hides and wool production). In the Middle Ages, this had already resulted in a flourishing trade in wool, linen, cotton, etc. With the industrial revolution many synthetic alternatives to the natural fibres were developed such as nylon, polyesters, polyacrylamides, etc., which became popular not only in the textile industry but also in the polymer world to produce plastics and composite materials (car interiors, tennis rackets, surfboards, window frames).

Nowadays lots of efforts are made to apply natural fibres as reinforcement to composite materials. Composites are now being reinforced using carbon, aramide and glass fibers which make the composites quite expensive. On the other hand, the pressure on composite manufacturers to develop biodegradable polymer materials has led to research into developing composites reinforced with natural fibres. Of course, there is no competition between the strength of a glass fibre (~3500 MPa) and for example, a technical flax fibre (~760 MPa). For many applications, where the manufactured parts do not undergo much strain, the use of flax fibres could be very advantageous because of the aforementioned increased bio-degradability and the reduced weight of the parts. The reduced weight results from the low density of flax fibres (1.48 g/cm^3) compared to glass fibres (2.6 g/cm^3), which is lower or equal to the density of the polymeric matrix. However, the reinforcement of composites with natural fibres (flax, hemp, sisal, jute, etc.) still poses some fundamental problems. The biggest problem concerns the adhesion between the natural fibre and the polymeric matrix and the impregnation of the fibre by the matrix. Other disadvantages of natural fibres are the elevated uptake of moisture, the degradation under the influence of sunlight and the non-constant quality of the fibres.

In order to improve the adhesion between the fibre and the matrix, different physical (stretch, heat treatment, electric discharge) and chemical modifications of fibres have been evaluated using coupling reagents such as γ-glycidoxypropyltrimethoxysilane and diisocyanates. In this way, a covalent link is being made between the natural fibre and the synthetic matrix. The adhesion is also strongly influenced by the amount of pectins and hemicellulosis in the fibre. Another application in which natural fibres are playing an essential role is in the development of modified fibre mats in order to clean up oil spills in seas and oceans. Recently, a chemically modified softwood fibre was developed which can take up 25 times its own weight in oil and which keeps floating on the water surface. Another recent application is the use of biodegradable mats to temporarily stabilize soils, roadsides or riverbanks before the natural vegetation takes over the stabilization. Much work is also in progress on the genetic modification of plants, e.g. cotton plants, in order to develop new crops which could produce bioplastics such as polyhydroxy-butanoates by the introduction of the genes necessary to produce the reserve polymer, i.e. β-ketothiolase, acetoacetyl reductase and PHA synthase.

Utilization of Waste Material

Utilization of wood and lignocellulosic residues

With the exception of pulp, wood is mostly used for material purposes: building, furniture, valuable objects, etc. The species and the quality of wood used in every case depend on the added value of the final object. When the finest parts have been selected, the wood residues can still find an application in a triturated form in the particleboard industry. Nevertheless, huge amounts of wood and lignocellulosic waste are still found, which can be very diverse in quality. It is important nowadays to take advantage of this source of renewable raw material.

The combustion of wood and other lignocellulosic waste has been utilized since human beings looked for shelter and tried to cook food. Direct burning is still used on both a reduced (home) scale and a significant industrial level, for instance for the production of steam by burning sugarcane bagasse in boilers.

Indirect energetic valorization of wood and lignocellulosics occurs through the production of combustibles or fuels with specific properties, for instance: charcoal, obtained by slow and incomplete combustion of wood; or ethanol, obtained by hydrolysis of polysaccharides and fermentation of the obtained sugars. Combustion and other indirect energetic uses of biomass are treated in more detail in Chapter 5.

The chemical valorization of wood tends to compete with energetic valorization since they use the same raw material. Moreover, the indirect energetic valorization and the chemical valorization overlap frequently in many applications. Here are some examples:

- Charcoal is appropriately considered as a chemical in metallurgy (reducing agent in steel fabrication) and in adsorption processes (activated carbon).
- Synthetic gas (*syngas*), constituted mainly of H_2 and CO obtained by wood gasification, serves not only as a combustible but also as a raw material for the synthesis of organic compounds (urea, ethylene, acetone).
- The pure chemical valorization of wood (xylochemistry) includes the transformations in which degradation by temperature is not an important factor, for instance hydrolysis (glucose and furfural production) or hydrogenation (phenolics).

Germany, Russia, Japan and France were, for a long time, at the head of the chemical valorization of wood, but in the last decade Japan took the pole position. Brazil is frequently pointed out as an example of the valorization of biomass, especially in the ethanol programme for motor purposes. Such a programme has recently enfeebled however.

Here the main chemical valorizations of wood will be presented briefly. Some of them can be applied to other lignocellulosics:

The *pyrolysis* of wood is a thermal decomposition carried out in a closed furnace with very limited air input (less than 10% stoichiometric). At the beginning of the process, the reactions are endothermic and a heat source is necessary. When the temperature reaches more than 250 °C, the reactions that take place are exothermic and the process is self-maintained up to 600 °C. Wood pyrolysis produces mainly a solid product (*charcoal*) but the following products are also obtained in important quantities: a highly viscous liquid (*tar*, from which phenol, benzol, guaiacol, cresol and other disinfectants can be obtained); a liquid product resulting from the condensation of vapours (*pyroligneous liquor*, constituted mostly of acetic acid, methanol and acetone); and a gas by-product (CO and CO_2 mainly). Vegetable charcoal competes with mineral coal in metallurgic applications. On the other hand, it is a preferred raw material for the production of activated carbon, of which approximately 70 000 t are produced annually in France.

The *gasification* of wood is analogous to pyrolysis. However, the thermal degradation is carried out in the presence of a limited quantity of air (20–80% stoicheometric). Gasification is achieved faster than pyrolysis. The main product is *syngas*, composed of CO and H_2, which can be of high quality if oxygen is used instead of air. Methanol can easily be produced by the reaction $CO + 2H_2 \rightarrow CH_3OH$. Other important industrial chemical transformations can also be performed (Figure 7.4).

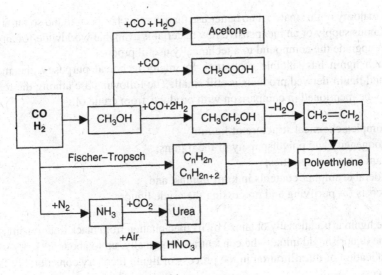

Figure 7.4 *Some industrial chemicals obtained from syngas*

The *solvolytic liquefaction* is the thermal degradation of wood in the presence of a solvent, which can also react with wood components. Such a solvent reagent may be phenol or a polyhydric alcohol. The process forms a liquid product composed of low-molecular-weight compounds, which bear phenolic groups or hydroxyl groups according to the solvent used. The number of different molecules is extremely high and their structure is only partially known. However, the liquefied wood is used as a raw material for the production of polymers. If phenol is used as a solvent, the liquefied wood can react with formaldehyde to produce resol or novolak-type resins. If a polyol is used, the liquefied wood can react with a diisocyanate to produce polyurethane in expanded (foam) or condensed form, rigid or elastic, according to the reagents used.

Utilization of technical lignins

In the process of wood digestion, all organic compounds, excluding cellulose, are dissolved generating the digesting liquor (about 50%). The solution obtained from the above process is called post-digesting liquor and is a waste product of the cellulose production.

   The current annual global production of lignin in the pulp industry amounts to some 50×10^6 t, which approaches the whole production of the plastic industry. The field of lignin utilization can be divided into four general groups (Fengel and Wegener, 1984):

1. lignin as fuel;
2. lignin as a remaining component in mechanical and unbleached chemical pulps;
3. lignin as polymeric product; and
4. lignin as a source of low-molecular-weight chemicals.

For the majority of this material no better use than burning has been found so far. Therefore, an enormous supply of an inexpensive raw material is available worldwide for any process that can upgrade the compound to a technically useful product.

Though lignin has suitable properties for many technical purposes, the market for lignin and lignin-derived products is still small. The following are among the restrictions that may be mentioned in comparison with products from crude oil:

- the complex chemical structure of lignin;
- inhomogeneity and polydispersity of the lignins;
- high amounts of impurities;
- considerable sulphur contents in kraft lignins; and
- high costs for purifying and processing the crude liquors.

Sulphate lignin is traditionally obtained by its precipitation from black liquor using sulphuric acid. The liquor should initially be concentrated thermally, up to about 50% dry substance. The application of ultrafiltration in the process of lignin recovery considerably facilitates this process.

The technical lignins and modified lignins have found a wide field of application. Crude lignin sulphonates have found considerable application in gravel-road surface stabilization, in soil stabilization, as mineral binder and as binding agent for animal feed pellets.

Applications of lignin sulphonates involve their utilization as tanning agents in combination with chrome tanning agents or as protective colloids for preventing scale formation in a steam boiler. Due to the ionic character of the lignins they can be processed to ion-exchange resins. One of the most important applications for the future will be the use of the adhesive properties of lignin for thermosetting resins and plastics.

Lignins can be modified either by further degradation and polymerization of the resulting oligomers, or by direct chemical modification of the polymer fragments using small molecules, or by grafting to form other large molecules. Sulphur lignin and electrically conducting lignins have been studied intensively in the last two decades. An attractive feature of lignins is their chemical composition, with a large amount of relatively easily accessible hydroxyl units that can be used for the attachment of a variety of functional groups. Hydroxyalkylation and carboxymethylation of lignin have been shown to produce a hydrophilic functionality that enhances surface activity. Alkenyl succinic anhydrides are well known in the pulp industry as paper-sizing agents.

Preparation of graft copolymers of lignin into main polymer backbones appears to be a suitable method to introduce a new class of engineering plastics. Furthermore, grafting some components into the main polymer backbone may enhance the degradability. For example, styrene graft copolymers of lignin are readily degraded by white-rot fungi whilst the polystyrene homopolymer is not degradable.

Lignin has also been modified with enzymes (laccases or peroxidases) for the production of new compounds. At present, the following approaches have been studied: (1) *in situ* polymerization of lignin for the production of particleboard; (2) enzymatic copolymerization of lignin and alkenes, because in the presence of organic hydroperoxides laccases catalyze the reaction between lignin and olefins; and (3) enzymatic activation of the middle-lamella lignin of wood fibres for the production of wood composites (Huttermann, Mai and Kharazipour, 2001).

The degradation of lignin is expensive, giving low yields of pure chemicals. Therefore, the production of low-molecular-weight compounds from lignins is restricted by economical and technological considerations. Additionally, the fractionation and purification processes are necessary because the technical lignins contain solubilized carbohydrates and considerable amounts of inorganic material, especially sulphur.

The most important product with respect to commercial utilization is vanillin, a flavouring agent. The softwood lignin sulphonate is heated for 2–12 h at 100–165 °C in the presence of sodium hydroxide, and the vanillin formed is extracted and purified by vacuum distillation and recrystallization. At present, apart from vanillin, only dimethylsulphide (DMS) and dimethylsulphoxide (DMSO) can be recovered from this process.

Utilization of hemicelluloses and furfural

Hemicelluloses represent 25–35% of the dry weight of hardwoods and 15–25% of the softwoods. The composition of monosaccharides in spent sulphite varies according to wood species. The spent liquor for spruce mainly contains mannose (11%), while for birch it mainly contains xylose (21.1%) and glucuronic acid (30.8%), and for aspen mainly xylose (24.3%).

Ethanolic fermentation of spent sulphite liquor with the baker yeast *Saccharomyces cerevisiae* is incomplete because this yeast cannot ferment the pentose sugars in the liquor. This results in poor ethanol yields and a residue problem. Using pentose and hexose fermenting yeasts such as *Pichia stipitis*, *Candida shehatae* and recombinants of these, the spent sulphite liquor can be nearly completely fermented with favourable results and increased ethanol production. The spent sulphite liquors are also used to produce fodder yeast, with *Candida* yeast or fungi *Paecilomyces varioti* as cattle fodder.

Furfural is the most important product of the dehydrogenation of xylose and is used as an industrial solvent, disinfectant, preservative or as a raw material in nylon synthesis and others. Furfuryl alcohol obtained by hydrogenation of furfural is used to produce the so-called furan resins. These liquids are applied in the production of reinforced plastics.

7.2.2 Starch

Starch is one of the most widespread polysaccharides in the world of plants; it functions as the reserve polysaccharide of the vast majority of the plants and contains two polymers – amylose and amylopectin. Amylose consists of α-D-glucose units which are connected through 1,4-bonds, whereas amylopectin is a branched polymer built up by 1,4-bonded α-D-glucose units with the branches formed by 1,6-bonds. The most important crops for starch production in Europe are wheat, corn and potato. Corn is the most important crop worldwide.

The use of starch for non-food applications is quite high, and it is estimated that 46% of the starch produced worldwide is used for non-food applications. The advantages for the use of starch in non-food applications are associated with the relatively easy isolation of starch from the starch granules by extraction with water after grinding of the part of the plant that contains starch. The industrial use of starch is mostly associated with its ability to change the viscosity of solutions, its dispersing ability, gel formation and glue characteristics.

Native Starch

The most important application of native starch is as a paper glue in the corrugated cardboard industry. In 1995 approximately 1.3×10^6 t of starch were used to manufacture 74×10^6 t of paper in Western Europe (18 kg of starch per tonne of paper). Almost three quarters of this is used for surface sizing (gives an improvement of the paper stiffness, the strength and a controlled ink receptivity), 16% for wet-end application and 11% for coating applications. In 2000 the use of starch for papermaking and corrugating was already 1.9×10^6 t.

Further applications of starch are in the area of slow release of biologically active components. Starch can be used as a capsule for the application of atrazine in corn fields in order to prevent atrazine from leaching into the water. In this way, the atrazine is less mobile and migrates less deeply into the soil (40 cm) than non-encapsulated atrazine (60 cm). Since the release of atrazine is slower, the activity period is also longer while using a similar amount of atrazine.

The same type of application is used in the encapsulation of ureum in order to get a more performant fertilizer system in rice production. Low-density polyethylene or polypropylene in combination with starch is used as coating material in order to improve the photodegradation of the polyethylene and in this way to control the speed of the release of the fertilizer. Composite materials can also be manufactured by extrusion of starch and a synthetic polymer. In this way materials can be made containing up to 50% starch.

Recently, Goodyear even produced a tyre using an organic filler based on corn starch, replacing 25% of the traditionally used silica and soot. This makes a tyre 100 g lighter and lowers the roll resistance by 20%. In Belgium for example, 70×10^6 kg of tyres are used yearly, with 10% of the weight being disposed of into the environment due to wear. High value-added applications of starch include its use as a binder or tabletting agent in the pharmaceutical industry and as a source for the production of sorbitol, vitamin C, etc.

Modified Starches

Modification of starch can be by physical, chemical or enzymatic means. The physical modification mainly consists of pregelatinization. Starch is easier to modify than cellulose, since modifications with a low degree of substitution already change the functional properties considerably.

The range of modified starches is large and all sorts of starches are being produced for typical applications. Among them are oxidized, esterified, etherified, cross-linked and hydrolyzed starches for applications as thickener, binder, cobuilder, thermoplastic, complexing agent, flocculating agent and coating. In the scope of this chapter it is impossible to give a complete overview of all applications, therefore only some specific applications will be dealt with.

- Cationic starch (etherified starch) is mainly produced by reaction of starch with quarternary alkyl ammonium salts (mostly via epoxide ring opening reactions) and are used for surface treatment of paper.
- Oxidized starches are mainly used in the textile industry for the strengthening of yarns, to prevent breakage during weaving. The anionic starches also result in better printability of the tissues.

- Oxidized starches are also an important target as builder or co-builder in washing powders because of their anti-redeposition features. The strongly oxidized starches, soluble in cold water, have a good complexation ability for calcium and magnesium ions which makes them a valuable alternative for polyacrylates (not very biodegradable) and phosphates (causing eutrophication in rivers and lakes). Since the washing process is in essence an operation in basic medium, the starch derivatives are stable during the washing and are hydrolyzed once they are liberated into the environment.
- Carboxymethylated starch, produced by reaction of starch with the sodium salt of chloroacetic acid, is also applied as anti-redeposition agent in the washing powder industry. However, the best results are obtained using ent-copolymers in which synthetic side chains, e.g. polyacrylates, are grafted onto the starch backbone using special catalysts.

7.2.3 Chitin and Chitosan

Chitosan is the biopolymer obtained by partial deacetylation of chitin, which is not only a major component of the exoskeleton of insects and crustaceae, but is also present in the cell wall of fungi, yeasts and algae. The name chitosan represents a group of acid-soluble heteropolysaccharides which mainly consist of $\beta(1\text{–}4)$-N-acetyl-β,D-glucosamine and $\beta(1\text{–}4)$-β,D-glucosamine (Scheme 7.2).

Chitin is a very interesting biopolymer since it is the most abundant polysaccharide in the world after cellulose. Since it was mostly considered as waste and available in large amounts, a lot of research has been done to evaluate the properties of chitin and its deacylated derivatives, including chitosan.

Chitosan can be produced by extraction of chitin from the shells of shrimps, lobsters, prawns and crabs, followed by deacetylation with strong base (40–45% NaOH solution). The production scheme for chitosan is depicted in Figure 7.5. The deacetylation can also be performed by chitin deacetylases present in fungi and insects. This procedure leads to the production of better-defined oligomers and polymers (Tsigos *et al.*, 2000).

Alternatively, chitosan can be produced by a number of yeasts such as *Absidia artrospora*, *Absidia coerulea*, *Absidea blakesleeana*, *Gongronella butleri*, *Lentinus edodes*, *Mucor rouxii*, *Phycomyces blakesleeanus*, etc. In spite of the fact that the chitosan obtained has mostly better-controlled physio-chemical properties, the production method is much more expensive.

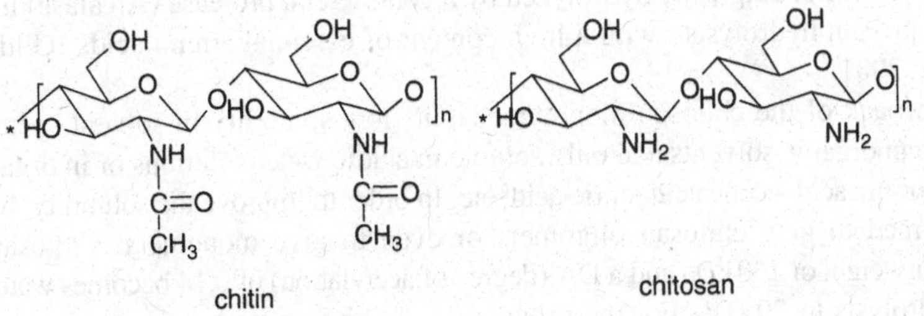

Scheme 7.2   *Structure of chitin and chitosan*

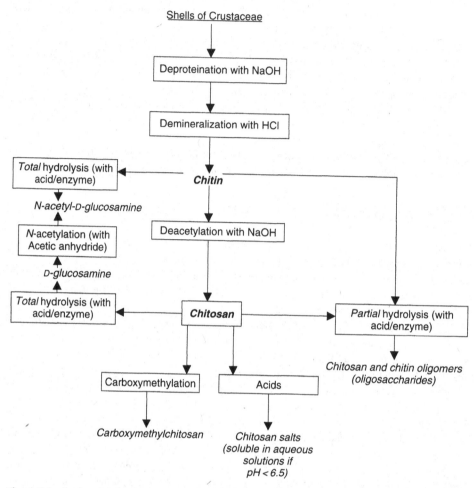

Figure 7.5   *Production scheme and derivatives of chitosan*

In view of the integral valorization of renewable resources, the use of chitosan for industrial applications is a very good example. Not only the chitin from the shell waste can be used, efforts are also being made to isolate the proteins from the shells. The shell proteins can, for example, be hydrolyzed by a commercial protease (Alcalase) in order to obtain a protein hydrolysate with a high content of essential amino acids (Gildberg and Stenberg, 2001).

A drawback of the chitosan biopolymer is its low solubility in solvent systems. It is insoluble in organic solvents and only soluble in acidic water solutions or in organic acids such as formic acid, acetic acid, citric acid, etc. In order to improve the solubility, hydrolysis is performed to give chitosan oligomers or even to give monomers. Chitosan with a molecular weight of 150 kDa and a DA (degree of acetylation) of 0.14 becomes water soluble after hydrolysis to 29 kDa. For the production of chitosan with a low molecular weight many methods have been described, for example degradation with hydrogen peroxide, hydrolysis with hydrochloric acid, phosphoric acid, enzymatic hydrolysis by *Streptomyces kurssanovii*, *Aspergillus* sp., *Mucor rouxii*, etc.

In this section, a number of examples will be given on the use and the advantages of chitosan derivatives for non-food applications. The number of examples of the use of chitin, chitosans and modified chitosans is very large and an overview of all the different applications is beyond the scope of this work. More detailed overviews are given in the specialized literature (Kubota and Kikuchi, 1998; Muzzarelli and Muzzarelli, 1998; Jenkins and Hudson, 2001).

A lot of applications of chitosans and chitosan derivatives are linked to the control of microorganisms since chitosans are well known for their anti-microbial (bactericidal and fungicidal) activity (Rabea *et al.*, 2003). However, there is still a lot of uncertainty concerning the mode of action of chitosan. One theory states that the complexation of ions by the chitosan is responsible for the action, others believe that it influences the cell wall.

Among the most important chitin derivatives is the *O*-carboxymethyl chitin, of which the sodium salt is soluble in water. The carboxymethylation preferentially occurs at the C6 position, although the reaction occurs at C3 as well. These derivatives are used for complexation of cations and the presence of the acetamido group at the 2-position delivers a high affinity to calcium ions. The reaction product of *O*-carboxymethyl chitin and mitomycin C (an antitumour agent) gives good results *in vitro* since the active agent is slowly released, which could not be achieved using starch or peptides.

Chitosan and its derivatives are also being used as wound dressing for third-degree burns and as material for the reconstruction of soft tissue (tissue engineering) (Fwu-Long *et al.*, 2001). The derivatives have haemostatic properties and allow regeneration of the tissue. These features are of course associated with the biocompatibility, low toxicity, and bacteriostatic and fungistatic activity of the derivatives (Yong Lee and Mooney, 2001). Methylpyrrolidone-derivatized chitosan has been used to enhance bone formation.

7.2.4 Fructans

Fructans are polysaccharides formed from fructose monomers. They are also formed as reserve polysaccharides in plants and in microorganisms. In the literature some confusion exists about the nomenclature, however, the fructose polymers in plants are most often called fructans whereas those formed in mircroorganisms are called levans. Fructans are very widespread and are present in monocotyles, dicotyles and green or blue-green algae. Inulin was the first fructan discovered (in 1804 from the roots of *Inula helenium*) and since then numerous fructans have been described from a multitude of plants such as *Dahlia variabilis*, *Helianthus tuberosus* and different kinds of agave.

Within the group of the fructans three types can be distinguished: (i) the inulin group, in which the fructose units are interconnected by a β(2,1′) glycosidic bond (occur mostly in dicotyles such as *Malpighiaceae*, *Primulaceae*, *Boraginaceae*, *Violaceae*, etc.). Inulin consists of fructose polymers and occasionally a glucose unit at the end of the fructose chain (Scheme 7.3); (ii) the phlein group with a β(2,6′) glycosidic bond (occur mostly in monocotyles such *as Liliaceae*, *Agavaceae*, *Amaryllidaceae*, *Iridaceae*, *Poaceae*, etc.); and (iii) the branched fructans with β (2,1′) glycosidic and β (2,6′) glycosidic interconnections (in green algae such as *Dasycladales*, *Acetabularia*, etc.).

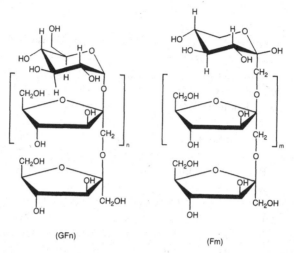

(GFn)

(Fm)

Scheme 7.3 *Structure of inulin*

In addition to all the modifications that are performed on starch derivatives, a lot of research has been done recently on the chemical modification of inulin-type polysaccharides (Stevens, Meriggi and Booten, 2001a). This interest is based on the intermediate molecular weight of the inulin derivatives (highest DP ~60) compared to the high molecular weight of starch, which can lead to readily soluble inulin derivatives that are valuable in the chemical industry. The difficulties associated with modified starch derivatives are mostly related to solubility problems.

Among all the typical chemical modifications of polysaccharides (oxidation, esterification, etherification, cross-linking, etc.), the most important modifications are carboxymethylation and carbamoylation.

Carboxymethylated inulin, formed by reaction with the sodium salt of chloroacetic acid in aqueous medium in the presence of sodium hydroxide, is commercially available as an inhibitor for calcium carbonate crystallization in closed water circuits. Addition of ppm amounts of the carboxymethylated inulin prevents scaling, which is extremely pernicious to heat transfer in closed water circuits.

A second promising modification of inulin consists of the carbamoylation of inulin by reaction with alkylisocyanates (Stevens *et al.*, 2001b; Stevens, Meriggi and Booten, 2001c). The grafting of the hydrophilic saccharide backbone with the hydrophobic side chains gives the end product an amphiphilic character which is of interest to the detergent industry. The molecules have great potential for use as emulsion stabilizers in cosmetics and pesticide formulations. The performance of the inulin carbamates is especially interesting since it gives very stable emulsions in applications with high salt concentrations, in which many other emulsion stabilizers fail to function.

7.3   Use of Oligosaccharides

7.3.1   Maltodextrins

Maltodextrins are the products formed by hydrolysis of starch and are divided into subgroups by their DE (dextrose equivalent) value. The maltodextrins consist mostly of

8–25 glucose units. The applications of the maltodextrins are situated between those of the polysaccharides and those of the disaccharides and monosaccharides. They are mostly used in food applications (sweetening, etc); specific non-food applications are not always well defined since they are mostly covered in the patent literature of the saccharides in general. However, sometimes specific characteristics can be attributed to these derivatives. The reaction of maltodextrins (DE from 1 to 47) with alkylisocyanates gives products (DS (degree of substitution) from 0.01 to 2) with excellent tensio-active properties in combination with good biodegradability (Stevens, Meriggi and Booten, 2001c). The grafting of the hydrophilic saccharide backbone with lipophilic side chains can be controlled to obtain tensio-active agents with a desired HLB.

7.3.2 Cyclodextrins

Cyclodextrins have, since their discovery approximately 100 years ago, been a laboratory curiosum for a long time, but are now becoming more and more valuable for industry and for very specific applications.

Cyclodextrins are cyclic oligosaccharides derived from starch by a number of microorganisms. The microorganisms that are now being used for the commercial production of cyclodextrins are *Bacillus macerans* and *B. circulans*, which contain cyclomaltodextrin transferase. Because the enzymes can transform only unbranched oligosaccharides and due to the fact that the enzymes are inhibited by 40 mg/l of cyclodextrins, only 50–60% of starch can be transformed into cyclodextrins, which makes production quite expensive.

The most important cyclodextrins are circular oligoglucoses containing six glucose units (α-cyclodextrin), seven glucose units (β-cyclodextrin) and eight glucose units (γ-cyclodextrin). The feature of the cyclodextrins that makes these molecules very interesting is the fact that the outside of the molecule is hydrophilic due to the hydroxy groups of the glucose units that are oriented to the outside, and that the inside of the torus is lipophilic due to the carbon atoms and the hydrogen atoms that are oriented to the inside. Therefore, the interior space of the cyclodextrins is ideal to host an organic molecule which fits into the cavity of the α,β,γ-cyclodextrin (cavity: 4.7–5.3×10^{-10} m (α); 6.0–6.5×10^{-10} m (β); 7.5–8.3×10^{-10} m (γ)).

Due to the possibility of complexation, cyclodextrins can be used for:

- stabilization of labile compounds against chemical influences such as oxidation or radiation;
- reducing the volatility of compounds;
- suppressing undesired smell or taste;
- improving the water solubility of compounds;
- mixing (in principle) non-miscible solvents;
- improving the availability of compounds in biological systems; and
- the separation and analysis of mixtures of isomers.

Industrial applications of native cyclodextrins are known in sprays (aqueous solution), to use on textiles (Fébrèze) in order to remove bad odour (e.g. from curtains), or in the application for the removal of cholesterol (e.g. from butter). An aqueous cyclodextrin solution is mixed with the fat at 70 °C during which the cholesterol is transferred from the lipophilic phase to the aqueous phase with complexation to the cyclodextrin and crystallization of the complex. After separation of the three phases a fat can be obtained which is nearly free of cholesterol. However, small amounts of cyclodextrin stay in the fat.

Since cyclodextrins have a haemolytic effect on red blood cells during intravenous applications, a lot of effort has been devoted to developing cyclodextrin derivatives which do not have a haemolytic effect in order to develop drug delivery applications in the human body. Up to now, the hydroxypropyl-β-cyclodextrin is much less haemolytic and is the compound that is being used for non-oral medical applications.

The most important application of cyclodextrins is the use of mostly permethylated cyclodextrins as stationary phase for chiral gas chromatographic separations. The different interaction of enantiomers with the cavity of the cyclodextrin molecule allows the separation of the enantiomers. Several types of chiral GC columns have been based on this special feature of the cyclodextrins.

On top of all these possibilities, big efforts are being made to use the cyclodextrin system as a moiety to develop synthetic enzymes, since they can act as a mimic of the active site of an enzyme which can complex with a substrate molecule (Breslow and Dong, 1998).

Although cyclodextrins are still quite expensive, the development of new applications in medicine, environmental sciences, food chemistry and pharmaceutical sciences will certainly push production technology to provide these compounds at reasonable prices.

7.4 Use of Disaccharides

7.4.1 Sucrose

Sucrose, formed by α-D-glucose and β-D-fructose is mainly produced from sugar cane in tropical and semi-tropical regions and from sugar beet in regions with a moderate climate. Sugar cane (*Saccharum officinarum*; yearly production of 750×10^6 t), which cannot tolerate any frost, is one of the plants able to transform most solar energy into chemical form (2%). Since the plant can be completely used for different applications, it is a very valuable crop for developing countries. Next to sugar production, the bagasse (which remains after the extraction of the sugar) can be dried (in developing countries in the sun) and used as fuel for the production of energy. Apart from the energy needed for sugar production, the sugar cane factories have 20–30% bagasse left to produce energy (up to 100 kwh/t) for other industries. Bagasse is also the most important source material for the production of paper and cardboard in sugar producing countries, since the bagasse cannot be readily used as feedstock due to the high fibre content which makes it hard for animals to digest. If bagasse is used as feedstock, it needs to be treated with sodium hydroxide (partial hydrolysis) or else it needs to undergo a steam treatment.

In the production of sugar beet, the complete crop can also be used. Besides the sugar, the sugar beet pulp can be used for feed and the molasses (extraction concentrate from which no sugar can be crystallized in an economical way) can be used as carbon source for microbial fermentations. For these two crops, the concept of integral crop use is well developed and this should always be the aim in view of the concept of sustainable development.

Sucrose can be extracted easily from the crop (as well from sugar cane as from sugar beet) using water. Attention should be paid in order to prevent hydrolysis of the disaccharide during the extraction process (with the formation of fructose and glucose, also called invert sugar). This is mostly done by addition of calcium oxide to precipitate the acids in the extraction mixture and to basify the medium, preventing hydrolysis. The excess of calcium oxide is removed later on in the process by the addition of carbon dioxide with the formation of calcium carbonate.

The most important non-food use of sucrose is certainly the production of ethanol, which can be used as biofuel (e.g. in Brazil). The concept of the production of biofuels is discussed in Chapter 5. Since ethanol can be dehydrated to ethene, sucrose could form the connection between a renewable resource-based industry and a fossil fuel-based industry. However, the price of sucrose (even in molasses) is still too high to compete with the actual prices of ethene.

Further, as mentioned above, sucrose is often used as carbon source during fermentation. Sucrose is becoming more and more important in the production of sucrose esters (mono- to octa-esters) by reaction with fatty acids (e.g. Nebraska–Snell process) or with triglycerides (Tate and Lyle process), with *in situ* generation of the fatty acids. As well as the food use of these esters as fat replacer (which is currently the subject of a lot of debate), the sucrose esters can be used as detergents. Despite the fact that the surface active properties are not as good as those of some synthetic detergents, the raw reaction mixtures are definitively a lot cheaper than the synthetic agents and are therefore very suitable for use in less developed countries which could produce easily their own detergents. For this reason, the production of sugar cane in developing countries can certainly facilitate their sustainable development.

Sucrose can also be used in the foam industry for the production of rigid foams. Reaction of sucrose with diisocyanates gives rise to the production of carbamate polymers with the formation of rigid foams. However, sucrose itself results in the formation of very brittle foams, therefore sucrose is first transformed to polyhydroxypropylsucrose and then reacted with diisocyanates and other foam-forming ingredients. The sucrose derivative gives strength as well as flexibility to the foams. In Europe, the market for the use of sucrose in polyurethanes is estimated at up to 10 000 t/yr.

Allylsucrose, formed by reaction of sucrose with allyl halides, is being used in film-forming materials, paints and coatings since it can be polymerized under thermal or oxidative conditions. Polysucrose, the polymer of sucrose formed by cross-linking of sucrose with epichlorohydrine, is a neutral, highly branched polymer which is very soluble. It can form aqueous solutions with very low viscosity and high specific density. It can therefore be used as a medium for cell separations with a density gradient.

Industrial Uses of Chemicals from Molasses

The molasses are mainly used as a fermentation source. Some minor components from molasses are also isolated for special applications. *N*-trimethylglycine (betaine) is present in molasses at about 3–5% and is used as a vitamin in chicken feed and as a phytoalexine-stimulating agent for the growth of certain crops. Oxalic acid can also be produced by oxidation of the remaining sucrose in the molasses by a vanadium pentoxide catalyst.

7.4.2 Maltose

The use of maltose, obtained by partial hydrolysis of starch, for non-food applications is rather limited. It is also used as a nutrient in pharmaceutical formulations for intravenous use and as carbon source for the biotechnological production of special antibiotics.

7.5   Use of Monosaccharides

7.5.1   Glucose

Glucose is the primary product of the photosynthesis of plants and approximately 200×10^9 t are produced yearly. However, 95% is further metabolized to other sugars or polysaccharides (sucrose, starch, cellulose, etc.).

The hydrolysis of starch into glucose syrups is very important in the food sector because of their application as viscosity modifier, moisture stabilizer, inhibitor of crystallization, freeze-point-lowering agent, sweetener, etc. Glucose is becoming more and more important for non-food applications and for the synthesis of chemicals using biotechnological routes. An overview of some important modifications is given in Scheme 7.4.

Dehydration

Dehydration leads to the formation of hydroxymethylfurfural (HMF), which has been recognized as an important intermediate in a renewable resource-based industry since it is readily produced by heating glucose and fructose in acidic medium, it can be easily stored and it has several functionalities which can be exploited for various applications (Kunz, 1993). HMF could be used for the production of polymers and resins, however the price still limits the usage of HMF for several applications.

Hydrocracking

Hydrocracking is a petrochemical technique that can also be used on glucose – leading to glycerol, glycol and 1,2-propanediol. The best results up to now are those leading to a mixture of 48% glycerol, 16% glycol and 20% 1,2-propanediol. However, because of the very low prices of these products when produced from fossil fuel, the processes are not being exploited at the moment.

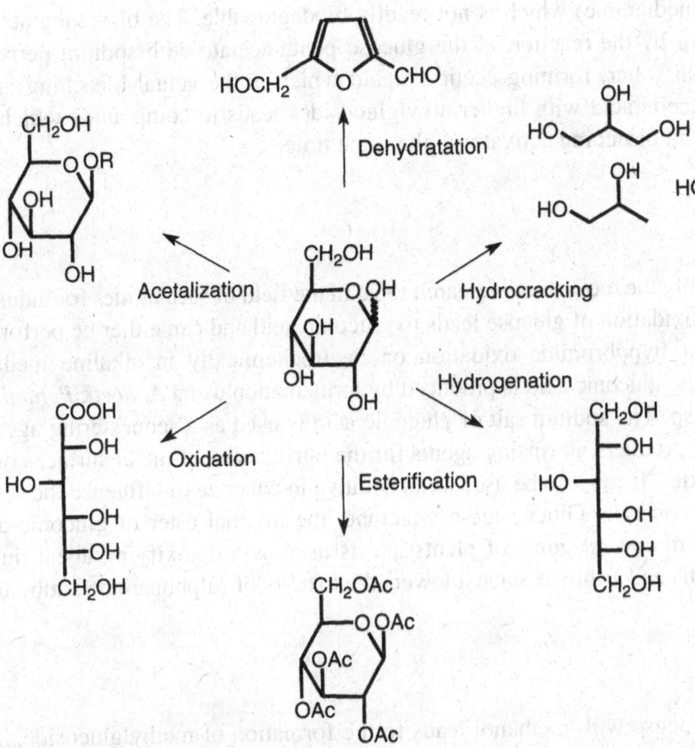

Scheme 7.4 *Modifications of glucose for industrial use*

Catalytic Hydrogenation

Catalytic hydrogenation of glucose leads to the production of sorbitol, which is mainly used (±25%) for the production of vitamin C by the Reichstein–Grüssner process. In this process, sorbitol is first oxidized to sorbose by *Acetobacter suboxydans* and then further transformed to ascorbic acid (vitamin C). Next to this important application, sorbitol is mainly used for its moisture regulating, softening and plastifying properties in adhesives, textile and paper applications, and in cosmetics and toothpaste (20%) because of its physico-chemical properties. The esterification of sorbitol with fatty acids leads to a whole set of industrially very important tensio-active agents (TWEEN and SPAN series) and represents approximately 5% of sorbitol production.

Esterification

Esterification is, as for many renewable resources, an important modification. The ester-ification of glucose with acetic acid to produce glucose penta-acetate is the most important and is now being used as a peroxide activator in detergent formulations requiring bleaching activity at low temperatures. In these applications, it is used as replacement for TAED (tetraacetyl ethylenediamine) which is not readily biodegradable. The bleaching activity is generated *in situ* by the reaction of the glucose penta-acetate with sodium perborate upon dissolution in water, forming acetic peracid which is the actual bleaching agent. Esterification of acetic acid with higher alkylglucosides leads to compounds which can act as co-builder and bleaching activator at the same time.

Oxidation

Oxidation is probably the most active research topic in the field of saccharides for industrial applications. The oxidation of glucose leads to gluconic acid and can either be performed by hypochlorite or hypobromide oxidation or electrochemically in alkaline medium. Actually, most of the gluconic acid is produced by fermentation using *A. aceti, Penicillium* sp. or *Aspergillus* sp. The sodium salt of gluconic acid is used as a sequestering agent in industrial cleaning products, in rinsing agents for the purification of metal surfaces or for the cleaning of bottles. It can also be used as an additive in concrete to influence the setting parameters of the concrete. Glucuronic-6,3-lactone, the internal ester of gluconic acid, can also be found in several gums of plants and is used as a detoxifying agent during treatment of hepatitis and arthrosis since it lowers the toxicity of sulphonamide antibiotics.

Acetalization

Acetalization of glucose with methanol leads to the formation of methylglucoside. This compound has been used extensively in the US as an additive in phenol resins which are utilized as glue for chipboard, and also in melamine resins. More important nowadays is the reaction of fatty alcohols with glucose (or maltodextrins) in acidic medium for the production of alkylglucosides (AGs) or alkylpolyglucosides (APGs). This market was developed by Rohm and Haas during the 1970s – who were the first producers of these compounds on industrial scale. Although the C_8–C_{10} polyglucosides were not ideal as a surfactant, Akzo Nobel and Cognis modified the products and these are now on the market. The derivatives with longer alkyl chains, e.g. the C_{12}–C_{16} polyglucosides, were developed in the 1980s, mostly by Procter & Gamble, and are used as co-surfactants in detergents and in personal care products.

Derivatives with shorter alkyl chains are mostly used in agrochemical applications as wetting agents because of their penetrating features. A very important characteristic of these compounds is that they have no cloud point as surface active agent and that they are highly tolerant of high electrolyte concentrations, much more than other non-ionic surfactants. It is also possible to make solutions with high concentrations of active compound (up to 60%).

The AGs and APGs are very environmentally friendly and are the first non-ionic surface active agents completely based on renewable resources.

7.5.2 Fructose

The use of fructose for non-food applications is very limited and it can only be used for the production of hydroxymethylfurfural and for the production of levulinic acid.

7.5.3 Xylose and Arabinose

These aldopentoses do not appear as free sugars in nature, but they are building blocks of the xylans and the arabinans which are present in hemicelluloses and pectins. Only the xyloses and arabinoses are of industrial importance and are isolated after acid hydrolysis of the corresponding pentosans. Xylans are present in the wood of angiosperms and in tissues of annual crops (e.g. stover, bagasse, straw, etc.). Arabinans are found in sugar beet pulp. Enriched solutions of these pentosans can be utilized for the production of hydroxymethylfurfural. Purified arabinose and xylose can also be hydrogenated to arabitol and xylitol for use as non-cariogenic sweetener in the bakery industry and in the pharmaceutical industry.

7.6 Conclusion

In this chapter, an overview has been given on the use of carbohydrates as a renewable resource for industrial products. First, the morphology and use of wood and natural fibres were discussed since the use of these sources is well known and obvious in daily life. The use of wood as an energy carrier, as a source of cellulose in paper production (with the different pulping processes) and as a source of fibres has been illustrated. Further, the importance of other natural fibres (e.g. cotton) for the textile industry has been highlighted and also the increasing importance of natural fibres in the composite industry. In the context of the integral use of natural resources, the use of by-products (lignin) from the different processes has been discussed.

The use of natural polysaccharides (starch, chitin, fructans) was then discussed in view of industrial usage related to the specific properties of the different polymers. Starch is being used primarily in the paper and corrugated paper industry, whereas chitin is primarily used for its bacteriostatic and fungicidal properties and in tissue engineering. The non-food use of inulin is situated in the field of anti-scaling products. Cyclodextrins evolved from a laboratory curiosum are useful products in separation technology of cations and organic compounds and in controlled release applications. The monosaccharide glucose is a very versatile compound for a lot of applications. Modification and chemical transformation of glucose leads to substances which are basic sources for detergents, emulsifiers, foams, vitamins, etc.

In general, the use of saccharides as a renewable source of energy and industrial products has great potential for the future. Many applications are technically relevant but cannot compete economically at this time with synthetic products derived from fossil fuels. However, research and application work has to be continued in preparation for the future generation when fossil fuels become scarce.

References

Breslow, R. and Dong, S.D. (1998). Biomimetic reactions catalyzed by cyclodextrins and their derivatives. *Chem. Rev.*, **98**, 1997–2011.

Fengel, D. and Wegener, G. (1984). *Wood: Chemistry, Ultrastructure, Reactions* (ed. W. De Gruyter), Berlin, New York.

Fwu-Long, M., Shin-Shing, S., Yu-Bey, W., Sung-Tao, L., Jen-Yeu, S. and Rong-Nan, H. (2001). Fabrication and characterization of a sponge-like asymmetric chitosan membrane as a wound dressing. *Biomaterials*, **22**, 165–173.

Gellerstedt, G. (2001). Pulping chemistry. In: *Wood and Cellulosic Chemistry* (eds D.N.-S. Hon and N. Shiraishi), pp. 859–905, M. Dekker, New York, Basel.

Gildberg, A. and Stenberg, E. (2001). A new process for advanced utilisation of shrimp waste. *Process Biochem.*, **36**, 809–812.

Huttermann, A., Mai, C. and Kharazipour, A. (2001). Modification of lignin for the production of new materials. *Appl. Microb. Biotechnol.*, **55**, 387–394.

Jenkins, D.W. and Hudson, S.M. (2001). Review of vinyl graft copolymerization featuring recent advances toward controlled radical-based reactions and illustrated with chitin/chitosan trunk polymers. *Chem. Rev.*, **101**, 3245–3273.

Johnson, T.F.N. (2001). Current and future market trends. In: *Regenerated Cellulose Fibres* (ed. C. Woodings), pp. 273–289. CRC Press, Boca Raton, Boston, New York, Washington DC.

Klemm, D., Philipp, B., Heinze, T., Heinze, U. and Wagenknecht, W. (1998). *Future Developments in Cellulose Chemistry – An Outlook in Comprehensive Cellulose Chemistry*, Wiley-VCH, Weinheim, **2**, 315–325.

Kubota, N. and Kikuchi, Y. (1998). Macromolecular complexes of chitosan. In: *Polysaccharides, Structural Diversity and Functional Versatility* (ed. S. Dimitriu), pp. 595–628, M. Dekker, Inc., New York.

Kunz, M. (1993). Hydroxymethylfurfural, a possible basic chemical for industrial intermediates. In: *Inulin and Inulin Containing Crops* (ed. A. Fuchs), pp. 149–160, Elsevier Science Publishers B.V.

Muzzarelli, R.A.A. and Muzzarelli, B. (1998). Structural and functional versatility of chitins. In: *Polysaccharides, Structural Diversity and Functional Versatility* (ed. S. Dimitriu), pp. 569–594, M. Dekker, Inc., New York.

Nicolau, K.C. and Mitchell, H.J. (2001). Adventures in carbohydrate chemistry: new synthetic technologies, chemical synthesis, molecular design, and chemical biology. *Angew. Chem. Int. Ed.*, **40**, 1576–1624.

Rabea, E.I., Badawy, M.E.T., Stevens, C.V., Smagghe, G. and Steurbaut, W. (2003). Chitosan as Antimicrobial Agent. Applications and mode of action. *Biomacromolecules*, **4**, 1457–1465.

Röper, H. (2002). Renewable raw material in Europe – Industrial utilisation of starch and sugar. *Starch*, **54**, 89–99.

Rosenau, T., Potthast, A., Sixta, H. and Kosma, P. (2001). The chemistry of side reactions and byproduct formation in the system NMMO/cellulose (Lyocell process). *Progr. Polym. Sci.*, **26**, 1763–1837.

Sjöström, E. (1981). *Pulping Chemistry in Wood Chemistry, Fundamentals and Applications*, pp. 104–124. Academic Press Inc., Orlando.

Stevens, C.V., Meriggi, A. and Booten, K. (2001a). Chemical modification of inulin, a valuable renewable resource, and its industrial applications. *Biomacromolecules*, **2**, 1–16.

Stevens, C.V., Meriggi, A., Peristeropoulou, M., Christov, P.P., Booten, K., Levecke, B., Vandamme, A., Pittevils, N. and Tadros, T.F. (2001b). Polymeric surfactants based on inulin, a polysaccharide extracted from chicory. 1. Synthesis and interfacial properties. *Biomacromolecules*, **2**, 1256–1259.

Stevens, C.V., Meriggi, A. and Booten, K. (2001c). *PCT WO 01/44303 A1* (C08B 31/4, CO8G 18/71, C11D 1/66), 21/06/2001. Tensio-active glucoside urethanes, p. 23.

Targonski, Z., Rogalski, J. and Leonowicz, A. (1992). Biological transformation of lignocellulose as potential source of food resources. *Biotechnology*, **2**, 56–65.

Tsigos, I., Martinou, A., Kafetzopoulos, D. and Bouriotis, V. (2000). Chitin deacetylases: new, versatile tools in biotechnology. *Trends in Biotechnology*, **18**, 305–312.

Yong Lee, K. and Mooney, D.J. (2001). Hydrogels for tissue engineering. *Chem. Rev.*, **101**, 1869–1879.

8

Occurrence, Functions and Biosynthesis of Non-Carbohydrate Biopolymers

Christian O. Brämer and Alexander Steinbüchel

8.1 Introduction

In previous chapters the use of natural products (mostly of plants) has been discussed as energy source and as substrate for the production of interesting chemicals. However, in many cases interesting compounds are produced in nature by microorganisms. Biopolymers produced by bacterial strains are a class of compounds that have attracted more and more interest since techniques became available to make the bacteria produce them in large quantities.

Biopolymers and their derivatives are diverse, abundant, important to life, exhibit fascinating properties and are of increasing interest for various applications. The biodegradability of these polymers is an important feature for industry to investigate further, in order to develop their production and potential applications.

In this chapter several aspects of the use of biopolymers are discussed:

- The different classes of biopolymers, their function and their synthesis in nature.
- A more detailed description of the recent development of three groups of biopolymers: the polyhydroxyalkanoates; the poly(thioesters); and the polyamides (including cyanophycin, poly(glutamic acid), poly(lysine) and poly(aspartic acid)).
- Further, the microbial production of monomers is described. These monomers may be chemically polymerized to yield biodegradable polymers for the packaging industry.

Renewable Bioresources: Scope and Modification for Non-food Applications. Edited by C.V. Stevens and R. Verhé
© 2004 John Wiley & Sons, Ltd   ISBNs: 0-470-85446-4 (HB); 0-470-85447-2 (PB)

8.2   Biopolymers: Functions and Synthesis in Nature

Biopolymers are classified into: (i) living matter such as ribonucleic acids and deoxy-ribonucleic acids; (ii) polyamides such as proteins and poly(amino acids); (iii) polysaccharides such as cellulose, starch and xanthan; (iv) organic polyoxoesters such as poly(hydroxyal-kanoic acids), poly(malic acid) and cutin; (v) polythioesters, which were reported only recently; (vi) polyanhydrides, with poly-phosphate as the only example; (vii) polyisopre-noides such as natural rubber or Gutta Percha; and (viii) polyphenols such as lignin or humic acids (Table 8.1).

Biopolymers occur in any organism and they contribute the major fraction to the cellular dry matter in most organisms. Biopolymers fulfil a wide range of quite different essential or beneficial functions for the organisms, such as: conservation and expression of genetic information; catalysis of reactions; storage of carbon, nitrogen, phosphorus and other nutrients as well as energy; defence and protection against the attack of other cells or hazardous environmental or intrinsic factors; sensor of biotic and abiotic factors; communication with the environment and other organisms; mediator for adhesion to the surfaces of other organisms or of non-living matter; and many more. In addition, many biopolymers are structural components of cells, tissues and entire organisms. To accomplish all of these different functions, biopolymers must exhibit rather diverse properties. They must interact very specifically with a large variety of different substances, components and materials, and often they must have extraordinarily high affinities for them. Finally, they must have a high strength. Some of these properties are utilized directly or indirectly for various applications. This and the possibility of producing them from renewable resources or CO_2, like living matter is always doing, makes biopolymers interesting candidates for the industry.

All organisms synthesizing biopolymers share general principles concerning (i) location of the biosynthetic processes; (ii) dependency on templates; and (iii) substrates for polymerizing enzymes. First, all biopolymers are synthesized by enzymatic processes, though the localization of these enzymes may differ, e.g. cytoplasm, compartments or organelles of the cell, cell wall or extracellularly. Thereby synthesis of a biopolymer is

Table 8.1 Classification of biopolymers (Steinbüchel, 2001. Reproduced by permission of Wiley-VCH.)

Class	Example	Template-dependent synthesis	Substrate of the polymerase	Synthesis in	
				Prokaryotes	Eukaryotes
Polynucleotides	Nucleic acids	Yes	dNTPs, NTPs	Yes	Yes
Polyamides	Proteins	Yes	Aminoacyl-tRNAs	Yes	Yes
	Poly(amino acids)	No	Amino acids	Yes	Yes
Polysaccharides	Xanthan, dextran	No	Sugar-NDP, sucrose	Yes	Yes
Polyoxoesters	PHB	No	Hydroxyacyl-CoA	Yes	(No)
Polythioesters	Poly(3HB-co-3MB)	No	Mercaptoacyl-CoA	Yes	No
Polyanhydrides	Polyphosphate	No	ATP	Yes	Yes
Polyisoprenoids	Natural rubber	No	Isopentenyl-pyrophosphate	No	Plants, some fungi
Polyphenols	Lignin	No	Radical intermediates	No	Only plants

not limited only to the location where it is initiated, as the polymer can also be transported into another compartment of the cell where synthesis is completed. Secondly, biopolymers are synthesized by either template-dependent (nucleic acids or proteins) or template-independent (e.g. polysaccharide, polyoxoester or polyisoprenoid) processes. Template-dependent biopolymers are monodisperse, i.e. all individual molecules have the same molecular weight, and the primary structure of these polymers is highly conserved. In contrast, template-independent biopolymers are polydisperse, i.e. they share only an average molecular weight. Finally, in most cases biopolymers are not synthesized by direct polymerization of their building blocks. Often activated substrates are used by the polymerizing enzymes, e.g. ADP- or UDP-glucose for glycogen synthesis or hydroxyacyl-CoA thioester for PHA synthesis, or the polymerization reaction is driven by hydrolysis of ATP as, for example, during biosynthesis of poly(γ-D-glutamate) (Kunioka, 1997).

Production of biopolymers can be realized in different ways in order to make them available for technical applications. First, biopolymers occurring abundantly in nature, e.g. cellulose or starch, can be isolated directly from plants or algae. Secondly, biotechnological production of biopolymers can occur intra- and extracellularly with different consequences due to limitations of the production or downstream processes which must be employed to purify the polymer. By intracellular accumulation of water-insoluble biopolymers like PHA (Anderson and Dawes, 1990) or cyanophycin (Oppermann-Sanio *et al.*, 1999), the amount is limited by the space in the cytoplasm of the cells. Therefore, during fermentative production of biopolymers using microorganisms, the yield per volume is limited by the cell density and by the fraction the polymer can contribute to the biomass. A further problem is the extraction of the product from cells or tissues by disintegration, which may be effective, but which leads to a concomitant release of cell constituents that have to be separated from the biopolymer. With extracellular accumulation of water-soluble polymers, like in case of the polysaccharides xanthan (Katzbauer, 1998) or dextran, which are accumulated by *Xanthomonas campestris* or *Leuconostoc mesenteroides* respectively, the problems mentioned above do not occur. Unfortunately, biotechnological processes can rarely take advantage of this, since the presence of these biopolymers in the medium causes a high viscosity resulting in rheological problems during fermentation processes. Thirdly, an alternative strategy for the production of biopolymers is the use of cell-free systems, for example the application of *in vitro* synthesis of biopolymers employing isolated enzymes, e.g. the application of heat-stable DNA polymerases in the polymerase chain reaction (PCR), or production of polymer constituents as monomers by fermentative processes and polymerization of them by a solely chemical process – as shown for the production of polylactic acid. Finally, generation of transgenic plants for the production of certain polymers.

Basic and applied research has revealed much knowledge on the enzyme systems catalysing biosynthesis, degradation and modification of biopolymers, as well as on the properties of biopolymers. This has also resulted in an increased interest in biopolymers for various applications in industry, medicine, pharmacy, agriculture, electronics and many other areas. However, the knowledge is still scarce. The genes for the biosynthesis pathways of many biopolymers are still not available, or were identified relatively recently (mainly) due to increased availability of genome sequencing data; many new biopolymers have been described only recently, and from just a minor fraction of the biopolymers the biological, chemical, physical and material properties have been investigated. Often promising biopolymers are not available in sufficient amounts for that purpose. Nevertheless, polymer chemists, engineers and material scientists in academia and industry have discovered biopolymers that are suitable chemicals and materials for many new applications, or they are considering biopolymers as models to design novel synthetic polymers.

Due to the complexity of the topic this chapter will focus on three of the technically relevant biopolymer classes, without neglecting other important biopolymers completely: the biosynthesis of microbial (i) *polyesters*; (ii) *polythioesters*; and (iii) *polyamides* will be covered. Traditional and novel processes for their production will be shown. In addition, biotechnological processes for the production of monomers suitable for subsequent polymer synthesis are presented.

8.3   Polyhydroxyalkanoates

Living systems are capable of synthesizing a wide range of different polyesters. Most of them are synthesized by plants as structural components of the cuticle that covers the aerial parts of plants, such as cutin and suberin, or by prokaryotic microorganisms as intracellular storage compounds. The latter are referred to as polyhydroxyalkanoic acids (PHAs). These polyesters from plants and bacteria are water-insoluble. Besides these, the water-soluble polyester poly(malic acid) is synthesized by a few eukaryotic organisms. In addition, polymers, e.g. 3-hydroxybutyrate, exhibiting a rather low degree of polymerization were detected, complexed with other biopolymers such as calcium polyphosphate or proteins. The latter were found in almost any organism looked at; however, the physiological function(s) have not been revealed yet.

The chemical composition of insoluble cytoplasmic inclusions in the Gram-positive bacterium *Bacillus megaterium* was already identified in 1926 as poly(3-hydroxybutyric acid) (Lemoigne, 1926). By the end of the 1950s, enough evidence had been accumulated from physiological studies to suggest that this biopolymer functions as an intracellular reserve for carbon and energy. Meanwhile, it is known that this or structurally related storage polyesters are synthesized by members of almost any phylogenetic taxon of prokaryotes.

An overwhelming number of different PHAs, comprising approximately 150 different hydroxyalkanoic acids as constituents of these polyesters, have been isolated from bacteria during the last 20 years (Steinbüchel and Valentin, 1995). In 1974 the identification of 3-hydroxyalkanoates, other than 3-hydroxybutyrate, such as 3-hydroxyvalerate and 3-hydroxyhexanoate, was reported in chloroform extracts of activated sludge. Since the 1980s, many bacteria were demonstrated to synthesize various types of polyesters containing 3-, 4- and 5-hydroxyalkanoate units. However, most of these PHAs are obtained only if precursor substrates, which are structurally more or less related to the constituents to be incorporated into PHAs, are provided as carbon source to the bacteria. Only a few PHAs can be obtained from the simple carbon sources that are available in large amounts from agriculture or CO_2 from the atmosphere. The thorough knowledge of the key enzyme, PHA synthase, and of the enzymes involved in the biosynthesis of hydroxyacyl coenzyme thioesters, as well as the availability of the corresponding genes, allows engineering of the metabolism to obtain new PHA biosynthesis pathways. PHA synthases can be divided into four classes comprising differences in their molecular composition as well as in their substrate specificity: (i) PHA synthases of class-I consist of only one subunit (PhaC) and incorporate short chain length (C_3–C_5) hydroxyacyl-CoA thioester into the polyester; (ii) PHA synthases of class-II also consist of one subunit (PhaC) but they use medium chain length (C_6–C_{14}) hydroxyacyl-CoA thioesters as substrates; (iii) PHA synthases of class-III consist of two classes of subunits (PhaC and PhaE). These enzymes use short chain (C_3–C_5), as well as medium (C_4–C_8), hydroxyacyl-CoA thioester as substrates; and (iv) PHA synthases of class-IV have been identified only in *B. megaterium* and is composed of two different subunits (PhaC and PhaR). This enzyme has substrate specificity for short chain (C_3–C_5) hydroxyacyl-CoA thioesters.

So far, little has been done to engineer metabolic pathways *in vitro* by combining isolated PHA synthases with additional enzymes, which allows the synthesis of hydroxyacyl-CoA thioesters and the recycling of coenzyme A. In this way often non-natural pathways were engineered. It may be demonstrated that such *in vitro* engineered pathways can also be established *in vivo* if the corresponding genes are expressed in the respective bacterium (Liu and Steinbüchel, 2000; Lütke-Eversloh *et al.*, 2002). Most studies on the engineering of PHA biosynthesis pathways have been done *in vivo*. *In vivo* metabolic engineering is of particular interest if pathways are engineered which mediate between central intermediates of the metabolism on one side and the hydroxyacyl-CoA thioesters, as substrates of PHA synthases, on the other side. This is because such studies could yield recombinant bacteria or transgenic plants which are capable of producing a wider range of PHAs from simple and cheap carbon sources or CO_2.

Despite the detection of so many different PHAs the use of precursor substrates is still a tool to obtain novel, hitherto unknown PHAs. The onset of the molecular biology revolution during the late 1970s provided new tools for biological research, which were successfully used to decipher genetic information and to further understand the principles behind PHA biosynthesis at the molecular level. By the end of the 1980s, the genes coding for enzymes involved in PHA biosynthesis were cloned from *Ralstonia eutropha* (formerly known as *Alcaligenes eutrophus*), and the genes were also shown to be functionally active in *Escherichia coli*. To date, about 60 PHA synthase structural genes have been cloned from different bacteria (Rehm and Steinbüchel, 2001). In addition, many genes encoding enzymes and proteins relating to PHA biosynthesis were cloned and characterized at a molecular level. Thereby a class of proteins, the amphiphilic phasins, was identified, which layer the PHA granules and play an important role during formation and structuring of granules. This stimulated research and provided new perspectives for biotechnological production of PHAs. This knowledge has been utilized to establish PHA biosynthesis in many prokaryotic organisms and plants, such as *Arabidopsis thaliana, Brassicus napus* and *Zea mays*. The methodology of metabolic engineering was successfully applied for effective production of various PHAs, by fermentation biotechnology or agriculture, in economically feasible processes. In particular, transgenic plants expressing PHA biosynthesis pathways may provide potential producers of PHAs in the future. One important aspect is the large-scale biotechnological production of PHAs by fermentative processes and by agriculture from renewable carbon sources and CO_2 respectively.

The PHA family of polyesters is thermoplastic with biodegradable and biocompatible properties. Many of these water-insoluble polyesters can be thermoformed by using conventional extrusion and moulding equipment into various types of products such as bottles, films and fibres – much like established petrochemical-based thermoplastics.

Some PHAs, most notably poly(3HB) and poly(3HB-*co*-3HV), have become commercially attractive as novel biodegradable materials for applications in various areas such as the packaging industry, fishing industry and medicine. In medicine, PHA found applications in cardiovascular products such as pericardial patches (Bowald and Johansson-Ruden, 1990, 1997; Malm *et al.*, 1992a,b), artery augmentation (Malm *et al.*, 1994), cardiovascular stents (Schmitz and Behrend, 1997; Unverdorben *et al.*, 1998), vascular grafts (Noisshiki and Komatsuzaki, 1995) and heart valves (Stock *et al.*, 2000), in the field of guided bone or tissue regeneration (Kostopoulos and Karring, 1994; Leenstra, Maltha and Kuijpers-Jagtman, 1995), in drug delivery as implants, tablets or microparticulate carriers (Akhtar *et al.*, 1989; Cargill *et al.*, 1989; Chen *et al.*, 2000), in human and animal nutrition (Peoples *et al.*, 1999; Martin, Peoples and Williams, 2000; Veech, 2000), as well as in orthopaedics and wound management. Applications are also known for the water-soluble polyester poly(malic acid). This explains the interest of the industry in PHAs and other polyesters as large-scale biotechnological products. The physical and mechanical properties can be regulated by varying the composition of the polyesters. As a result, PHA can be made into a wide variety of polymeric materials, from hard crystalline plastics to very elastic rubber. Besides thermoplasticity, one of the most important characteristics of PHA products is their biodegradability (Jendrossek, Schirmer and Schlegel, 1996). PHA products such as films and fibres are degraded in soil, sludge, seawater and also compost. Under optimum conditions the degradation rate is extremely fast. Many prokaryotic and eukaryotic micro-organisms excrete extracellular PHA depolymerases that hydrolyze PHA products, and they utilize the decomposed compounds as nutrients. Also these genes have been cloned and characterized at a molecular level. Today, interdisciplinary research and development of biological polyesters are rapidly expanding in both the biological and polymer sciences. Concerted multidisciplinary scientific approaches have been directed to elucidate various new aspects of PHA. One important impact of studying and introducing natural polyesters was that efforts to establish new synthetic biodegradable materials were stimulated very much. As a consequence, many new biodegradable packaging materials were developed by the chemical industry, and production processes for already-existing synthetic polyesters were extremely optimized. For example, polylactides, which were formerly affordable only for medical applications, will now also become available for bulk applications.

8.4   Poly(thioesters)

The possibilities recently created by genetic engineering can also lead to the development of new biopolymers in the cells. Therefore a variety of the polyoxoesters described above can be formed as PTEs, which contain thioester bonds instead of oxoester bonds in the polymer backbone. They belong to the organic biopolymers containing sulphur. Initial studies on synthetic PTEs from dithiols, adipoyl chlorides or terephthalyl chlorides were reported 50 years ago (Marvel and Kotch, 1951), but recently the production of PTEs by microorganisms was described (Lütke-Eversloh *et al.*, 2001). 3-Mercaptobutyrate (3MB), 3-mercaptopropionate (3MP) and 3-mercaptovalerate (3MV) were identified as constituents of co- or homopolymers when cells of the wild-type strain *R. eutropha* or recombinant *E. coli* strains were cultivated in the presence of the corresponding 3-mercaptoalkanoic acid (3MA) as precursors. With regard to the chemical structure of these substrates, the β-position of the sulphhydryl group is required for incorporation into the polymer. As most of these precursors are not commercially available, with the exception of 3-mercaptopropionic acid, these 3-mercaptoalkanoic acids must be chemically synthesized via the corresponding acetylmercaptoalkanoic acid, which can be synthesized by the addition of thioacetic acid to 2-alkenoic acid with subsequent alkaline cleavage (Schjånberg, 1941).

For biotechnological production of copolyester consisting of either 3-hydroxybutyrate (3HB) and 3MP (poly(3HB-*co*-3MP)) or 3HB and 3MB (poly(3HB-*co*-3MB)), the PHA-accumulating bacterium *R. eutropha* was employed. As this strain is unable to use 3MA as a carbon and energy source, a second carbon source for growth, e.g. gluconate or fructose, must be applied. Due to the toxic effect of 3MAs, this substrate has to be added in suitable portions during fed-batch fermentation. The accumulation of these copolyesters in *R. eutropha* most likely occurs by the well-known metabolic routes comprising the activation of the 3MA to its thioester and incorporation into the polymer (Figure 8.1a). A possible route of production of homopolyesters of 3MA compounds entails the use of recombinant

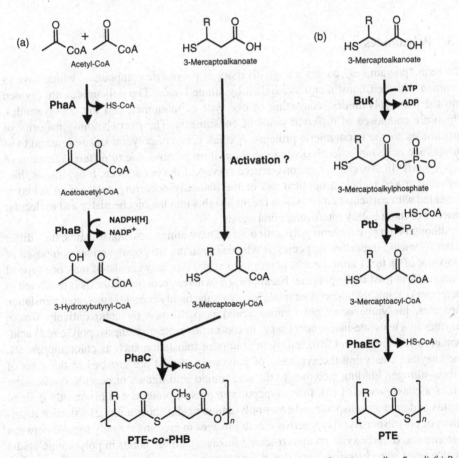

Figure 8.1 *Putative metabolic pathways for PTE biosynthesis in R. eutropha (a) as well as E. coli (b) Buk, butyrate kinase; Ptb, phosphotransbutyrylase; PhaEC, PHA synthase; PhaA, β-ketothiolase, PhaB, acetoacetyl-CoA reductase; PhaC, PHB synthase*

strains of non-polymer-accumulating bacteria, e.g. *E. coli*, if the genes for polymer production are expressed in these strains. A recombinant *E. coli* strain harbouring the butyrate kinase (*buk*) and phospho-transbutyrylase (*ptb*) genes of *Clostridium acetobutylicum*, as well as the two PHA synthase genes (*phaE* and *phaC*) of *Thiocapsa pfennigii* (Liu and Steinbüchel, 2000; Lütke-Eversloh *et al.*, 2002), was employed for the production of homopolyesters of 3MP, 3MB and 3MV by feeding the corresponding monomers as precursor substrates. The precursors are first phosphorylated by the *ptb* gene product, to the 3MA-phosphate, and afterwards activated to the CoA-thioester by Buk. These 3MA-CoA are the substrates for the polymerization reaction catalysed by the PHA-synthase of *T. pfennigii* (Figure 8.1b).

Although PTEs seem to be analogous to polyoxoesters, the physical characterization of these polymers revealed a few differences, and the microstructures of PTEs are highly non-homogenous as shown by nuclear magnetic resonance (NMR) spectroscopy and thermal analysis. The discovery of PTEs by fermentative processes makes it possible to evaluate their potential for technical applications.

8.5   Polyamides

The term "polyamides" covers a huge diversity of polymeric compounds, which have in common that their constituents are linked by amide bonds. The polyamides are divided into the homopolyamides, consisting of one type of monomer, and the copolyamides, which are composed of different kinds of constituents. The overwhelming majority of polyamides are the copolymeric proteins. A small group of polyamides are referred to as poly(amino acids) in order to distinguish them from proteins due to different features of biosynthesis. In this chapter, a comparative survey of the occurrence, biosynthesis, biodegradation and technical applications of the naturally occurring poly(amino acids) is presented with particular emphasis on recent insights into the biochemistry and molecular genetics of these highly interesting compounds.

Although both proteins and poly(amino acids) have amino acid constituents, they differ in the following important respects: (i) whereas proteins are copolyamides composed of a mixture of up to 21 amino acids, poly(amino acids) typically consist of only one type of amino acid, at least in the polymer backbone; (ii) whereas protein synthesis is catalysed in a template (i.e. mRNA)-dependent mode involving the highly complex ribosomal translation apparatus, biosynthesis of poly(amino acids) is performed by comparatively simple enzymes in a template-independent way. In contrast to protein synthesis, poly(amino acid) formation is therefore not affected by inhibitors of translation such as chloramphenicol. The enzymes catalysing the synthesis of poly(amino acids) are grouped in the class of carbon–nitrogen binding enzymes to the acid-amino acid ligases or peptide synthetases; (iii) as a consequence of (ii), the polypeptide strands of proteins exhibit exactly defined lengths and are monodisperse, whereas poly(amino acids) show a remarkable size distribution or polydispersity; (iv) whereas amide linkages in proteins are only formed between α-amino and α-carboxylic groups (α-amide linkages), amide bonds in poly(amino acids) involve other side chain functions (i.e. β- and γ-carboxylic and ε-amino groups).

Only three different poly(amino acids) are known to occur in nature – poly(glutamic acid), poly(lysine), and cyanophycin (Figure 8.2). However, there may be other poly

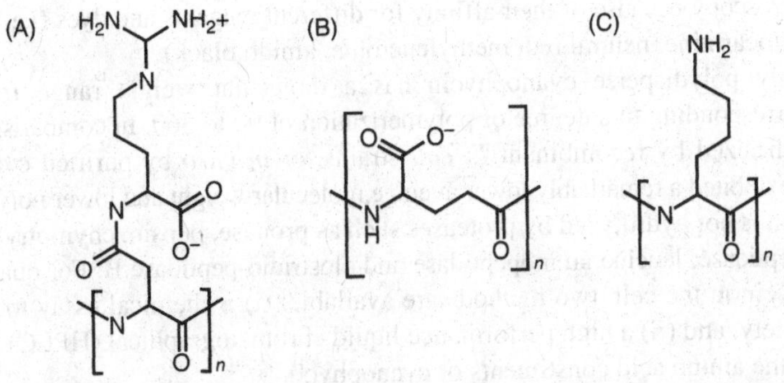

Figure 8.2 *Chemical structures of (A) cyanophycin [n = 90–360], (B) poly(glutamic acid) [n = 700–7000] and (C) poly(lysine) [n = 25–30]*

(amino acids) that have not been found yet. In poly(glutamic acid), the amide linkages are formed between the α-amino group and the γ-carboxyl group in the polymer backbone, whereas in poly(lysine) the α-carboxyl group is linked to the ε-amino group of lysine. In contrast to these two poly(amino acids), the constituents of the third poly(amino acid), cyanophycin, are α-aspartic acid residues containing pendent arginine residues linked to the β-carboxyl group. As all three poly(amino acids) are completely biodegradable they may be considered as a basis for environmentally friendly polyamides in various technical applications such as sustained-release material, thickener, in bioremediation processes or as a lubricant.

8.5.1 Cyanophycin

In contrast to poly(glutamic acid) and poly(lysine), cyanophycin is located only in the cytoplasm of producing cells. Cyanophycin, which is also referred to as cyanophycin granule polypeptide (CGP), was discovered microscopically as highly refractile cell inclusions more than 100 years ago (Borzi, 1887). The granules, which vary in size and shape, occur in all cyanobacterial groups. Under these conditions cyanophycin synthesis and accumulation can be promoted by addition of translational or transcriptional inhibitors, and deprivation of sulphur or phosphorus. It was first concluded that cyanophycin is limited to cyanobacteria, but recently a gene encoding a cyanophycin synthetase (*cphA*) was identified in the non-cyanobacterial strain *Acinetobacter* sp. ADP1 and its functionality as cyanophycin synthetase was shown. In addition, genes homologous to *cphA* of *Synechocystis* sp. strain PCC 6803 were found in species of the genera *Bordetella*, *Clostridium*, *Desulfitobacterium* and *Nitrosomonas* (Krehenbrink, Oppermann-Sanio and Steinbüchel, 2002). In non-heterocyst-forming cyanobacteria, the cyanophycin granules are distributed in the protoplasts. In heterocyst-forming cyanobacteria, the cyanophycin granules are frequently present in the heterocysts, which are specialized for the fixation of molecular nitrogen. The cyanophycin granules represent only one type of a variety of cell inclusions that can be found in cyanobacterial cells. Among the other simple inclusions, which consist of polyglucans, poly(3-hydroxybutyrate), lipids or polyphosphate and the more complex carboxysomes, phycobilisomes and gas vacuoles, the cyanophycin granules can be recognized by light microscopy because of their affinity for different reagents and dyes (i.e. Sakagushi reagent, acetocarmine, neutral red, methylene blue, amido black).

The highly polydisperse cyanophycin has a molecular weight range from 25 to 100 kDa, corresponding to a degree of polymerization of 90 to 360. In comparison, cyanophycin synthesized by recombinant *E. coli* strains, or *in vitro* by purified cyanophycin synthetase, exhibited a remarkably lower average molecular weight and lower polydispersity. Cyanophycin is not hydrolyzed by proteases such as pronase, pepsin, chymotrypsin, α- or β-carboxypeptidase, leucine aminopeptidase and clostridio-peptidase B. For quantification of cyanophycin in the cell, two methods are available: (i) a chemical assay to detect the arginine moiety; and (ii) a high performance liquid chromatographical (HPLC) method to detect both the amino acid constituents of cyanophycin.

In their natural aquatic habitat, cyanobacteria are often exposed to limitation or depletion of nitrogen. Therefore, cyanophycin works as a temporary nitrogen-storage compound, as it is accumulated during transition from the exponential to the stationary growth phase and disappears when balanced growth conditions resume. Thereby it fulfils the criteria of a very good nitrogen-storage compound, as five nitrogen atoms are deposited for every building block of the polymer and cyanophycin is insoluble in the aquatic cell environment under physiological pH and ionic strength. In comparison, poly(arginine) would store more nitrogen as cyanophycin, but this polycationic polypeptide is soluble under physiological conditions and would therefore disturb the internal cell milieu by binding to polyanionic DNA and by changing the osmolarity.

Biosynthesis of cyanophycin is catalysed by one single enzyme, the cyanophycin synthetase, which polymerizes *in vitro* arginine and aspartic acid, when ATP, Mg^{2+}, K^+, a sulphhydryl reagent and a small amount of cyanophycin as primer are available. The primer is elongated at its C-terminus by stepwise incorporation of the amino acid substrate (Figure 8.3).

First, the poly(aspartic acid) backbone is phosphorylated at the α-carboxylic acid group in an ATP-dependent reaction, followed by linkage to the α-amino group of an aspartic acid residue. In a second reaction step, the β-carboxylic acid group of the poly (aspartic acid) chain is phosphorylated and linked to the α-amino group of an incoming arginine, resulting in the formation of cyanophycin. As cyanophycin is a storage compound, enzymes also exist to mobilize this transient nitrogen storage compound. These

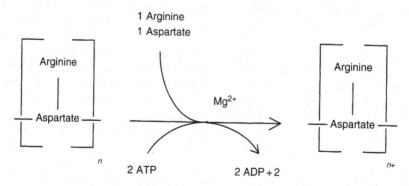

Figure 8.3   *Reaction of cyanophycin synthetase*

cyanophycinases are hydrolases and are located intracellularly and, as recently shown, also extracellularly (Obst *et al.*, 2002). The reaction products are dipeptides consisting of arginine and aspartic acid, which have to be cleaved from the amino acids in a second step.

Cyanophycin has attracted the interest of the chemical industry as it can be modified chemically to a derivative with a reduced amount of arginine, which can be employed in technical processes as a biodegradable substitute for poly(acrylate). But the natural producer, cyanobacteria, are unsuitable tools for biotechnological production of cyanophycin as the cell yields and cyanophycin contents are low. They need a long cultivation time and highly sophisticated fermentation processes. Therefore, cyanophycin synthesis has been established on other bacterial host strains such as *E. coli*, *Corynebacterium glutamicum* or *R. eutropha* to achieve an economically profitable production.

8.5.2   Poly(Glutamic Acid)

Poly(glutamic acid), PGA, occurs as a cell-capsule component of strains of the Gram-positive genus *Bacillus* and was first detected 60 years ago in *B. anthracis*, which causes the disease anthrax. The physiological function of PGA in *B. anthracis* depends on its weak immunogenity, preventing the production of antibodies and thereby increasing the virulence of this strain (Roelants, Senyk and Goodman, 1969). In addition to these endospore-forming bacilli, two halophilic eubacteria (*Sporosarcina halophila* and *Planococcus halophila*), the halophilic archaea (*Natrialba aegyptica*) as well as the eukaryotic organism (*Cnidaria*) are also capable of producing PGA. The function of excreted PGA is not known with certainty, but besides its immobilizing function this polymer constitutes an external carbon and nitrogen reserve. Furthermore, PGA-containing slime, excreted by some halophilic bacteria, increases the availability of water in hypersaline environments.

Nutritional requirements for PGA production were studied with *B. licheniformis* and revealed that a high content of carbon sources is needed. In addition, other *Bacillus* strains need glutamic acid in the medium to produce PGA. Under optimized conditions this polymer can be produced in amounts ranging from 20 to 50 g/l by strains of *B. licheniformis* or *B. subtilis*. Purified PGA is a water-soluble, highly hygroscopic white powder and chelating agent, leading to highly viscous solutions even at low concentrations.

The precursor for biosynthesis of PGA is the tricarboxylic acid cycle intermediate, 2-oxoglutarate, which can be directly transaminated to L-glutamic acid. *B. licheniformis* expresses two L-glutamic acid-synthesizing enzymes – glutamate synthase and glutamate dehydrogenase. These enzymes are relatively insensitive towards end-product inhibition and lead to high intracellular concentrations of L-glutamic acid, which is directed to PGA synthesis. All PGA-producing bacteria are able to metabolize this polymer as their sole source of carbon and nitrogen. During growth on this polyamide a depolymerase is expressed which seems to be physically associated to the cell surrounding PGA.

The polyanionic character of PGA makes it a suitable environmentally sound substitute for chemosynthetic polyacrylate in several technical fields, for example as a sustained-release material, drug carrier in food, thickener (aging prevention of food, viscosity enhancement for drinks) or humectant (use for skin-care cosmetics), or in bioremediation processes as flocculant or metal absorbent. In addition, attempts have been made recently to employ PGA-producing bacteria for conversion of ammonium nitrogen in liquid manure to a transient depot form of nitrogen (Pötter, Oppermann-Sanio and Steinbüchel, 2001). Chemically synthesized derivates of PGA are of great interest as they reveal improved material properties. Cross-linking of PGA by γ-irradiation or chemical reagents leads to the formation of the so-called hydrogel, which might serve as implants for drug-delivery or protein-slow-release systems in biomedical applications (Fan, Gonzales and Sevoian, 1996). They also show a fantastic water absorption capability corresponding to specific water contents (mass of absorbed water/mass of dry polymer) in the range of 800–3500 depending on the dose of irradiation applied during the synthesis.

8.5.3 Poly(Lysine)

The polyamide poly(lysine), PL, is only known to be produced by the Gram-positive bacterium *Streptomyces albulus* ssp. *lysinopolymerus*, which was isolated from Japanese soil during the screening for a Dragendorff-positive substance (i.e. alkaloid or quaternary nitrogen compound). By employing the antimetabolite method, a mutant of this strain was isolated capable of excreting four times more poly(lysine). This polymer has a cationic character at neutral pH and consists only of 25–30 constituents. It can be isolated from the culture filtrate by ionic exchange chromatography, treatment with active charcoal and finally precipitation. PL exhibits a growth inhibitory effect on Gram-positive as well as Gram-negative bacteria, at concentrations of 1–8 mg/l, and might therefore be employed as a preservative in animal or human feed such as boiled rice in lunchboxes in Japan. In addition, hydrogels with high water-absorbing capacity can be prepared from this polymer for use in the fields of agriculture, food processing and medicine. Biotechnologically produced PL is offered by the Chisso Corporation (Japan).

8.5.4 Poly(Aspartic Acid)

A fourth polyamide, poly(aspartic acid), which is chemically synthesized by pre-reaction of maleic acid anhydride with ammonia and subsequent thermal polymerization, is produced in a volume of 2000 t/yr. This process, which is employed by the BAYER AG for production of poly(aspartic acid), is nearly wasteless as by-products are recovered in the production process. Furthermore, this polyamide is biodegradable as shown by the Zahn-Wellens-test (OECD 302 B). Poly(aspartic acid) is sold as Baypure® DS 100 and can be applied as a dispersant, water softener or as an anti-corrosion agent. Due to its structural relationship to poly(acrylic acid) (Figure 8.4), poly(aspartic acid) is suitable to replace this persistent

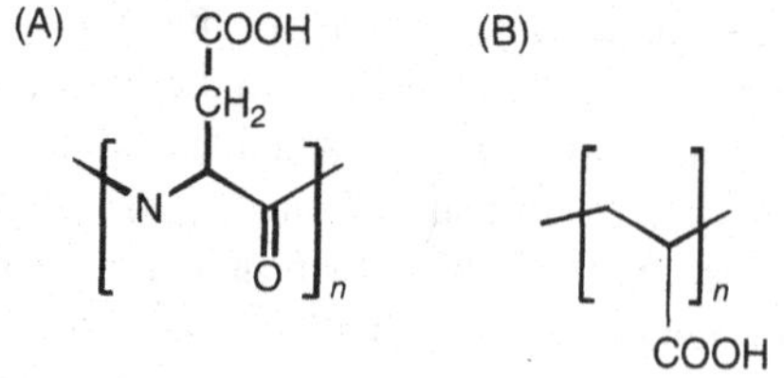

Figure 8.4 *Chemical structures of (A) poly(aspartic acid) and (B) poly(acrylic acid)*

substance in many technical applications. Examples include: as a scale inhibitor in cooling water circuits, by their adsorption at the surface of microscopically small salt crystals; as an environmentally sound detergent; as a corrosion inhibitor; as a calcium binder in textile processes; and finally in the liquefaction of mineral slurries. In future, this polyamide might be produced by a biotechnological process from renewable resources relying on the metabolism of cyanophycin.

8.6   Production of Monomers

In the previous sections the biosynthesis, biotechnological production and applications of naturally occurring biopolymers were described. Another way to achieve polymers is the polymerization of monomeric building blocks, which can be prepared by either chemical or biological methods. Naturally occurring metabolites such as amino acids, hydroxy acids and diols or dicarboxylic acids are often used as building blocks for chemical polymerization processes. These polymers have the advantage that they show biodegradability and biocompatibility and are therefore interesting for a variety of industrial applications. Until now the costs for biotechnological production of these monomers were too high, with the exception of lactic acid, in comparison to petrochemical-derived monomers, therefore they were not competitive. In the last decades, a lot of endeavours have been made to improve biotechnological and enzyme-catalysed synthetic processes, aiming at a reduction of costs. Production processes of some characteristic representatives of each class of monomers (amino acids, hydroxy acids and diols as well as dicarboxylic acids) are presented below.

8.6.1   Amino Acids

All 21 amino acids might be produced as primary metabolites by fermentative processes, but their biosynthetic pathways are often strictly regulated and hence most wild type strains are unable to produce L-amino acids. Therefore, strains were constructed employing different strategies, e.g. metabolic engineering, that are able to excrete amino acids. For the production of L-glutamic acid, L-lysine and L-phenylalanine fermentative processes exist, which make their production competitive with other methods like chemical or biocatalytic synthesis.

In 1957 *Corynebacterium glutamicum* was discovered as a strain which excretes large amounts of "L-glutamic acid" during cultivation. L-glutamic acid represents the major fraction of industrially available amino acids, and due to its two carboxylic groups, it acts as a precursor for two different poly-glutamic acids – poly(γ-glutamic acid) and poly(L-glutamic acid). Many different strategies were developed to reach a high yield of glutamic acid, i.e. strain development by random or rational mutagenesis and metabolic engineering, control of permeability of the cell membrane by adding surfactants or penicillin, and optimization of ammonium supply (Eggeling and Sahm, 1999).

A further commercially important amino acid, "L-lysine", which is produced by fermentation of *C. glutamicum* (Nakayama, 1985), is employed as feed additive and as the monomeric unit of poly(ε-lysine) or poly(L-lysine). Strains of *C. glutamicum*, which are

Table 8.2 Enzymatic production of amino acids (summarized from Leuchtenberger and Plocker, 1990. Reproduced by permission of Wiley-VCH.)

Enzyme	Substrates	Product
Aminoacylase	N-acetyl-D-, L-amino acids	Methionine, valine, alanine, tryptophan
α-Chymotrypsin	Esters of D-, L-amino acids	Aromatic amino acids
Amidase	Amide of D-, L-amino acids	Valine, phenylalanine, tryptophan, lysine
Dehydrogenase	D-, L-α-hydroxycarboxylic acid	Alanine, leucine, valine, phenylalanine
Aspartase, phenylalanine ammonialyase	α-, β-unsaturated compounds	Aspartic acid, phenylalanine
Dehydrogenase	α-keto acids	Alanine, leucine, valine, phenylalanine
Transaminase	α-keto acids	Phenylalanine, tyrosine, tryptophan, methionine, serine, cysteine, alanine, leucine, isoleucine, valine

employed for fermentative production of L-lysine, were obtained by undirected mutagenesis. Enhanced L-lysine production was also achieved by metabolic engineering strategies such as reducing the growth rate, increasing the availability of phosphoenolpyruvate as a precursor for L-lysine biosynthesis and enhancing expression of key enzymes of the L-lysine biosynthetic pathway.

An alternative process for the fermentative production of amino acids represents the enzymatic synthesis, which reveals several advantages including low cost of raw materials, high product purity, high enantioselectivity and regioselectivity as well as mild conditions. Enzymatic production is employed for the amino acids L-aspartic acid, L-alanine, L-methionine and L-valine (Leuchtenberger, 1996). A summary of enzymatic production of amino acids is shown in Table 8.2.

8.6.2 Dicarboxylic Acids

Dicarboxylic acids such as fumaric, itaconic and adipic acids are produced by biological processes or subsequent chemical conversion of biological products. The condensation of these acids results in the formation of different polyesters.

The unsaturated dicarboxylic acid *itaconic acid* (IA) can easily be incorporated into polymers and has received attention as a substitute for chemically synthesized acrylic acid and methacrylic acid. IA is produced by fungal fermentation of carbohydrates employing *Aspergillus terreus, Ustilago, Candida* or *Rhodotorula* sp., whereas the metabolic pathway is similar to that of citric acid production (Bonnarme *et al.*, 1995). However, the current production costs for IA by fermentative processes are approximately US$4 per kg, which makes this substance unsuitable for a wide range of applications.

Fumaric acid (FA), an intermediate of the tricarboxylic acid cycle (TCA), has found applications as an acidulant in food and beverage products. Furthermore, it represents a monomer for polymer synthesis. Currently, FA is chemically synthesized by isomerization of maleic acid, a derivative of the carcinogenic benzene. Hence new metabolic pathways for biotechnological production of FA from renewable resources are under investigation. Several factors are known to influence the FA production using a submerged fermentation of *Rhizopus*, such as nitrogen limitation under the condition of a high carbon-to-nitrogen ratio, a pH of 5.5 and the presence of $CaCO_3$. In *Rhizopus*, pyruvate carboxylase (pyruvate + CO_2 → oxaloacetate), fumarase and malate dehydrogenase are involved in the synthesis of this dicarboxylic acid. The reduction of production costs by applying cheap carbon sources, e.g. cassava bagasse hydrolysate (Carta *et al.*, 1999), or repeated batch processes employing cells of *Rhizopus* immobilized in polyurethane sponges (Petruccioli, Angiani and Federici, 1996) is essential for the biotechnological production of FA.

For the biotechnological production of "succinic acid", an intermediate of the TCA cycle, strains of *Anaerobiospirillum succiniciproducens* and *Actinobacillus succinogenes* are most efficient with a maximum volumetric productivity of 6.1 g of succinic acid $L^{-1} h^{-1}$. Furthermore, metabolic engineering of *E. coli* was employed in order to enhance the succinic acid production by establishing a new metabolic route converting phospho-enolpyruvate to malic acid. Besides other uses, succinic acid might provide an interesting feedstock for 1,4-butanediol and for the production of adipic acid and nylon.

Industrial production of *adipic acid*, an intermediate in the manufacture of nylon, is carried out by a multi-step chemical synthesis. Adipic acid may also be produced by fermentative processes, employing strains of *Acinetobacter*, *Nocardia* and *Pseudomonas*, and using cyclohexanol, hexanoate or derivatives of it as substrate, but the costs are too high for their commercial use. A genetically modified strain of *E. coli*, which harbours a plasmid containing the 3-dehydroshikimate dehydratase and protocatechuate decarboxylase genes of *Klebsiella pneumoniae* as well as the *Acinetobacter calcoaceticus* 1,2-dioxygenase gene, was able to convert several carbon sources to adipic acid via catechol and *cis, cis*-muconic acid. But until recently, the production of adipic acid was only based on chemical synthesis, rather than biological production, due to insufficient yields of adipic acid by fermentative processes and the requirement for expensive carbon sources.

8.6.3 Hydroxy Acids

Hydroxy acids such as lactic acid, lactones and cyclic esters contain a carboxylic acid as well as an alcohol group, which are necessary to build an ester bond in the polymer. Therefore these monomers represent suitable precursors for polyester synthesis.

The biotechnological production of the very important product *lactic acid* (LA) has raised several problems, such as the efficiency and economics of separation/purification steps, the requirement for complex nutrients due to the limited ability of *Lactobacilli* to synthesize B vitamins and amino acids, as well as low LA yields during continuous fermentations causing high production costs. To overcome these problems much attention has been focused on the optimization of several factors, including the employed production strains, the carbon/nitrogen sources, pH, temperature and various culture methods, in order to achieve an efficient production of LA. LA is currently used for the production of poly(lactide) (PLA), by the company Cargill-Dow (Minnetonka, Minnesota, USA), which might find application as sustainable packaging material due to a new combination of attributes like stiffness, clarity, deadfold and twist retention, as well as flavour and aroma barrier characteristics. A further application of PLA is in fibres and non-wovens – fabrics containing PLA are considered for sportswear as well as outerwear because this polymer shows several beneficial properties such as good moisture management, good drape/hand-feel and excellent crease resistance. In 1998 Kanebo Inc. introduced a PLA fibre under the trade name LACTRON™ at the Nagano Winter Olympics.

8.6.4   Diols

The polycondensation of diols such as 1,2-propanediol, 1,3-propanediol or 1,4-butanediol with dicarboxylic acids results in the formation of polyesters. These diols can be produced either by biological methods or by subsequent chemical conversion of biological products.

The chiral *1,2-propanediol*, which can be used for the synthesis of chiral polymers, is produced by *Thermoanaerobacterium thermosaccharolyticum* using glucose, xylose, mannose or cellobiose as substrate. Furthermore, strains of *E. coli* have been generated by metabolic engineering that are able to produce 1,2-propanediol by supplying the required enzymes and/or reducing power.

Many bacteria are able to produce the anaerobic fermentation product *1,3-propanediol* when glycerol is applied as substrate. The two facultative strains *K. pneumoniae* and *Citrobacter freundii* are selected for their ease of handling. As, in many countries, glucose and other sugars are considerably cheaper than glycerol various strategies have been applied to construct strains capable of producing 1,3-propanediol from these sugars. An application of 1,3-propanediol is its polycondensation with terephthalic acid, resulting in the formation of poly(propylene terephthalate) which can be used in carpeting and textiles with good recyclability. This process was developed by DuPont (Wilmington, DE, USA).

The production of *1,4-butanediol* is carried out by chemical transformation of various chemicals such as 1,4-butynediol, succinic acid, succinic anhydride, maleic acid or maleic anhydride. Biological production has also been reported, including ω-oxidation of *n*-butanol by *Pseudomonas* strains as well as enzymatic conversion of 1,4-butanediol diester employing a carboxylesterase from *Brevibacterium linens* IFO 12171. This monomer can be polymerized to thermoplastic polyesters with low melting points by condensation with aliphatic dicarboxylic acids.

8.7   Conclusion

In this chapter several aspects of the biosynthesis, occurrence and application of biopolymers as well as of monomers, produced by microbial processes, have been discussed.

1. *Polyhydroxyalkanoates (PHA)*   More than 150 different constituents of these intracellular storage compounds are known. The composition of these water-insoluble polymers depends on the precursor substrate and the microbial strain employed for biosynthesis. To enhance the spectrum of constituents, genetic engineering was employed to establish new metabolic pathways in microorganisms or plants, or to channel metabolites into PHA biosynthesis. The key enzymes of PHA biosynthesis are the PHA synthases, which differ in their substrate specificity (short or medium chain length hydroxyacyl-CoA thioesters) and molecular composition (one or two subunits). One important property of PHAs, besides their thermoplasticity, is their biodegradability and biocompatibility – conditional upon intra- and extracellular PHA depolymerases. PHA, especially poly(3HB) and poly(3HB-*co*-3HV), has found numerous applications in the packaging industry, fishing industry and medicine.
2. *Polyamides*   This class of biopolymers is characterized by amide bond linkage of their constituents (amino acids), with homo- and copolyamides (i.e. proteins) being

distinguished. In nature, only three different poly(amino acids) occur – poly(glutamic acid), poly(lysine) and cyanophycin. Their biodegradability makes these polyamides an environmentally friendly alternative for technical applications employed in agriculture, food processing or medicine.

3. *Monomers* Another way to achieve biopolymers is the polymerization of monomeric building blocks (i.e. amino acids, dicarboxylic acids, hydroxy acids or diols), which may be prepared by chemical or biological methods. These polymers also show bio-degradability and biocompatibility, and are therefore interesting for several industrial applications such as packaging materials or fibres. Furthermore, the monomers are employed for applications in the food and beverage industry as additives or acidulants.

References

Akhtar, S., Pouton, C.W., Notarianni, L.J. and Gould, P.L. (1989). A study of the mechanism of drug release from solvent evaporated carrier systems of biodegradable P(HB-HV) polyesters. *The Journal of Pharmacy and Pharmacology*, **41**, 5.

Anderson, A.J. and Dawes, E.A. (1990). Occurrence, metabolism, metabolic role and industrial uses of bacterial polyhydroxyalkanoates. *Microbiology Reviews*, **54**, 450–472.

Bonnarme, P., Gillet, B., Sepulchre, A.M., Role, C., Beloeil, J.C. and Ducrocq, C. (1995). Itaconate biosynthesis in *Aspergillus terreus*. *Journal of Bacteriology*, **177**, 3573–3578.

Borzi, A. (1887). Le communicazioni intracellulari delle Nostochinee. *Malphigia*, **1**, 74–203.

Bowald, S.F. and Johansson-Ruden, E.G. (1990). A novel surgical material. *European Patent Application* No. 0,349,505 A2.

Bowald, S.F. and Johansson-Ruden, E.G. (1997). A novel surgical material. *European Patent Application* No. 0,754,476 A1.

Cargill, R., Engle, K., Gardner, C.R., Porter, P., Sparer, R.V. and Fix, J.A. (1989). Controlled gastric emptying. II. *In vitro* erosion and gastric residence times of an erodible device in Beagle dogs. *Pharmaceutical Research*, **6**, 506–509.

Carta, F.S., Soccol, C.R., Ramos, L.P. and Fontana, J.D. (1999). Production of fumaric acid by fermentation of enzymatic hydrolysates derived from cassava bagasse. *Bioresource Technology*, **68**, 23–28.

Chen, J.-H., Chen, Z.-L., Hou, L.-B. and Liu, S.-T. (2000). Preparation and characterization of diazepam-polyhydroxybutyrate microspheres. *Gongneng Gaofenzi Xuebao*, **13**, 61–64.

Eggeling, L. and Sahm, H. (1999). Amino acid production: principles of metabolic engineering. In: *Metabolic Engineering* (eds S.Y. Lee and E.T. Papoutsakis), Marcel Dekker, New York.

Fan, K., Gonzales, D. and Sevoian, M. (1996). Hydrolytic and enzymatic degradation of poly (γ-glutamic acid) hydrogels and their application in slow-release systems for proteins. *Journal of Environmental Polymer Degradation*, **4**, 253–260.

Jendrossek, D., Schirmer, A. and Schlegel, H.G. (1996). Biodegradation of polyhydroxyalkanoic acids. *Applied Microbiology and Biotechnology*, **46**, 451–463.

Katzbauer, B. (1998). Properties and applications of xanthan gum. *Polymer Degradation and Stability*, **59**, 81–84.

Kostopoulos, L. and Karring, T. (1994). Guided bone regeneration in mandibular defects in rats using a bioresorbable polymer. *Clinical Oral Implants Research*, **5**, 66–74.

Krehenbrink, M., Oppermann-Sanio, F.B. and Steinbüchel, A. (2002). Evaluation of non-cyanobacterial genome sequences for occurrence of genes encoding proteins homologous to cyanophycin synthetase and cloning of an active cyanophycin synthetase from *Acinetobacter* sp. strain DSM 587. *Archives of Microbiology*, **177**, 371–380.

Kunioka, M. (1997). Biosynthesis and chemical reactions of poly(amino acids) from microorganisms. *Applied Microbiology and Biotechnology*, **47**, 469–475.

Leenstra, T.S., Maltha, J.C. and Kuijpers-Jagtman, A.M. (1995). Biodegradation of non-porous films after submucoperiosteal implantation on the palate of Beagle dogs. *Journal of Materials Science. Materials in Medicine*, **6**, 445–450.

Lemoigne, M. (1926). Produits de deshydration et de polymerization de lácide β-oxobutyrique'. *Bulletin de la Societe Chimique de France*, **8**, 770–782.

Leuchtenberger, W. (1996). Products of primary metabolism. *Biotechnology*, **6**, 455–502.

Leuchtenberger, W. and Plocker, U. (1990). Industrial uses of enzymes. In: *Enzymes in Industry* (ed. E. Gerhartz), Wiley-VCH, Weinheim.

Liu, S.J. and Steinbüchel, A. (2000). A novel genetically engineered pathway for synthesis of poly(hydroxyalkanoic acids) in *Escherichia coli'*. *Applied Environmental Microbiology*, **66**, 739–743.

Lütke-Eversloh, T., Bergander, K., Luftmann, H. and Steinbüchel, A. (2001). Identification of a new class of biopolymer: bacterial synthesis of a sulfur-containing polymer with thioester linkages. *Microbiology*, **147**, 11–19.

Lütke-Eversloh, T., Fischer, A., Remminghorst, U., Kawada, J., Marchessault, R.H., Bögershausen, A., Kalwei, M., Eckert, H., Reichelt, R., Liu, S.-J. and Steinbüchel, A. (2002). Biosynthesis of novel thermoplastic polythioesters by engineered *Escherichia coli*. *Nature Materials*, **1**, 236–240.

Malm, T., Bowald, S., Bylock, A., Saldeen, T. and Busch, C. (1992a). Regeneration of pericardial tissue on absorbable polymer patches implanted into the pericardial sac. An immunohistochemical, ultrasound and biochemical study in sheep. *Scandinavian Journal of Thoracic and Cardiovascular Surgery*, **26**, 15–21.

Malm, T., Bowald, S., Bylock, A. and Busch, C. (1992b). Prevention of postoperative pericardial adhesions by closure of the pericardium with absorbable polymer patches. An experimental study. *Scandinavian Journal of Thoracic and Cardiovascular Surgery*, **104**, 600–607.

Malm, T., Bowald, S., Bylock, A., Busch, C. and Saldeen, T. (1994). Enlargement of the right ventricular putflow tract and the pulmonary artery with a new biodegradable patch in transannular position. *European Surgical Research*, **26**, 298–308.

Martin, D.P., Peoples, O.P. and Williams, S.F. (2000). Nutritional and therapeutic uses of 3-hydroxy-alkanoate oligomers. *PCT Patent Application* No. WO 00/04895.

Marvel, C.S. and Kotch, A. (1951). Polythioesters. *Journal of the American Chemical Society*, **73**, 1100–1102.

Nakayama, K. (1985). Lysine. In: *Comprehensive Biotechnology*, Vol. 3 (eds H.W. Blanch, S. Drew and D.I.C. Wang), Pergamon Press, New York.

Noisshiki, Y. and Komatsuzaki, S. (1995). Medical materials for soft tissue use. *Japanese Patent Application* No. JP7,275,344 A2.

Obst, M., Oppermann-Sanio, F.B., Luftmann, H. and Steinbüchel, A. (2002). Isolation of cyanophycin-degrading bacteria, cloning and characterization of an extracellular cyanophycinase gene (*cphE*) from *Pseudomonas anguilliseptica* strain BI. The *cphE* gene from *P. anguilliseptica* BI encodes a cyanophycinhydrolyzing enzyme. *Journal of Biological Chemistry*, **277**, 25096–25105.

Oppermann-Sanio, F.B., Hai, T., Aboulmagd, E., Hezayen, F.F., Jossek, S. and Steinbüchel, A. (1999). Biochemistry of microbial polyamides metabolism. In: *Biochemical Principles and Mechanisms of Biosynthesis and Biodegradation of Polymers* (ed. A. Steinbüchel), Wiley-VCH, Weinheim.

Peoples, O.P., Saunders, C., Nichols, S. and Beach, L. (1999). Animal nutrition compositions. *PCT Patent Applications* No. WO 99/34687.

Petruccioli, M., Angiani, E. and Federici, F. (1996). Semi-continuous fumaric acid production by *Rhizopus arrhizus* immobilized in polyurethane sponge. *Process Biochemistry*, **31**, 463–469.

Pötter, M., Oppermann-Sanio, F.B. and Steinbüchel, A. (2001). Cultivation of bacteria producing polyamino acids with liquid manure as carbon and nitrogen source. *Applied and Environmental Biology*, **67**, 617–622.

Rehm, B.H.A. and Steinbüchel, A. (2001). PHA synthases the key enzymes of PHA synthesis. In: *Biopolymers: Polyesters* (eds I.A. Steinbüchel and Y. Doi), Wiley-VCH, Weinheim.

Roelants, G.E., Senyk, G. and Goodman, J.W. (1969). Immunochemical studies on the poly-γ-glutamyl capsule of *Bacillus anthracis*. V. The *in vivo* fate and distribution in rabbits of the polypeptide in immunogenic and nonimmunogenic forms. *Israel Journal of Medicine Sciences*, **5**, 196–208.

Schjånberg, E. (1941). Pentensäure und Thioessigsäure. *Chem. Ber.*, **74**, 1751–1759.

Schmitz, K.-P. and Behrend, D. (1997). Method of manufacturing intraluminal stents made of polymer material. *European Patent Application* No. 0,770,401 A2.

Steinbüchel, A. (2001). Perspectives for biotechnological production and utilization of biopolymers: metabolic engineering of polyhydroxyalkanoate biosynthesis pathways as a successful example. *Macromolecular Bioscience*, **1**, 1–14.

Steinbüchel, A. and Valentin, H.E. (1995). Diversity of microbial polyhydroxyalkanoic acids. *FEMS Microbiology Letters*, **128**, 219–228.

Stock, U.A., Sakamoto, T., Hatsuoka, S., Martin, D.P., Nagashima, M., Moran, A.M., Moses, M.A., Khalil, P.N., Schoen, F.J., Vacanti, J.P. and Mayer, J.E. (2000). Patch augmentation of the pulmonary artery with bioabsorbable polymers and autologous cell seeding. *The Journal of Thoracic and Cardiovascular Surgery*, **120**, 1158–1168.

Unverdorben, M., Schywalsky, M., Labahn, D., Hartwig, S., Laenger, F., Lootz, D., Behrend, D., Schmitz, K., Schaldach, M. and Vallbracht, C. (1998). Polyhydroxybutyrate (PHB) stent-experience in the rabbit. *American Journal of Cardiac Transcatheter Cardiovascular Therapeutics*, Abstract TCT-11, 5S.

Veech, R.L. (2000). Therapeutic compositions. *PCT Patent Application* No. WO 00/15216.

9

Industrial Products from Lipids and Proteins

Roland Verhé

In co-authorship with *Martin Mittelbach, Sandrine Mateo, Valérie Eychenne, Pascale De Caro, Zéphirin Mouloungui, Christian V. Stevens*

9.1   Introduction

Oils and fats of natural origin have been widely used for lighting and as lubricants, long before derivatives from petroleum and mineral oils became generally available during the second half of the 19th century. Currently, a marked return to oils and fats of vegetable and animal origin to replace petroleum-based and synthetic products can be observed due to a number of beneficial forces and it is foreseen that these long-term trends are favourable for the oleochemical industry.

The driving forces for this trend are mainly environmental and ecological. Oils and fats are renewable resources and do not lead to an increase in CO_2 production. In addition, their biodegradability is less complicated in comparison with the synthetic analogues.

Oleochemistry has several beneficial factors for agriculture, the chemical industry and ecology, and can influence consumer behaviour and perception.

Most of the oils/fats are consumed as human food (80%) or animal feed (5%), leaving only 15% for industrial use. The chemical industry has introduced the concept of sustainability and the efficient use of resources is extremely important for the chemical industry where raw materials, energy, plant and processing costs are the main cost factors.

The consumer acceptance of and demand for products derived from natural and renewable resources have increased in the last decennium.

Renewable Bioresources: Scope and Modification for Non-food Applications. Edited by C.V. Stevens and R. Verhé
© 2004 John Wiley & Sons, Ltd   ISBNs: 0-470-85446-4 (HB); 0-470-85447-2 (PB)

The aim of this chapter is to provide an overview of the use of lipids of natural origin, including:

- the chemical structure and properties of lipids;
- the main sources of lipids for the oleochemical industry;
- the applied technology; and
- the products and their application of oleochemical compounds.

9.2   Chemical Structure of Oils and Fats

(Gunstone and Padley, 1997; Akoh and Min, 1998; Gunstone and Hamilton, 2001).

9.2.1   Composition of Lipids

The major lipids, which mainly accumulate as fine droplets (in particular animal and vegetable tissues), consist mainly of triacylglycerols (formerly known as triglycerides) together with minor components as di- and monoacylglycerols and phosphoglycerides. The glycerol esters, esters between glycerol and fatty acids, are called the saponifiable fraction due to the alkaline cleavage of the ester bond giving rise to soap and glycerol. The unsaponifiable fraction mainly consists of minor compounds such as sterols, tocopherols (vitamin E), squalene and waxes. However, this latter fraction is not that important today for the oleochemical industry and will be partially removed during refining (Scheme 9.1).

9.2.2   Structure of Fatty Acids

Due to the fact that the physical properties of lipids are largely dependent on the structure of the fatty acids in the acylglycerols, the chemical functionality of fatty acids is commented on in this section.

Other types of lipids include esters with other alcohols, sterols (sterol esters) and sugars (sugar esters). Although there is a systematic nomenclature based upon IUPAC rules,

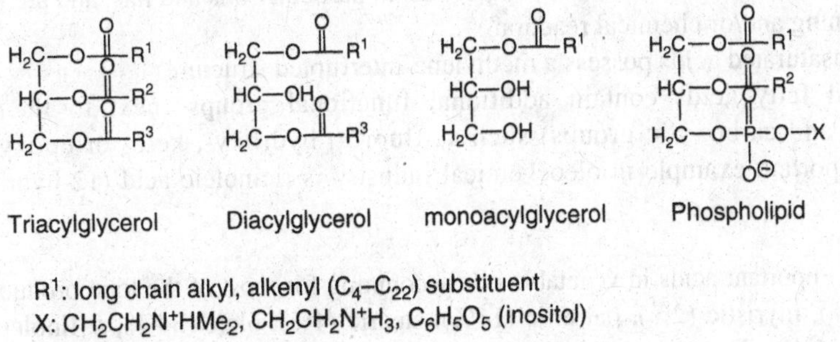

| Triacylglycerol | Diacylglycerol | monoacylglycerol | Phospholipid |

R^1: long chain alkyl, alkenyl (C_4–C_{22}) substituent
X: $CH_2CH_2N^+HMe_2$, $CH_2CH_2N^+H_3$, $C_6H_5O_5$ (inositol)

Scheme 9.1   *Composition of glycerides*

Table 9.1 Systematic and trivial names of fatty acids

Chain length	Systematic name	Trivial name	Position double bond
12:0	Dodecanoic	Lauric	–
14:0	Tetradecanoic	Myristic	–
16:0	Hexadecanoic	Palmitic	–
18:0	Octadecanoic	Stearic	–
18:1	9-Octadecenoic	Oleic	9 (n-9)
18:2	9,12-Octadecadeinoic	Linoleic	9,12 (n-6)
18:3	6,9,12-Octadecatreinoic	γ-linolenic	6,9,12 (n-6)
18:3	9,12,15-Octadecatrienoic	α-linolenic	9,12,15 (n-3)
20:0	Eicosanoic	Arachidic	–
20:1	Eicosaenoic	–	9 (n-11)
20:4	Eicosatetraenoic	Arachidonic	5,8,11,14 (n-6)
20:5	Eicosapentaenoic	EPA	5,8,11,14,17 (n-3)
22:0	Docosanoic	–	–
22:1	Docosenoic	Erucic	13 (n-9)
22:5	Docosapentaenoic	DPA	7,10,13,16,19 (n-3)
22:6	Docosahexaenoic	DHA	4,7,10,13,16,19 (n-3)

Notes:

The numbering along the carbon chain starts from the carboxylic acid carbon.
Numbering with respect to CH_3-end (often used in food science).

trivial names (which in many cases relate to the source of the acid) are still widely used, examples include palmitic acid from palm oil, oleic acid from olive oil, linoleic and linolenic acids from linseed oil, ricinoleic acid from castor oil (*Ricinus communus*) and arachidic acid from groundnut oil (*Arachnis hypogaea*).

In Table 9.1, the main fatty acids are summarised with their chain length, number and position of double bonds, and their trivial and systematic names.

The main structural features of the fatty acids are the following:

1. Natural fatty acids are straight-chain compounds with an even number of carbon atoms (uneven or branched fatty acids are minor components in particular lipids). The most common fatty acids have 12–22 carbon atoms.
2. Unsaturated acids (carbon–carbon double bond) mostly have the Z(*cis*)-configuration. Trans-fatty acids are also minor compounds, in particular oils and fats, and are the result of refining and/or chemical reaction.
3. Polyunsaturated acids possess a methylene-interrupted structure.
4. Special fatty acids contain additional functional groups (next to the olefinic bonds and carboxylic groups) such as fluoro-, hydroxy-, keto- or epoxygroups. An important example in oleochemical industry is ricinoleic acid (12-hydroxyoleic acid).

The most important acids in vegetable oils accounting for most of the total production are: lauric (4%), myristic (2%), palmitic (11%), stearic (4%), oleic (34%), α-linolenic (5%) and erucic (3%). The major acids in animal fats and fish oil are myristic, palmitic, stearic, oleic, eicosenoic, arachidonic, DPA and DHA.

The physical state of the lipids is mainly dependent not only the nature of the fatty acids but also on the distribution of the fatty acids along the positions in the glycerol moiety.

It is generally recognised that a fat is solid at 25 °C, while it is called an oil if the melting point is <25 °C. The melting point of a lipid is dependent upon the following parameters:

1. The higher the number of carbon atoms in the chain the higher the melting point.
2. The presence of a double bond in the fatty acid decreases the melting point in comparison with the saturated fatty acid with the same number of carbon atoms.
3. The higher the number of double bonds the lower the melting point for the same number of carbon atoms.
4. An E (*trans*) configuration increases the melting point in relation to the Z (*cis*) isomer.
5. Conjugation gives rise to higher melting points than the non-conjugated isomers.

9.3 Major Sources of Oils and Fats Used in Oleochemistry

The total amount of oils and fats produced worldwide increased from 53×10^6 t in 1980 to over 110×10^6 t in 2002, with an average growth of 3.3%/yr in the 1990s. From the 110×10^6 t, only 15.6×10^6 t have been transformed into chemical products of which 10×10^6 t are soap. About 80% is of plant origin and most of the total amount of oils and fats are used for human nutrition (Scheme 9.2).

Due to its climate, Asia is the main player for oils and fats used in the oleochemical industry. Malaysia, Indonesia and the Philippines are the main countries producing palm kernel and coconut oil, the two suppliers of lauric oil (C_{12}). Animal fats are produced mainly in the US and Europe. Mainly lauric oils and palm oils including palm stearin (vegetable oils), and mainly tallow and lard (animal fats) are the most used natural raw materials. Major growth occurred in the production of vegetable oils (86.6×10^6 t in 2000). The production of animal fats was 20.1×10^6 t in 2000. The current major

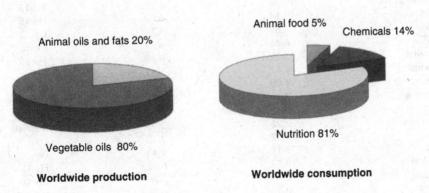

Animal oils and fats 20%

Animal food 5%

Chemicals 14%

Vegetable oils 80%

Nutrition 81%

Worldwide production

Worldwide consumption

Scheme 9.2 Production and use of oils and fats

growth in the production of vegetable oils is occurring in the palm oil industry (42%). Another oil which is becoming very important is rapeseed/canola. In 2000 Asia's share in the oil and fat production was about 44%, North America 16%, Europe 15% and South America 10%.

The rising world population also increases the demand for oils and fats. The production of oils and fats per capita on a worldwide basis continues to rise from 13.4 kg/yr/person in 1980 to 17.4 kg/yr/person in 2000. This overall growth in the production has been bigger than was estimated in 1990.

Of the 17 traded commodity oils, soyabean oil continues to be the major oil consumed, followed by palm oil, animal fats and rapeseed oil (Table 9.2). The fatty acids produced have a chain length of 8–22 carbon atoms, but most of the oleochemical industry relates to 12-, 14-, 16-, 18- and 22-carbon atom chain length materials (Renaud, 2002).

Most oils used for industrial purposes include soyabean, rapeseed/canola and high erucic rapeseed oil, palm (especially palm stearin), palm kernel, fish oil, linseed, castor and tallow.

One other important oleochemical feedstock, which has not been mentioned before in this chapter, does not have a triglyceride structure. Tall oil, a by-product of the wood pulp industry, is produced when pinewood is reacted with sodium hydroxide and sodium sulphate, converting the acids into sodium soaps. Tall oil is produced in the US and Scandinavia. Tall oil is not a triacylglyceride but a mixture of fatty acids (75–80% oleic–linoleic acids) together with resins. Also jojoba oil, which consists of esters of long chain fatty acids with long chain fatty alcohol (wax), possesses special properties suitable for cosmetics and will become more important in the near future.

Although small in percentage, the use of oils and fats as a source of raw material for the chemical industry can be increased if some important conditions are fulfilled:

Table 9.2 *Average annual production (million tonnes) of 17 commodity oils*

Oil	1996/2000	2001/2005	Main producer
Soyabean	22.84	26.52	US
Palm	17.93	23.53	Malaysia
Rapeseed/canola	12.56	15.29	EU-15
Sunflower	9.14	10.77	EU-15
Tallow	7.65	8.24	US
Lard	6.21	6.75	China
Butterfat	5.75	6.26	EU-15
Groundnut	4.62	5.03	China
Cottonseed	4.00	4.49	China
Coconut	3.10	3.47	Philippines
Olive	2.42	2.52	EU-15
Palm kernel	2.26	2.95	Malaysia
Corn	1.97	2.30	US
Fish	1.11	1.13	Peru
Linseed	0.73	0.83	EU-15
Sesame	0.70	0.76	China
Castor	0.47	0.56	India

- A higher concentration of one fatty acid in the raw material.
- Most of the vegetable oils contain too much C_{16}- and C_{18}-fatty acids. Sources for longer chains (more than C_{20}, fish oils, castor oil) and shorter chains (less than C_{14}, palm kernel, coconut) are limited.
- More oils with a higher degree of unsaturation would be preferable as most vegetable oils and animal fats contain too high percentage of saturated fatty acids, which is unfavourable for its chemical reactivity.
- The introduction of a higher degree of functionality (other than unsaturation) such as epoxy or hydroxyl functions would enhance its chemical derivatisation.
- The economic value of oils is dependent not only on the fatty acid content but also on the value of the cake as animal feed. The protein content and the absence of anti-nutritional factors are especially important.
- Introduction of new crops, e.g. crambe (erucic acid), cuphea (short chain fatty acids) and vernonia (epoxy fatty acids), and genetic engineering can enhance the production of fatty acids for the oleochemistry industry.

9.4   Technology of Oils and Fats

It is very rare that crude oils and fats derived from natural resources are used as such in oleochemistry. Lipids are first separated from the lipid-containing raw materials by mechanical crushing, pressing or by solvent extraction.

9.4.1   Extraction of Lipids

Before extraction, the oilseeds are cleaned, cracked, cooked and conditioned to an optimum humidity and temperature. The cleaning involves the removal of impurities such as sand, stones, stems, leaves and weed seeds and this first processing step is carried out with vibrating screens. Magnets remove iron particles.

De-hulling or de-cortication (removal of the protective outer shell of seeds and beans) involves cracking the oilseeds by bar or disk hullers, screening and aspiration or flotation. The objective is to remove as much hull as possible, to increase protein content of the defatted meal (soya) and to improve the cold stability (less waxes in crude oil). Oilseeds are usually flattened into thin flakes of a thickness of 0.3–0.4 mm in order to improve the extractability of the oil by rupture of the cell wall. Flaking is accomplished by passing the seeds through heavy steel rolls in parallel, revolving in opposite directions. The temperature can go up to 60 °C.

For most oilseeds, cooking is necessary especially for those that are too soft to withstand the pressure generated within a screw press. Cooking also ruptures oil cells and reduces the oil viscosity and inactivates the enzymes. Cooking occurs at 100–130 °C for about 15 min in stack cookers or horizontal cookers. Drying to a moisture of 2–3% is carried out in vented drying vessels in series.

Some oils also need a detoxification step. Gossypol (from cottonseed) is removed by cooking, whereby gossypol is bound to the cottonseed protein. Toxic compounds (ricine, ricinine) such as those in castor oil are removed by treatment with $Ca(OH)_2$, NaOH and

NaOCl, while aflatoxins (metabolites from *Aspergillus flavus*) are reduced by treatment with ammonia.

Mechanical Extraction

Crude oils are recovered from the flakes of raw material by applying a mechanical pressure by a screw press (expeller) consisting of a rapidly rotating Archimedean screw within a drained barrel. Screw presses can be applied in two ways: pre-pressing and full-pressing. In pre-pressing, only part of the oil is recovered and the partially de-oiled meal (cake with 18–20% oil) is further treated by solvent extraction. Combined pre-pressing and solvent extraction is commonly applied for oilseeds with high oil content (30–40%, rapeseed, sunflower). Pre-pressing generally has a beneficial effect on the solvent extractability of the remaining oil (low residual oil content). Full-pressing requires 95 000 kPa to squeeze out as much oil as possible, preferably up to 3–5% residual fat for animal material. Full-pressing can also be carried out in a pre-press and a final press. During pressing, the product temperature increases due to friction, which can have a potentially negative effect on the quality of the oil and the meal. In these cases cold pressing is preferred.

Palm fruit – palm kernel

Palm fruit arrives at the mill containing 20–25% moisture. The first step is the sterilisation of the fruit to inhibit the enzyme systems. They then pass into a digester that breaks the skin but leaves the kernel intact. Pressing produces a liquor of 50% oil, 40% water and 10% solids, which, after settling and centrifugation, gives the raw palm oil. The kernel from the press solids are separated by pneumatic separation, centrifugation and via a hydrocyclone. The oil is recovered from the kernel by high-pressure pressing.

Animal lipids

The separation of fat from animal residues (rendering) involves water removal, pressing or solvent extraction. Fats can also be removed through dry rendering by boiling the tissues and drainage of the free fat. Fish oil is produced in a similar way.

Solvent Extraction

Solvent extraction involves different unit operations: extraction of the oil from the oilseeds using hexane as a solvent; evaporation of the solvent; distillation of the oil–hexane mixture (called miscella); and toasting of the de-oiled meal. In special cases, other solvents can be used: halogenated solvents (mostly dichloromethane), acetone, ethanol or isopropanol. Supercritical extraction can also be performed using CO_2.

The extraction most commonly used today is percolation extraction with a countercurrent flow where the fresh solvent comes in contact with the exhausted material and fresh feed is mixed with the solvent that already contains some oil. In this way, the countercurrent procedure produces high yields with a minimum use of solvent.

There are several types of extractors classified by the way the solids are transported: perforated belt, rectangular loop, circular basket and sliding. The most widely used system is the perforated belt extractor (Extraction DeSmet, Belgium) in which the prepared seeds (pre-pressed or not) are transported on a perforated belt through the horizontal body of the unit. Solvent is introduced at the discharge of the extractor and sprayed over the bed of the product. Solvent percolates through the product and extracts the oil. The oil–solvent mixture is then collected in a hopper and pumped to a sprayer located next to the previous one, towards the extractor inlet. The solvent is recovered by distillation in three steps. In the first-stage evaporator, the oil in the miscella is concentrated from an average of 25–70% in an "economiser", where hot-solvent vapours that were boiled from the solids condense against the first-stage evaporator and remove solvent from the miscella. In the second stage, steam stripping (3–4 bar) concentrates the oil to 95% (indirect heating). In the final stage, a stripping column removes the last traces of hexane.

9.4.2 Refining

The majority of oils and fats after extraction are not suitable or ready for human consumption or for industrial use. A number of unwanted compounds need to be removed in the refining process, and this consists of several steps. The undesirable products are non-triacylglycerol compounds: free fatty acids (FFAs), mono- and diacylglycerols, phospholipids, sterols, tocopherols, pigments, oxidation products, trace elements such as metals, protein-degradation products, waxes, squalene, hydrocarbons, moisture, dirt, contaminants such as pesticide residues, polycyclic aromatic compounds, foreign organic compounds (dioxin), etc. Two refining routes are used: chemical and physical refining, which refer to the way FFA are removed. In the chemical process, they are neutralised by the addition of an alkaline solution (neutralisation with formation of soaps). In the physical process, FFAs are discarded by vacuum distillation (in a deodouriser). The physical-refining method has the advantage of being more environmentally friendly than the chemical-refining process because less wastewater is produced. In the chemical-refining process, an excess of water is used for neutralisation, followed by acidification, so as to avoid soap stock.

In the degumming process, the hydratable phosphatides are removed by water, whereas the non-hydratable ones are removed by treatment with citric or phosphoric acids at 60–85 °C in special mixes. During the neutralisation step (only in the chemical refining), caustic soda will not only neutralise the FFAs but also react with the triacylglycerols causing loss of neutral oil by saponification. In addition, during this process gums and pigments are partially removed. Neutralisation is carried out continuously using three separators followed by washing out traces of soap and drying (80–100 °C, 20–50 torr).

Bleaching is primarily used to remove colour pigments but it can also remove traces of soap and phospholipids and can decompose oxidation products. Bleaching is an adsorption process using natural clays (bentonite and montmorrillonite) activated by acid treatment. In combination with active carbon, organic compounds such as pesticide residues and environmental pollutants can also be removed. In this decolourisation step, bleaching earth is added to the oil at 90–120 °C in agitated vessels followed by filtration. Deodourisation consists of a vacuum-steam distillation at 200–210/1–9 mbar in order to remove FFAs (distillation), undesirable odour and flavour (deodourisation), pigments (degradation, decolourisation), pesticide residues and polyaromatic hydrocarbons. However, deodourisation can cause undesirable effects by the loss of valuable components (tocopherols, sterols), isomerisation (*cis* → *trans*) and by fat degradation due to polymerisation.

In Europe, physical-refining deodourisation is carried out at slightly higher temperatures using more purge steam and a longer residence time than in chemical refining. Deodourisation can be carried out in a vertical or horizontal design or in a packed column. The process of deodourisation involves several stages: de-aeration, pre-heating, high temperature heating (indirect with thermal fluids), steam stripping, heat recovery, chelation of trace metals with citric acid and polish filtration by a bag or cartridge filters. In some cases, oil refining involves winterisation. This is a dry fractionation (a fractional crystallisation) in which the more solid saturated triacylglycerol fraction (stearin) is filtered off from the more liquid unsaturated triacylglycerols fraction (olein). In this process, the fats are fractionated to produce oils with no deposits at low temperatures. If the solid phase that is removed consists of a wax, the process is called "dewaxing" and no cloudy material is formed in the oils during storage at low temperature.

9.4.3   Fractionation

Fractionation is a fractional crystallisation process in which the oil is crystallised and the crystals are separated from the liquid phase due to the different solubilities of the triacylglycerols, depending on the molecular weight and the degree of unsaturation of the fatty acids present. Fractionation is carried out for the removal of unwanted compounds, the enrichment of a desirable triglyceride, the separation into two or more fractions with higher potentials for a specific application and as an alternative for hydrogenation.

The fats are heated to approximately 20 °C above the melting points and are cooled slowly (1–3 °C/h), sometimes with gentle agitation. The remaining oil is removed from the solid fat by vacuum filtration, centrifugal separation or by pressing. The fractionation can be improved substantially by using an organic solvent. Fractionation is often used for palm oil, palm kernel, coconut oils and beef tallow.

9.4.4   Interesterification

One way of changing the physical properties of a lipid without changing the fatty acid composition is interesterification. This is a catalytic process involving the exchange of fatty acid chains between the existing glycerol esters in order to form new esters. Because of this rearrangement, the physical and chemical properties can be altered and therefore it is (besides fractionation and hydrogenation) an important technique for modifying the properties of lipids. The catalysts are alkali-metal derivatives (alcoholates) and enzymes (lipases). The latter have the advantage of specificity and milder reaction conditions. An example is given for an interesterification of a 50/50 mixture.

9.5   Oleochemistry

9.5.1   Chemical Reactivity of Lipids

The chemical structure and the sites of chemical reactivity of lipids can be divided into two parts: the glycerol skeleton and the acyl chain of which the length, the presence of double bonds and the substituents provide the specificity and the chemical properties of the lipid. On the ester part, chemical reactions such as hydrolysis, transesterification, saponification and amidation can take place. On the unsaturated fatty acyl chain, numerous reactions can occur: dimerisation, hydrogenation, oxidation, epoxidation, metathesis, addition, halogenation, sulphation and sulphonation.

9.5.2   Production of Basic Oleochemicals

There are four classes of basic oleochemicals: fatty acids (FAs), methyl (or other alkyl) fatty acid esters (FAMEs), fatty alcohol and fatty amines. Glycerol can be added to this list, as it is a by-product of many oleochemical reactions.

Triglycerides are converted into fatty acids by hydrolysis (fat splitting) with steam at 240 °C/20–60 bar in a continuous uncatalysed countercurrent process. Fatty acids can also be produced through saponification with a strong aqueous base (sodium or potassium hydroxide) followed by acidification of the corresponding soaps.

Fatty acid esters are mainly produced through a transesterification reaction, mostly with methanol (methanolysis) in the presence of catalytic amounts of sodium methoxide or by a lipase.

The catalytic reduction of fatty acid esters at 200 °C/200–300 bar (with a copper containing catalyst and zinc chromide) is an important route for the production of respectively saturated and unsaturated fatty alcohols.

Fatty amines are produced from fatty acids by using cyanides in the presence of ammonia at 280–350 °C (via the intermediate fatty amides [dehydrative cyanation]). The nitriles are hydrogenated to primary or secondary amines (Nickel or Cobalt catalyst at 120–180 °C). According to the reaction conditions, primary amines are produced at a pressure of 20–40 bar in the presence of ammonia and secondary amines at lower pressure and with the beneficial removal of ammonia. The amines can be transformed through a catalytic reaction with formaldehyde to the tertiary amines (reductive methylation). Quaternary ammonium salts are produced from fatty amines by alkylation (methyl or benzyl chloride) (Scheme 9.3).

9.5.3   Industrial Chemical Reactions on Lipid Derivatives

The objective of this section is to give a general overview of the chemical reactions applied to lipids by industry. The classification is based upon the type of reaction and not on the description of the products produced.

Hardening

Of all the chemical processes, hardening is by far the most important. The use of liquid oils requires the modification of two properties: oxidative stability and melting point.

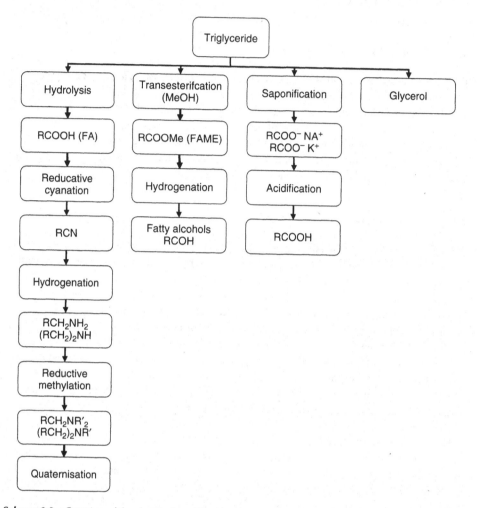

Scheme 9.3 *Overview of the chemical modification of triglycerides*

Three pathways are available for this transformation: hydrogenation, fractionation and interesterification. Hardening is widely used in spite of some negative aspects connected to the *cis–trans* isomerisation of the double bonds in the acyl chains. Basically, the hardening process is a catalytic reaction between a gas and a liquid, and hydrogen must be transported from the gas into the fluid phase where it has to reach the catalyst particles. The most critical point is the absorption of the hydrogen by the fluid, and agitation is a key component in hydrogen solubilisation. Ni-catalysts, either medium pore or wide pore, are used in tank reactors (well-stirred autoclaves or loop reactors, bubble columns and fixed-bed reactors).

For fats and fatty acids having different chain lengths and degrees of saturation (and which can easily isomerise during the hardening), selectivity is of considerable importance. The most common technical reaction is given in Scheme 9.4.

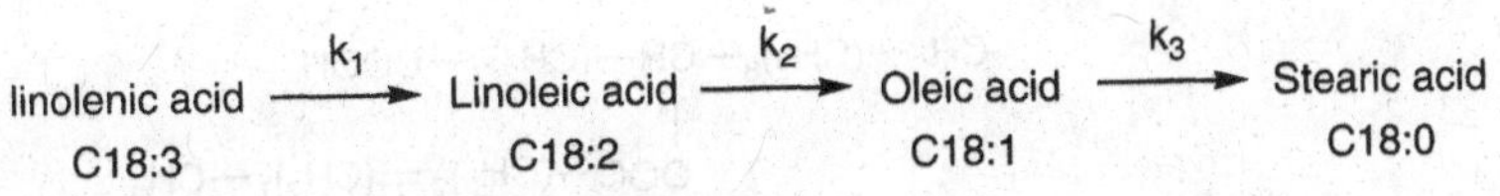

linolenic acid $\xrightarrow{k_1}$ Linoleic acid $\xrightarrow{k_2}$ Oleic acid $\xrightarrow{k_3}$ Stearic acid

C18:3   C18:2   C18:1   C18:0

Scheme 9.4 *Hydrogenation of an unsaturated fatty acid*

For this consecutive reaction, selectivity towards the intermediate product is strongly determined by the limited diffusion of lipid molecules and the hydrogen adsorption within the pores. Important parameters for selectivity are the k_1/k_2 and the k_2/k_3 ratios, which are influenced to a great extent by the catalyst type and the reaction conditions.

Dimerisation

Dimeric fatty acids are formed in complex mixtures after unsaturated acids are heated at 230 °C in the presence of natural clays or cationic catalysts. Distillation produces, next to some monomer, 80% of the dimer, 20% of a trimer and a further polymerised mixture. Oleic acid gives mainly C_{36} acrylic and monocyclic dimers, while linoleic acid gives mainly mono- and bicyclic compounds via a Diels–Alder reaction. The C_{36} dibasic acids are commonly used in alkyd resins, lubricants, polyesters, polyamides and adhesives. Diels–Alder reactions with other dienophiles such as acrylic acid derivatives give cyclic dibasic acids. In this way, linolenic acid reacts smoothly with acrylic acid to give a mixture of two isomeric C_{21} dicarboxylic acids, filling the gap between the commercially available short-chained dicarboxylic acids (adipic, azelaic) and the high molecular C_{36} dimers. Dimerisation in the presence of 10% water forms estolides. These dimers contain an ester and an acid function (Schemes 9.5 and 9.6).

Treatment of fatty alcohols with potassium hydroxide or potassium alkoxide at 200–300 °C gives dimeric monofunctional alcohols named "Guerbet alcohols" which can be oxidised into Guerbet acids. Guerbet alcohols are used in cosmetics as plasticisers and lubricants. Di-Guerbet esters are also used as lubricants.

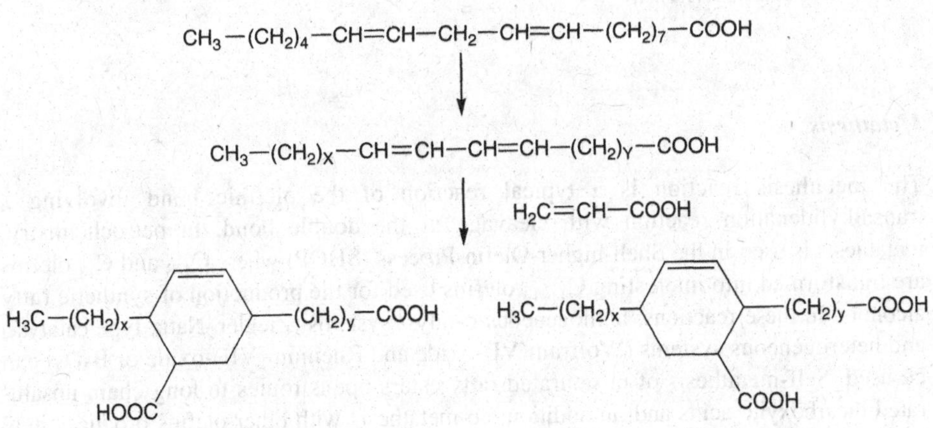

$CH_3-(CH_2)_4-CH=CH-CH_2-CH=CH-(CH_2)_7-COOH$

$\downarrow$

$CH_3-(CH_2)_x-CH=CH-CH=CH-(CH_2)_y-COOH$

$\downarrow$ $H_2C=CH-COOH$

Scheme 9.5 *Dimerisation of fatty acids*

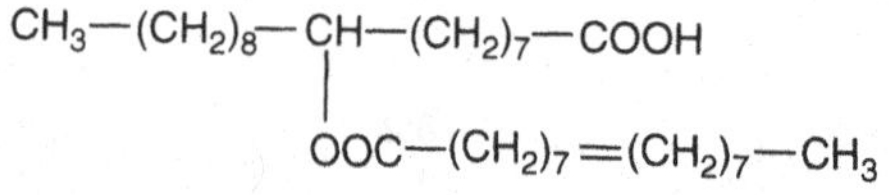

$$CH_3-(CH_2)_8-\underset{\underset{\displaystyle OOC-(CH_2)_7=(CH_2)_7-CH_3}{|}}{CH}-(CH_2)_7-COOH$$

dimeric estolide

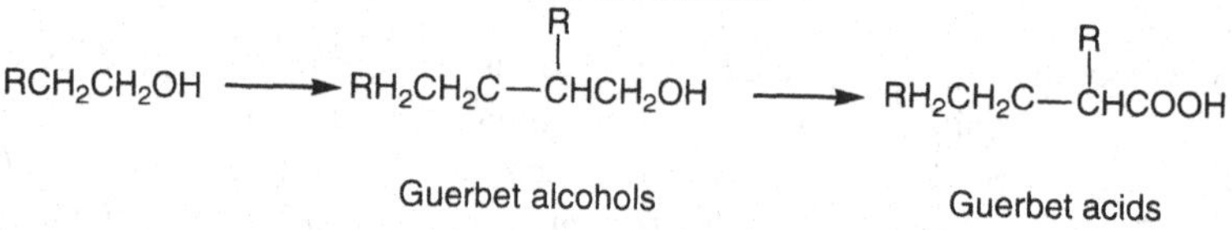

$$RCH_2CH_2OH \longrightarrow RH_2CH_2C-\underset{\underset{\displaystyle R}{|}}{C}HCH_2OH \longrightarrow RH_2CH_2C-\underset{\underset{\displaystyle R}{|}}{C}HCOOH$$

Guerbet alcohols   Guerbet acids

Scheme 9.6 *Formation of Guerbet alcohols*

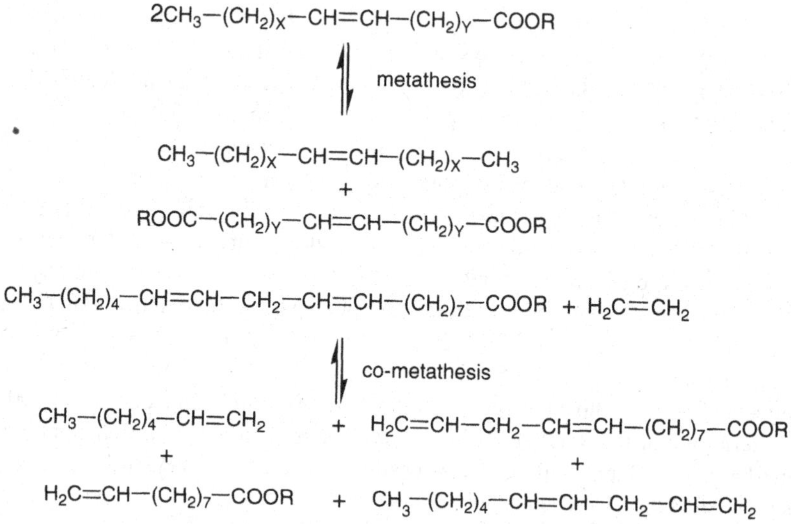

$$2CH_3-(CH_2)_X-CH=CH-(CH_2)_Y-COOR$$

metathesis

$$CH_3-(CH_2)_X-CH=CH-(CH_2)_X-CH_3$$
$$+$$
$$ROOC-(CH_2)_Y-CH=CH-(CH_2)_Y-COOR$$

$$CH_3-(CH_2)_4-CH=CH-CH_2-CH=CH-(CH_2)_7-COOR \ + \ H_2C=CH_2$$

co-metathesis

$$CH_3-(CH_2)_4-CH=CH_2 \quad + \quad H_2C=CH-CH_2-CH=CH-(CH_2)_7-COOR$$
$$+ \qquad\qquad\qquad +$$
$$H_2C=CH-(CH_2)_7-COOR \quad + \quad CH_3-(CH_2)_4-CH=CH-CH_2-CH=CH_2$$

Scheme 9.7 *Metathesis reactions*

Metathesis

The metathesis reaction is a typical reaction of the olefinic bond involving a transalkylidenation reaction with cleavage of the double bond. In petrochemistry, metathesis is used in the Shell-higher-Olefin-Process (SHOP) where C_{4-8} and C_{18} olefins are transformed into interesting C_{11-14} olefins used for the production of synthetic fatty alcohols. In these reactions, homogeneous catalyst systems (Ziegler–Natta type catalyst) and heterogeneous systems (Wolfram(VI)-oxide and Rhenium(VII)-oxide or B_2O_3) can be used. Self-metathesis of unsaturated fatty esters opens routes to long chain unsaturated dicarboxylic acids and, in addition, co-metathesis with other olefins produces new fatty acids (Scheme 9.7).

The possibilities for metathesis in fat chemistry are obvious when considering the composition of natural lipids against the demand of some fractions such as unsaturated C_{12-14} fatty acid derivatives.

Ozonolysis

Although the ozonolytic cleavage is a well-known process, there are only two actual industrial processes known: the cleavage of oleic acid into nonanoic acid and the dibasic azelaic acid (C_9), and erucic acid into nonanoic acid and brassylic acid (C_{13}). The reasons why industry is not eager to apply ozonolysis is that the generation of ozone requires high investments, needs much energy and is accompanied by often-explosive intermediate peroxides. The dibasic acids are used as starting materials for polyamides and polyesters while the di-esters from higher alcohols are used as plasticisers in PVC and as synthetic lubricants.

Cleavage of Ricinollic Acid Derivatives

Thermal cleavage of ricinollic methyl esters (methyl 12-hydroxy oleate) at 500 °C produces heptanal and methyl 10-undecyloate (C_{11}). The latter can be converted in 11-undecanoic acid as starting material for polyamides.

Sebacic acid (C_{10}) is obtained from castor oil or ricinollic acid through treatment with sodium hydroxide at 250–270 °C. Reaction with diamines produces 6,10-nylon fibres that can be used for various applications (Scheme 9.8).

Another reaction of ricinoleic acid involves dehydration, providing diene-enriched oils used as alternatives to drying oils as tung oil. Hydrogenation gives 12-hydroxystearate used in cosmetics, coatings and greases.

9.5.4 Functionalisation of Fatty Acids and Fatty Acid Esters

Two separate types of functionalisation are possible: functionalisation in the long C-Chain and functionalisation at the carbonyl function in the acid or the ester.

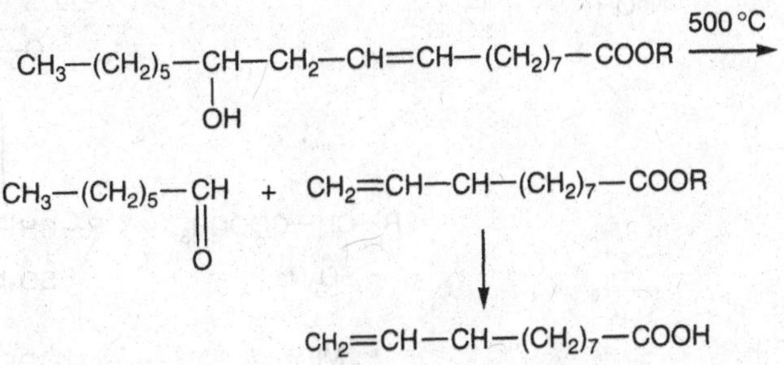

$$CH_3-(CH_2)_5-\underset{\underset{OH}{|}}{CH}-CH_2-CH=CH-(CH_2)_7-COOR \xrightarrow{\;500\,°C\;}$$

$$CH_3-(CH_2)_5-\underset{\underset{O}{\|}}{CH} \;+\; CH_2=CH-CH-(CH_2)_7-COOR$$

$$\downarrow$$

$$CH_2=CH-CH-(CH_2)_7-COOH$$

Scheme 9.8 *Thermal cleavage of ricinoleic methyl esters*

Functionalisation at the Carbon Chain

One of the most used industrial functionalisations of fatty acid esters involves the α-sulphonation process, giving rise to α-ester sulphonates that are widely used as anionic surfactant in laundry powders as an alternative for linear alkylbenzene sulphonates due to their greater biodegradability. The sulphonation reaction is assumed to proceed in two steps. In the first step the ester reacts with two equivalents of SO_3 leading to an intermediate anhydride. The latter acts as a sulphonating agent for the non-sulphonated methyl esters.

Sulphonated methyl esters and the di-sodium salts are obtained normally in the ratio of 80/10 (Scheme 9.9).

Due to the formation of mono- and dimethyl sulphate a diester (DSA) can be formed. Sulphation converts the hydroxyl function of ricinollic acid to a sulphate that was (apart from soap) the first anionic surfactant which is still used in textile processing, in leather and in hydraulic fluids. Epoxidation converts unsaturated fatty acid derivatives into epoxides that are easily converted into a series of products upon reaction with a variety of nucleophiles. Mainly, linseed oils, soja oils and palm oils are converted into epoxidised oils. Epoxidised oils are formed by reaction with H_2O_2, formic acid or acetic acid in the presence of sulphuric acid, with the formation of the corresponding peroxy acid. These oils are used as plasticisers and stabilisers of PVC as they are quenching the liberated HCl, from PVC by light or heat.

Functionalisation of fatty acids in order to produce polymerisable derivatives is used in the production of surface coatings. Especially castor oil and linseed oil with a high percentage of polyunsaturated acids are valuable resources for suitable and durable coatings, where a cross-linked structure is desirable. The use of oils for mechanically suitable and durable paints is based on the formation of a cross-linked structure. The latter structure is generated by the so-called overall drying process and involves the formation of hydroperoxides followed by break down into aldehydes and free radicals (mostly induced by cobalt salts) and subsequent free-radical linking with formation of intermolecular networks. Next to the dimerisation of unsaturated fatty acids (making the so-called stand oils), maleïnisation, reaction with monomers (styrene, acrylates) provided starting materials for

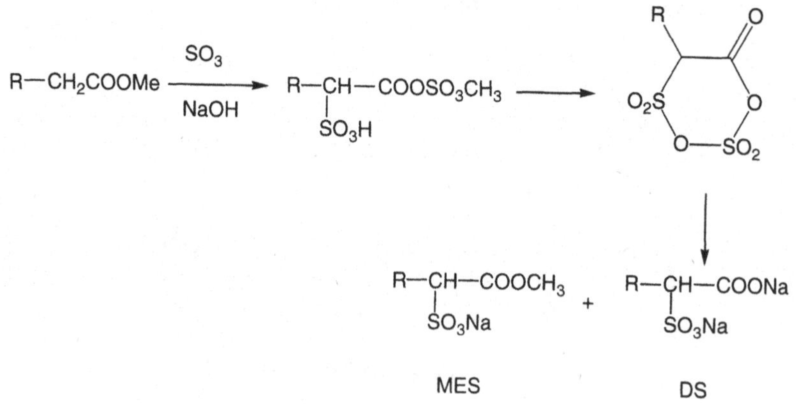

Scheme 9.9   *The sulphonation of FAME*

cross-linking. Maleïnisation is used to improve the water solubility and to increase the viscosity.

Functionalisation of the Acid/Ester Function

Esterification products from fatty acids with higher alcohols are widely used for industrial purposes.

Partial esters of glycerol with fatty acids are mostly prepared via partial hydrolysis in the presence of an alkaline catalyst and not by direct esterification of glycerol with fatty acids due to the low solubility of glycerol in fatty acids. Polyglycerol esters are directly prepared via esterification or by transesterification of triacylglycerides with polyglycerol. Tetraesters used as lubricants and oil additives are formed by esterification with erythritol.

Hexitol esters are formed via esterification of sorbitol with lauric, palmitic, stearic and oleic acids. Sorbitan esters are ethoxylated with ethylene oxide to form polysorbates which are extremely safe emulsifiers and disperging agents. Sorbitan esters are a mixture of mono-, di- and triesters and cyclised compounds. Esteramines, which are becoming more widely used, are typically obtained by the reaction of fatty acids with alkanolamines and ethoxylated fatty amines. Esterquats are used as specific fabric softeners. Alkyd resins are polyesters which are incorporating varying amounts of oil-derived fatty acids. The polyesters are mostly derived from *o*-phthalic anhydride by reaction with glycerol derivatives (Scheme 9.10).

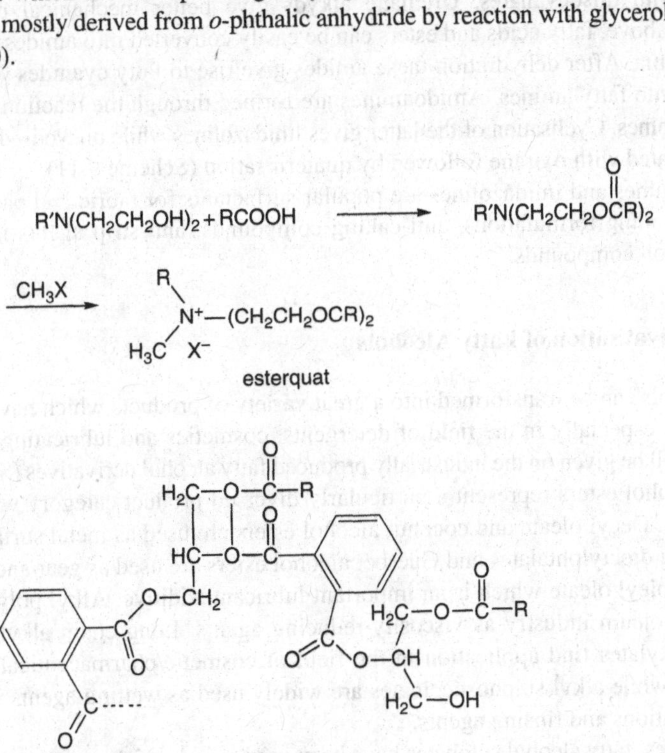

Scheme 9.10 Structure of an esterquat and alkyd resin

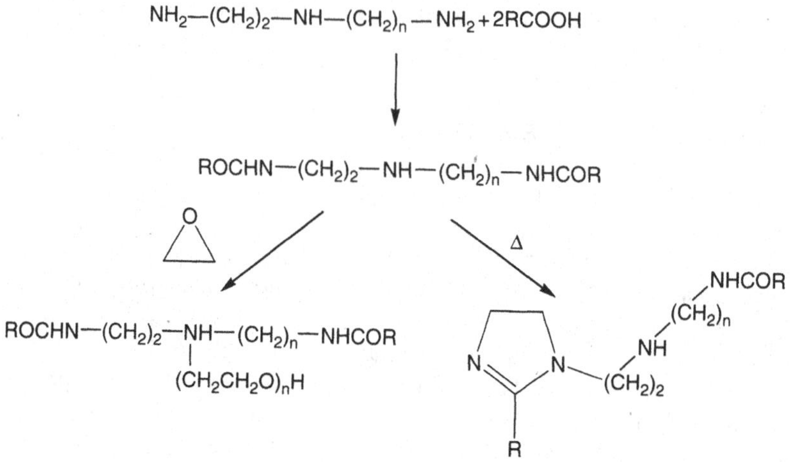

$$NH_2-(CH_2)_2-NH-(CH_2)_n-NH_2+2RCOOH$$

$$\downarrow$$

$$ROCHN-(CH_2)_2-NH-(CH_2)_n-NHCOR$$

Scheme 9.11   *Reactions with polyamines and fatty acids*

Other alkyd resins involved urethane functions via reactions of oil-derived glycerol structure with diisocyanates. Urethane alkyds give better mechanical properties. As mentioned above, fatty acids and esters can be easily converted into amides upon reaction with ammonia. After dehydration these amides give rise to fatty cyanides which are then converted into fatty amines. Amidoamines are formed through the reaction of fatty acids with polyamines. Cyclisation of the latter gives imidazolines while uncyclised amidoamines are ethoxylated with oxirane followed by quaternisation (Scheme 9.11).

Amidoamines and imidazolines are popular surfactants for fabric and paper softening, hair-conditioning formulations, anti-caking compounds, anti-strip agents in asphalt and anti-corrosion compounds.

9.5.5   Derivatisation of Fatty Alcohols

Fatty alcohols can be transformed into a great variety of products which have found large applications especially in the field of detergents, cosmetics and lubricating oils. A short overview will be given on the industrially produced fatty alcohol derivatives (Scheme 9.12).

Fatty alcohol esters represent a particularly diversed product category with numerous applications. Decyl oleate and coconut alcohol esters are used as metal surface treatment agents while dioctylphtalates and Guerbet alcohol esters are used as gear and motor oil in addition to oleyl oleate which is an important lubricant additive. Alkyl polyacrylates are used in petroleum industry as viscosity-reducing agents. Long chain alkyl lactates and alkyl carboxylates find application in the field of cosmetic/pharmaceutical applications and lotions while alkyl sulphosuccinates are widely used as wetting agents in shampoos, bath preparations and rinsing agents.

For decades, fatty alcohol sulphates have been used in industrial processes and consumer products due to this cleaning, foaming and emulsifying action. They are prepared by reaction of SO_3 in tubular reactors. The sulphates are more stable in alkali than in acid due to

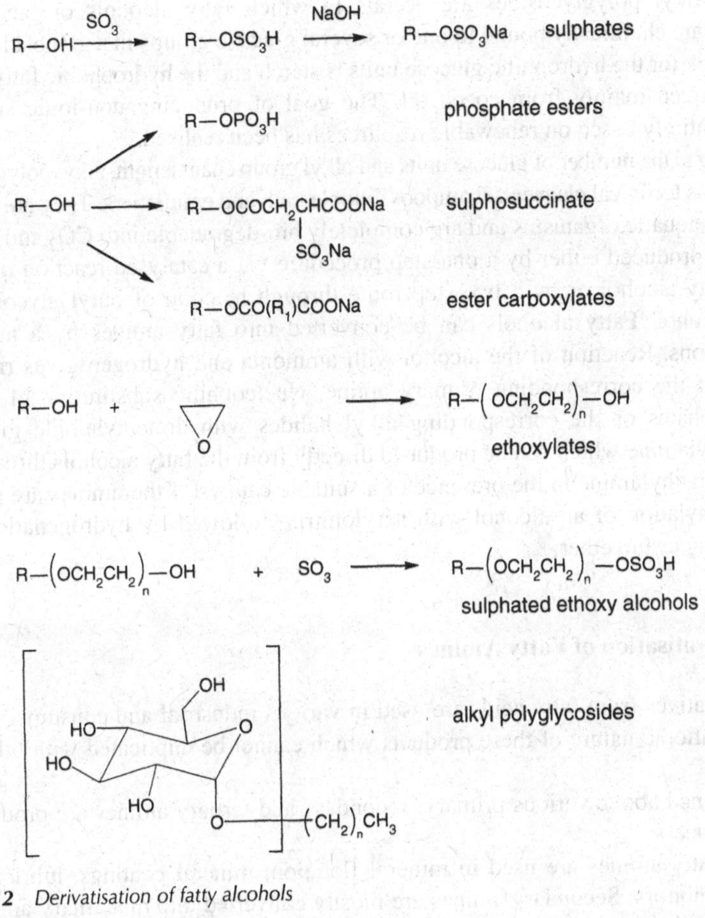

Scheme 9.12 *Derivatisation of fatty alcohols*

hydrolysis which releases sulphuric acid which catalyses the hydrolysis. Other anionic detergents such as phosphate esters, sulphosuccinates and ester carboxylates can be produced from fatty alcohols. During the manufacture of detergents, modification of the fatty alcohol in order to increase the water solubility is performed by reaction with ethylene oxide and/or propylene oxide giving rise to fatty alcohol ethoxylates.

Especially the group with a fatty alkyl group C_{12-14} and $n = 6–10$ produces excellent wetting agents and detergents which play an important role in washing at lower temperatures. These fatty alcohol ethoxylates can be derivatised as the single fatty alcohols into the corresponding sulphates, phosphates, ester carboxylates and succinates. The advantage of these ethoxylates is an increase in the viscosity of detergent formulations. Especially, the sulphated ethoxy alcohols (or alkyl ether sulphates) are one of the most used detergents derived from lipids.

A replacement of the hydroxyl group by a sulphate function at the end of the alkyl chain gives a high increase in the water solubility and provides stable foam, good wetting and excellent detergency if the alkyl group is C_{12-16}. In addition, the fatty alcohol ethoxylates can be converted into long chain ethers by reaction with alkyl halides according to Williamson's synthesis. Alkyl polyglycosides are acetals in which fatty alcohols or fatty alcohol ethoxylates are chemically bound to one or several glucose groups in a glycoside linkage. The feedstock for the hydrophilic glucose units is starch and the hydrophobic fatty alcohols may be sourced mainly from cocoa oil. The goal of producing non-ionic surfactants which are entirely based on renewable resources has been realised.

According to the number of glucose units and alkyl group chain length, alkyl polyglycosides can be used as technical cleaners, shampoos, bubble bath and emulsifiers. They are relatively less toxic to aquatic organisms and are completely bio-degradable into CO_2 and H_2O.

They are produced either by a one-step procedure via a catalysed reaction of glucose with the fatty alcohol or by a two-step route through reaction of butyl glycoside with the fatty alcohol. Fatty alcohols can be converted into fatty amines by a number of transformations. Reaction of the alcohol with ammonia and hydrogen gives rise to the formation of the corresponding primary amine. Nucleophilic substitution of the fatty alcohol sulphates or the corresponding alkyl halides with dimethylamine gives fatty alkyldimethylamine which can be produced directly from the fatty alcohols through reaction with dimethylamine in the presence of a suitable catalyst. Etheramines are produced via cyanoethylation of an alcohol with acrylonitrile followed by hydrogenation of the corresponding cyano ether.

9.5.6   Derivatisation of Fatty Amines

Amine derivatives from fatty acids are used in various industrial and consumer products due to the cationic nature of these products which cannot be duplicated with other ionic products.

As mentioned above various primary, secondary and tertiary amines are produced via the nitrile process.

Primary fatty amines are used in mineral flotation, mineral coatings, lubricants and corrosion inhibitors. Secondary amines are mostly converted into quaternary ammonium salts while tertiary amines are intermediates for quaternary ammonium salts and amine oxides, which are used in cleaning agents and biocides.

Alkyl polyamines are produced by the addition of primary amines to acrylonitrile followed by hydrogenation to diamine or triamine. N-alkyl-1,3-propanediamines are used as anticorrosive agents and lubricants.

Alkoxylation, in most cases with ethyleneoxide, gives alkoxylated amines which find applications as emulsifiers, such as agricultural adjuvants for herbicides and pesticides.

Fatty amines and derivatives can be converted into quaternaries $R_4N^+X^-$ by reaction with alkylating agents methyl chloride, benzyl chloride, dimethyl- and diethyl sulphate. Quaternaries are the main fatty-amine-based surfactants with a global consumption of 500 000 t/yr. They are used as fabric softening, as organoclays for rheology control of organic fluids and in personal care products for their properties of smoothness, softness and lubricity of the skin and hair and as disinfectants and sanitation agents.

Oxides of tertiary amines find use as laundry detergents, sanitisers, emulsifiers, personal care products, foam stabilisers and polymer stabilisers. They are produced by reaction with ozone, peracids and hydrogen peroxide. Another functionalisation of fatty amines involves the formation of betaines. Betaines offer good foaming and mildness working

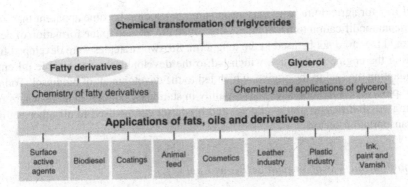

Scheme 9.13 Overview of the different application areas of oils and fats

well in personal care applications with the primary anionic surfactants to improve foam qualities and mildness.

9.6 Industrial Uses of Oils and Fats

The industrial uses of oils and fats are very broad and a selection of some important areas will be given in this section. The most important market for oils is the detergent industry with a consumption of 60% of oil derivatives produced, which correspond to 7×10^6 t. Other lipochemical activity consumes approximately 4×10^6 t which corresponds to 33% of the market. The other main applications are biodiesel, lubricants and hydraulic fluids (0.7×10^6 t), inks and paints (0.2×10^6 t) and phytochemical compounds with less than 0.2×10^6 t (Scheme 9.13).

9.6.1 Biodiesel

Due to the rising demand for fuel for transportation and due to the limited availability of mineral oil, the use of alternative fuels is increasingly important. In particular, the use of biofuels as renewable resources combines the advantages of almost unlimited availability and ecological benefits such as integration in a closed carbon cycle. The idea of using biofuels is not new. Even the inventor of the diesel engine, Rudolf Diesel, described the possible use of vegetable oils in his patent in 1897.

However, for a long period this idea was not developed further. Even during times of fuel shortage during the two World Wars, the use of vegetable oils was not considered because the material was needed for food purposes. The rapid development of the mineral oil industry and the sufficient supply of mineral fuel hindered the use of vegetable oil fuels. The energy crisis in the 1970s and agricultural overproduction in Europe, however, led to the discussion of utilising agricultural areas for the production of energy. Not only the use of biomass for primarily producing heat and electricity was started, but also the production of biofuels such as vegetable oils and ethanol from biomass and their use as substitutes for mineral fuel was initiated. Initial trials at using vegetable oils blended with diesel fuel for agriculture were very promising, but soon it became apparent that without significant modification to the engine, pure vegetable oils led to the formation of deposits and could therefore not be used in the long term. So, two strategies were developed further: adapting the engine to the fuel, which led to the development of vegetable oil engines; and adapting the fuel to the engine, which led to the development of biodiesel. Today, the use of pure vegetable oil makes sense mainly in stationary engines in areas where further refining possibilities are not available, and biodiesel can be used in all other engines in the transportation sector.

Potential for the Use of Vegetable Oils as Fuel

From figures on the overall demand for fuels and the availability of vegetable oils and animal fats, it is obvious that mineral fuel will never be completely substituted. Even the very ambitious goal of the European Commission to reach a market share of 5.75% biofuels by 2010 will be achieved only by including other biofuels such as biogas and ethanol. However, the use of vegetable oils, animal fats and other fat sources for the production of diesel fuel has currently reached a very high level of development, so that this application will play a major role in the future use of oils and fats.

As can be seen in Table 9.3, today's biodiesel production capacity in the EU-15 has reached 2×10^6 t/yr, which equals 1.7% of the diesel demand or 14% of the total production of fats and oils in the EU-15.

Definition of Biodiesel

Nowadays, biodiesel is defined as fatty acid methyl esters prepared from any kind of biological feedstock including vegetable oils, animal fat, single cell oils and waste material. Because of the current extensive application of rapeseed oil for biodiesel production, the terms rapeseed oil methyl esters (RME) or vegetable oil methyl esters (VOME) are also widely used. Basically, fatty acid ethyl esters could also be used and defined as biodiesel,

Table 9.3 *Biodiesel production capacities in Europe, 2003 (Bockey and Körbitz, 2002. Reproduced by permission of A. Munack and J. Krahl.)*

Country	Capacity [t/a]
Germany	1 109 000
France	440 000
Italy	350 000
Czech Republic	60 000
Denmark	60 000
Austria	45 000
Sweden	30 000
Great Britain	30 000
Sum	2 124 000
Diesel fuel consumption 1998 EU-15 (Bockey and Körlitz, 2002)	126 613 000
Production of oils, fats 2000 EU-15 (Oil World 2020, 1999)	15 400 000

but because of the relatively high price of ethanol compared with that of methanol, the use of ethyl esters has not so far been established.

Production of Biodiesel

Because of the vastly different fuel characteristics of vegetable oils and diesel fuel, especially the high viscosity and the low volatility of pure vegetable oils, chemical modification is necessary in order to split the large triglyceride molecule into smaller pieces. As triglycerides, the main components of vegetable oils and animal fats are esters of glycerol and fatty acids. A process called transesterification, which is the exchange of the alcohol or acid moiety of an ester, can achieve the preparation of fatty acid methyl esters. The transesterification of an ester with an alcohol is also called alcoholysis (methanolysis in the case of methanol).

Methanolysis of triglycerides is a three-step consecutive reaction with di- and monoglycerides as intermediates. As final products three molecules of fatty acid methyl esters and one molecule of glycerol are formed. If fatty acids are the starting material for biodiesel production, a simple esterification with methanol leads to the corresponding methyl esters. Esterification and transesterification reactions have to be catalysed in order to achieve fast reaction conversions under feasible reaction conditions. In Table 9.4, a selection of possible catalysts are listed.

Although there are numerous patents filed for the technical preparation of fatty acid methyl esters, mainly alkaline catalysts, especially sodium hydroxide and potassium hydroxide, are used for today's biodiesel production. This is due to the fact that very mild reaction conditions, such as ambient temperature and pressure, can be used without the need for final distillation of the end product. This is in contrast to the technical production of fatty acid methyl esters used in the oleochemical industry as intermediates for the preparation of fatty alcohols, where mainly alkali alcoholates are used at high temperature and pressure.

In common transesterification technologies for biodiesel production, an excess of methanol is necessary in order to achieve high conversion rates, even at temperatures between 20 and 40 °C. The catalyst is dissolved in methanol, and the reaction can be carried out in reactors either in batches or continuously. To achieve satisfactory conversion rates, the reaction should be carried out in two steps. In this way, after the reaction only the removal of excess methanol and different washing steps in order to remove soaps and traces of catalyst are necessary.

Table 9.4 *Possible catalysts for transesterification reactions*

1. Homogeneous Catalysts
   Alkalihydroxides, -alcoholates
   Acids: sulphuric acid, *p*-toluene sulphonic acid
2. Heterogeneous Catalysts
   Metal oxides (Mg, Ca, Al, Fe)
   Alkali carbonates
   Ion exchange resins (acidic, alkaline)
   Enzymes (lipases)
   Silicates

The driving force of the reaction under catalysis with hydroxides is the formation of alcoholates from alcohol and hydroxides and the separation of an insoluble glycerol layer during the reaction that removes glycerol from the equilibrium. So, the reaction will be forced into the direction of fatty acid methyl esters. Kinetic studies of the complicated multiphase reaction have shown that the first step of the reaction leading to the formation of diglycerides is the slowest, whereas the two other steps have higher reaction rates (Mittelbach and Trathnigg, 1990). Under mild reaction conditions and low water content of the starting material, side reactions like hydrolysis of glycerides can be largely suppressed.

One drawback of alkaline catalysis is the fact that during the reaction only glycerides are transformed into fatty acid methyl esters, whereas FFAs are not converted into methyl esters but into soaps, which have to be removed or would decrease the overall yield of conversion. Therefore mainly fully refined, deacidified oils are used for transesterification under alkaline conditions. But it is also possible to convert feedstocks with a maximum content of approximately 3% FFAs under alkaline catalysis. However, these FFAs will be lost during the reaction. In order to achieve an overall transformation rate of almost 100%, the FFAs can either be esterified in an additional step prior to the transesterification or re-esterified with methanol (Mittelbach and Koncar, 1994) after the main transesterification. The latter route is advantageous as soaps formed as side products by hydrolysis are finally transformed into methyl esters. For esterification of the free fatty acids, mainly strong acids like sulphuric acid, *p*-toluene sulphonic acid or strong acidic ion exchange resins are used. This can be done at temperatures near the boiling point of methanol, so that no pressure higher than atmospheric pressure is necessary.

A typical flow scheme for the production of biodiesel from oils and fats, using alkaline catalysts and re-esterification of fatty acids, is outlined in Figure 9.1.

In future, the use of heterogeneous catalysts under more drastic conditions will be increasingly important. The main advantages of such a process are the re-use of catalysts and the simpler purification of the glycerol layer which has to be released only from excess methanol. However, under these conditions the conversion of glycerides is not usually complete, so that the methyl ester has to be distilled under reduced pressure in order to obtain a high-quality end product. This, however, decreases the overall yield and therefore the economy of the process. As heterogeneous catalysts, enzymes, especially lipases, can be used (Mittelbach and Trathnigg, 1990). Though these enzymes normally catalyse the hydrolysis of fats and oils, they can also be used for the synthesis of fatty acid esters if there is almost no water available in the reaction mixture. For technical applications, the use of immobilised enzymes is recommended. Special lipases show high activities in alcoholysis and esterification reactions, but relatively high prices so far limit their use for the production of low-cost products like fuels.

Glycerol as a Co-product of Biodiesel Production

A very important co-product of biodiesel production is glycerol, which is yielded from the process in a quantity of approximately 10% (m/m), related to the vegetable oil of the starting material. So, a widespread utilisation of biodiesel will inevitably increase the availability of glycerol. Glycerol from biodiesel production plants can be further purified, including distillation in vacuum leading to pharmaceutically pure glycerol. However, in

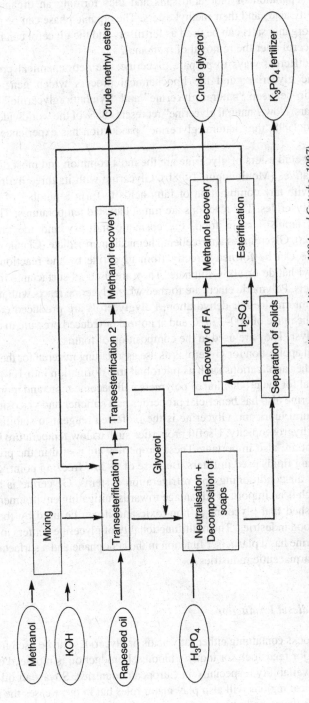

Figure 9.1 *Schematics of a typical biodiesel production process according to Mittelbach and Koncar, 1994 and Gutsche, 1997)*

most existing biodiesel production plants, glycerol is purified only by the addition of mineral acid, causing the separation of inorganic salts and thus forming an organic phase that mainly contains fatty acids and their methyl esters. The organic phase can be recycled into the process, the inorganic salts can be used as fertilisers and the glycerol can be put on the market as raw glycerol after the removal of methanol.

There are two different ways to prepare glycerine: the petrochemical process which produces synthetic glycerine; and the lipochemical process which generates natural glycerine. The ratio between "natural glycerine" and "synthetic glycerine" has evolved over the past 20 years. Now, "natural glycerine" represents 88% of the worldwide production. Between 1992 and 1998, the "natural glycerine" production has experienced an annual increase of 4.3%.

Long chain fatty acid esters of glycerine are the most common and most diverse group of glycerine derivatives (Mouloungui, 1998b). Glycerine with its three hydroxyl groups can be esterified with any combination of fatty acids to form a number of fatty esters, mono-, di- and triglycerides. The reactions are run at elevated temperatures. They are used as stabilisers and emulsifiers in food, the cosmetic industry and the phytoindustry (Mouloungui, 1998a). Glycerine is an excellent chemical intermediate (Claude et al., 2000). Ethers of glycerine can be formed directly from glycerine by the reaction of sodium glycerate with alkyl halide or alkyl sulphate. They are used as surfactants in detergents and washing products. Polymeric ethers are formed when glycerine reacts with itself to form polyglycerols via intermolecular dehydration. Polyglycerols are produced commercially at temperatures in the range of 200–275 °C and at normal or reduced pressure in the presence of an alkaline catalyst. They are used in the composition of foams.

Another technical application for glycerol is its use as a starting material for the preparation of explosives. Further applications such as microbial transformation into 1,3-propanediol as a starting material for the production of polymers will become more and more important in the future. Glycerine also has beneficial properties as a softener and viscosity modifier, so it is a good natural lubricant. Glycerine is the preferred reagent to stabilise moisture because of its high hygroscopicity. Useful properties such as low temperature fluidity and a high boiling point are also important. These properties are useful in the production of viscous dust-trapping fluids used in filters. Because of its low freezing point, glycerine is used in antifreeze, air-conditioning and refrigeration systems. Glycerine is also highly biodegradable which is an important advantage towards mitigating environmental effects. It is firmly established that glycerine is non-toxic and it can be used by the cosmetic, pharmaceutic and food industries. The multifunctionality of glycerine confers to it different applications. Glycerine has a plasticiser function in the cellophane and a surfactant function in cosmetic and pharmaceutic industries.

Feedstocks for Biodiesel Production

Basically, all feedstocks containing either fatty acids or glycerides can be used for biodiesel production. The major feedstock for today's biodiesel production is rapeseed oil because of its widespread availability, especially in European countries. Soyabean oil in the US and palm oil in tropical regions will also play major roles but in these cases the production of biodiesel is not as developed as it is in Europe.

Table 9.5 *Most important feedstocks for biodiesel production (Oil World, 2003)*

Fatty Acid [%]	Rapeseed Oil	Soyabean Oil	Olive Oil	Sunflower Oil	Palm Oil	Used Frying Oil	Tallow
C_{14}					1–2	1–3	2–8
C_{16}	3–6	8–12	7–16	5.5–8	40–48	13–25	25–38
C_{18}	1–2,5	3–5	2–4	2.5–6.5	4–5	5–12	15–28
C_{20}	<1	<0.5	0.5	<0.5			0.5
C_{22}	<0.5	traces	traces	0.5–1	0.5		
$C_{16:1}$	0.1–0.5	traces	traces	<0.5		0–4	2–3
$C_{18:1}$	52–66	18–25	64–86	14–34	37–46	43–52	40–50
$C_{18:2}$	17–25	49–57	4–15	55–73	9–11	7–22	1–5
$C_{18:3}$	8–11	6–11	0.5–1	<0.4	<1	1–3	<1
$C_{20:1}$	1.5–3.5	<0.5	0.5	<0.5		1	0.5–1
$C_{22:1}$	<2.5			<0.3			
Sum of saturated	4–9	11–17	9–20	8–16	44–53	19–40	40–64
Iodine number	96–115	121–143	78–94	127–142	35–57	80–91	35–48
World production [Mt/yr] (Jan./Dec. 2003)	12.49	31.36	2.81	8.96	27.52		8.66

Source: OIL WORLD, Hamburg/Germany, www.oilworld.biz. Reproduced by permission of ISTA Mielke GmbH.

On the one hand, the choice of a feedstock depends on its availability and therefore on the price of the product. On the other hand, the demands of the engine have to be considered as well. As can be seen in Table 9.5, the main difference between the various oils and fats is their fatty acid distribution. The use of methyl esters as fuel necessitates a low proportion of saturated fatty acids in order for the fuel to be usable even at cold temperatures. Here rapeseed oil and olive oil would be the best feedstocks. Palm oil and animal fat can be used only in warmer regions or as blends with other oils. However, the content of higher unsaturated fatty acids should also be limited because of the possible formation of deposits in engines produced by highly unsaturated products. So, the ideal feedstock can be a mixture of different oils and fats, depending on their availability and the climate of the region where the fuel is utilised. The utilisation of special feedstocks is also limited by national and international specifications for biodiesel which have to be met. Regarding the price of feedstocks, used frying oil and other waste materials are especially promising. In addition animal fats, which in the future may no longer be allowed for inclusion in animal feed, may become very important alternatives.

Fuel Characteristics and Specifications

Fatty acid methyl esters have very similar fuel properties as mineral diesel fuel. This is surprising because of the differences in chemical composition. The main difference is the almost uniform composition of biodiesel in contrast to mineral fuel, which is a mixture of a large number of different hydrocarbons with different structures and boiling points. Moreover, mineral fuel contains only carbon and hydrogen, whereas biodiesel has an additional oxygen content of approximately 10%. Therefore, the net calorific value of vegetable oil-based fuels is 13% lower than that of mineral fuel. This lower heating value is vastly compensated by the higher density of biodiesel, so compared by volume the difference is approximately only 8%.

It can be seen that the viscosities of mineral diesel and biodiesel are very similar, whereas the value for rapeseed oil is approximately 10 times higher. Because of the different composition of mineral diesel fuel, the boiling range lies between 180 and 350 °C, whereas biodiesel has a very close boiling range. The cetane number defines the quality of a diesel fuel relating to its combustion characteristics in a diesel engine. Because straight-chain hydrocarbons have superior burning characteristics than branched-chain isomers, n-hexadecane ("cetane") was once defined to have the cetane number 100. As all fatty acids have straight carbon chains, they all have relatively high cetane numbers. However, the cetane numbers of fatty acids depend on their degree of saturation, with saturated fatty acids having higher numbers than unsaturated acids. So the best biodiesel would consist only of saturated fatty acids, but this fuel would be solid at ambient temperature. The cold filter plugging point (CFPP) describes the temperature where a fuel begins to crystallise, leading to the plugging of the fuel filter. This value depends on the content of paraffins in mineral diesel fuel and the content of saturated fatty acids in biodiesel. The limit of CFPP values in the different specifications for mineral fuel depends on climate conditions in different countries.

Because of very similar properties, biodiesel and mineral fuel are soluble in each other and therefore can be blended in any ratio. However, because of the good solvation properties of esters, biodiesel is aggressive to paints and rubber parts like fittings and fuel lines, so that these materials have to be replaced by more stable polymers like viton, if higher quantities of biodiesel are used. No other changes are necessary within the engine and engine parts.

Quality of Biodiesel

During the early stages of biodiesel utilisation, it soon became apparent that the performance of biodiesel in an engine largely depends on the quality and purity of the fuel. It could be seen that incomplete conversion of triglycerides can lead to the formation of deposits in the engine. Elsewhere, traces of remaining catalysts as soaps can lead to serious problems within the injection pump. So, very early discussions on specifications were initiated. Austria was the first country to develop and publish a biodiesel standard ON C 1190 for RMEs as the basis for warranties given by many diesel engine companies. Several other countries followed, and finally in 2003 harmonised specifications for all the present and future members of the European Union were valid (prEN 14214).

The basis for this specification was the specification valid for mineral fuel (EN 590). However, additional oil-specific parameters had to be added. In particular, the content of impurities like methanol and glycerol, residues of catalysts (NaOH, KOH) and unreacted material (tri-glycerides, di-, monoglycerides) is limited. The question whether there should be any limitation of possible feedstocks was subject to fierce discussions during the development of European specifications. Finally, it was decided not to specify the raw material, so in principle all feedstocks can be used either as blend or as pure material provided that the specifications are met. However, feedstocks with high amounts of saturated fatty acids are almost entirely excluded because of the restrictions of CFPP, and the use of feedstocks with a high content of higher unsaturated fatty acids are also limited because of the restriction of the Iodine number. So 100% soyabean oil, sunflower oil, palm oil and animal fat cannot be used for the biodiesel production in Europe now, but special blends considering the different constraints will be possible.

Engine Performance and Emissions

Until 2003, several million tonnes of biodiesel have already been produced and utilised in different vehicles either as blends with mineral fuel or as 100% fuel in a wide variety of different diesel engines in passenger cars, tractors, heavy-duty vehicles and stationary engines. Also numerous results of test bench runnings and field tests have been published. In the following section some of these results will be summarised.

Due to the lower heating value of biodiesel its fuel consumption is slightly higher than that of mineral fuel. On a volumetric basis the fuel consumption is approximately 11% higher, whereas on a gravimetric basis the difference amounts to only 5–6% due to the higher density of biodiesel.

The thermal efficiencies of converting the chemical energy into mechanical energy are within the same range for both fuels, depending on engine conditions such as load. Some authors even report a slightly higher thermal efficiency for biodiesel, which could be explained by its oxygen content, which facilitates combustion.

Limitations on the emissions of passenger cars in Europe are in place for carbon monoxide (CO), hydrocarbons (HC), nitrogen oxides (NO_x) and particulate matter (PM), which are limited only to diesel engines. So far there are no limits for carbon dioxide (CO_2), which is the major product of combustion engines and which is the major greenhouse gas.

Using 100% biodiesel, a slight or even significant reduction of CO and HC emissions could be observed, depending on the type of engine and the choice of test cycle. For NO_x an increase between 7 and 20% could be observed in most cases. The lower HC and CO emissions can be explained by a more complete burning of the fuel due to the oxygen content of biodiesel, whereas for the increase of NO_x emissions several reasons are being considered. The better combustion characteristics of biodiesel can lead to higher combustion temperatures and therefore to higher amounts of NO_x. Another explanation could be the higher compression module of biodiesel, which causes an advance in the injection and ignition timing of the engines. In general, early ignition leads to high combustion pressure and maximum temperature, and thus results in increased NO_x emissions. Delaying the time of fuel injection, which can be done electronically, could reduce the increase of NO_x emissions. There is also the possibility of installing a sensor in the fuel supply which measures the composition of the fuel and automatically controls the optimum injection time. Non-limited gaseous emissions include benzene and aldehydes, for which no significant trend could be observed, as both increases and decreases in emissions have been reported.

In the case of PM, which consists of soot particles as well as adsorbed unburnt fuel molecules and polyaromatic hydrocarbons, a significant reduction of up to 70% can be observed in most measurements. However, there are also reports that show an increase in PM. Because of the very complex chemical composition of PM, it is better to compare the different chemical classes of compounds. So the amount of the very toxic polyaromatic hydrocarbons (PAH), which are adsorbed on the surface of soot particles, is significantly reduced with biodiesel, and also the amount of sulphates is significantly lower because of the very low content of sulphur in the fuel.

As diesel exhaust gases are classified as carcinogenic, it is important to compare the total carcinogenic potential of the exhaust gases. The mutagenic potential of the soluble part of the PM was measured by the Ames-test using *Salmonella typhimurium*. Here a significantly lower carcinogenic potential for biodiesel could be found, which might be explained by the relative low content of aromatics and sulphur compounds in biodiesel.

Summarising the emission data, it can be said that with the use of biodiesel there are significant reductions in the emission of HC, CO, aromatics and particulates, whereas there are slightly higher emissions of NO_x. HC, CO, aromatics and PM can be significantly in both cases with the help of a catalytical converter, which in the case of a diesel engine can only be an oxidative two-step converter fully oxidising unburnt carbon compounds. Because of the necessity of surplus oxygen for combustion in a diesel engine, simultaneous reduction of NO_x is not possible. This is one drawback of diesel engines, which can be overcome in the future by the after-treatment of exhaust gases, for instance with the help of urea. Furthermore, the use of particle filters for the removal of toxic particles will help to reduce local emissions to a very low level.

Environmental Aspects

The main advantages of biofuels from renewable resources are the saving of energy resources, the reduction of emissions of greenhouse gases and their almost unlimited availability in the future. However, the total ecological benefits can be analysed only by complete life cycle analyses, which have to distinguish between the different feedstocks and also the different uses of biodiesel. For instance, it is important to distinguish between the use of 100% biodiesel and the use of blends. It is also quite obvious that there must be a difference between the uses of neat vegetable oils, which can also be used for food purposes, and the use of waste material such as used frying oils, trap grease or animal fat. Moreover, one has to distinguish between local benefits, for instance the reduction of soot in urban areas, or global benefits such as the reduction of greenhouse gases. Finally, the secure supply of fuels in times of crises as well as socio-economic aspects have to be considered. In most of the literature available today, biodiesel from rapeseed oil is compared with fossil fuel, so there might be significant changes if other feedstocks are used.

One important aspect for the production of biofuels is the respective energy balance. This is the ratio of energy that has to be used for the production of the fuel and the energy which you can get out of the fuel (input: output ratio). For the production of biodiesel from rapeseed oil a range of 1:1.9 up to 1:3.2 is reported. This means that one gets up to three times more energy out of biodiesel that is needed for its production. This is in contrast to mineral fuel where the ratio is approximately 1:0.9 because of the high energy input for exploration, transport and raffination. The differences in the figures for rapeseed oil mainly come from the consideration of side products like straw and oil cake during vegetable oil production or glycerol during biodiesel production.

Global warming potential

One major aspect of life cycle assessments is the potential of global warming, expressed as CO_2-equivalents. CO_2 itself is produced during the whole production process of fuels. Because of the positive energy balance of biodiesel and the fact that biodiesel mainly consists of renewable material, one could expect a large saving of greenhouse gases compared to mineral fuel. This is indeed true for CO_2, but if you consider other greenhouse gases like N_2O and CH_4, which have an even higher global warming potential, the advantages for biodiesel are slightly diminished. Controversial statements can be found in the literature concerning the impact of N_2O from agricultural production, but all studies report a considerable advantage for RME over fossil diesel fuel. The relative savings of greenhouse gases for the use of RME are approximately 2.7 kg of saved CO_2 equivalents for every kg of substituted fossil diesel fuel. These figures should of course be even better if recycled products such as used frying oil and animal fat are used for biodiesel production.

Another aspect in life cycle analyses is the acidification potential, describing the emissions of ammonia, SO_x and NO_x, which can lead to the formation of acid rain. Though biodiesel contains only traces of sulphur and therefore emissions of SO_x are negligible, slightly higher emissions of NO_x and higher emissions of NH_3 during agricultural production contribute to a slight disadvantage of using biodiesel. In the case of the ozone depletion potential, biodiesel produced from rapeseed oil has a higher potential because of the formation of N_2O. Slightly higher figures are also reported for the tropospheric ozone formation potential.

Toxicological and eco-toxicological aspects are a major issue of life cycle analyses. Controversial results are reported in this context as well, especially if the toxic effects on humans are considered. However, the high biodegradability of biodiesel as well as the low eco-toxicity to aquatic organisms are significant advantages.

Franke and Reinhardt (2002) conclude that "an overall assessment of environmental impacts in favour of RME can be justified", even if remaining uncertainties will have to be clarified.

Further work has to be done on evaluating other feedstocks like waste material for biodiesel production, which will become more and more important in the future. By then it is likely that the results of life cycle analysis will be far more positive, because all aspects of agricultural production will be irrelevant. Table 9.6 gives an overview of the ecological properties of biodiesel.

9.5.2 Soaps

The most important utilisation of fatty acids, obtained by hydrolysis of triglycerides, is the synthesis of soap by saponification or neutralisation. The hydrolysis is carried out according to several continuous or discontinuous processes. The discontinuous processes seem to be gradually replaced by continuous processes. The saponification is divided into three main stages. The first stage is the reaction in which the oil and caustic soda are mixed to an emulsion, during the second stage the mixture is washed so that the water and glycerine are partially removed, and the last stage concerns the adjustment of the concentration of the soap. This last stage tends to disappear because it is a critical point and

Table 9.6 *Overview of ecological properties of biodiesel*

Energy input:output ratio[1]	1:1.9–3.2
Savings of CO_2-equivalents[1]	approx. 2.7 kg/kg biodiesel
Local emissions[1]	
PAH	up to 75% less
Particulates	up to 35% less
HC, CO	reduction of HC, CO
NO_x	slightly higher NO_x
Toxicity[2]	LD50 > 17000 mg/kg
Ecotoxicity[3]	80–90% less toxic to aquatic organisms
Biodegradability[4]	88% (28 d) mineral diesel: 26% (28 d)

[1] Scharmer (2001), Kaltschmitt, Reinhardt and Stelzer (1997), Reinhardt (1999), Krahl *et al.* (1996); [2] National Biodiesel Board (2002); [3] Rodinger (1998); [4] Zhang *et al.* (1998).

Table 9.7 *Main industrial consumption of fatty acids (Karleskind, 1992. Reproduced by permission of Lavoisier.)*

Fatty acids	Consumption (%)
Soaps	47.5
Fatty alcohols	13.5
Nitrogen compounds	9.0
Surface coatings	7.0
Dimers – Trimers	5.5
Metallic soaps	4.0
Plasticizers	2.5
Food	2.0
Rubber	2.0
Cosmetic	1.5
Others	4.5

chemists are still improving the first two stages. Finally, the soap is decanted and recovered. Although it seems to be an easy preparation, making a good soap requires quite a lot of experience.

The preparation of polymers and metallic soaps are applied in chemical processes. The other applications are as additives in the fields of cosmetics or surface coatings (Table 9.7).

9.5.3 Other Applications of Methyl Esters

Methyl esters can be produced by esterification of fatty acids or transesterification of triglycerides with methanol. The latter is currently the most profitable. Transesterification is an equilibrated reaction, but with an excess of methanol and with the removal of glycerol, the reaction can be pushed to completion.

Fatty acid methyl esters are used as intermediates for the synthesis of fatty alcohols, fatty amides and polyol esters. Methyl esters of fluid oils are excellent plasticisers of polymers, solvents for insecticides and active compounds.

The derivatives of fatty esters like polyethylene glycol esters, sorbitan esters and polyglycerol esters are good emulsifiers and dispersing agents for the pharmaceutical and cosmetic industries. They are also used for some specific technical reasons in the textile industry and phytopharmacy.

Methyl esters can be hydrogenated also by high processes using solid mixed metal oxide catalysts to give fatty alcohols. By using mixed catalysts, the reaction can be controlled so that the double bond in unsaturated fatty acid esters is not hydrogenated. Catalysts containing copper give saturated fatty alcohols, while special catalysts containing zinc preserve the double bonds in the starting compounds and give unsaturated fatty alcohols such as oleyl alcohol. Fatty alcohols are not only used by the detergent industry in this form or after chemical transformation, but they also find applications in other chemical industries such as the cosmetic industry and in the processing of mineral oils.

Two-thirds of the fatty alcohols produced in Western Europe are from natural sources. The world consumption of fatty alcohols totalled 925 000 tonnes in 1988. In Western Europe, companies such as Henkel, Condea and Oleofina produce natural fatty alcohols. In the US and in Asia, Procter & Gamble is the primary producer of natural fatty alcohols.

9.5.4 Surface Active Agents

One of the main application domains of oils and fats is the detergent sector. When a molecule has a hydrophilic part and a lipophilic part, it is called an amphiphile and has special properties. In particular, it lowers the surface tension between water and an organic phase. To minimise the energy, amphiphilic molecules are localised at the interface of the two phases, creating a thin layer with the hydrophilic part in the water and the lipophilic part in the organic phase. Depending on the nature of the two parts, the properties of the molecules can be different. The lipophilic part comes from fatty oils like copra and palm kernel for lauric and myristic moieties and from palm and rapeseed oil for palmitic and erucic moieties. The hydrophilic part can either be anionic, non-ionic or cationic. Many examples are known of surface-active agents that are used in household detergents and cleaning products (dishwashing and textile washing), and in cosmetic and body care formulations such as shampoo or bath soap.

9.5.5 Cosmetics

The use of fatty substances is the origin of the use of cosmetics since fatty compounds protect and moisturise the skin due to the affinity between the lipids of the skin and the natural lipids. The use of vegetable oils in cosmetics is numerous and their nature can be very different. Not only exotic oils like sweet almond oil, avocado oil, Jojoba oil, copra oil and castor oil, but also more usual oils such as palm oil are frequently used.

Fatty substances are used for two reasons:

1. Their emollient and nutritional properties which are needed in most cosmetic formulations; and
2. next to the surface-active activity, fatty acids also have foaming, wetting, cleaning and solubilising activity.

A large number of fatty substances are available, and the formulator will choose one in relation to the function of the final properties desired and the sensorial criterion of the required formulation.

Next to the fats used as softening agents, animal fats can be used as cosmetic additives. Lanolin (a mixture of esterified fatty acids and free alcohols) obtained from refined lamb's wool grease and squalene obtained from shark liver oil have been used extensively, but due to the movement in favour of the protection of animals and the environment, animal fats will be used less as softening agents in cosmetic preparations.

9.5.6   The Plastics Industry

In the plastics industry, fatty substances like metallic soaps, esters, amides and amines are used as additives. The main desired properties are antistatic, lubricant, heat stabilizer and plasticiser. Antistatic additives are used to remove static electricity and to lower the dust attraction that can lead to process and packaging problems. Other static additives are applied to lower the friction: they are called antiblock or slipping agents. The most important antiblock agent for the low-density polyethylene processing is erucamide, prepared from erucic rapeseed oil. In general, antistatic agents are quaternary ammonium salts or ethoxylated amines. Lubricants improve the rheological properties during polymer processing. Heat stabilisers lower thermal degradation, especially PVC degradation. The role of heat stabilisers is to neutralise the hydrochloric acid formed. Fatty heat stabilisers are mixtures of metallic soap. Epoxidised soyabean oil can also be used as a secondary stabiliser. The plasticisers are additives incorporated in the polymer to increase its plasticity. They are mainly esters, di- or triesters.

9.5.7   The Ink, Paint and Varnish Industries

In the ink, paint and varnish industries, fatty components are used as binders and as solvents. Oils can be used for this, but polyesters are mostly used because they are modern and effective binders.

The solvating power of hydrophobic compounds is the required property. Some oils are used without chemical transformation:

- Soyabean oil and sunflower oil as additives in the ink industry.
- Linseed oil and sunflower oil as reactive solvents in the paint industry.
- Different oils to help the penetration of insecticides and fungicides during wood treatment.

A paint formulation is normally composed of three parts: a pigment, a binder (e.g. an alkyd resin) and a solvent (today mainly white spirit). Among the chemical derivatives, alkyd resins are the most used and also the most efficient binder. Alkyd resin are polyesters made from polyols, polyacids and fatty acids. Each oil has its specific characteristics. Some are siccative or semisiccative, the others are non-siccatives.

9.5.8   Solvents

In many formulations (paints), in surface-cleaning or dry-cleaning agents, the main problem is the solvent that is mainly white spirit, an aromatic hydrocarbon considered eco-toxic as a VOC (Volatile Organic Compounds). Toxicological risks, health and environmental considerations have started to impose new legislation and a new solvent chemistry. In order to replace toxic solvents, research is progressing towards aqueous formulations, to organic formulations without aromatic solvents and to organic formulations with solvents derived from vegetable oils. There are a number of benefits in using vegetable oil instead of mineral oil. Contrary to mineral oils, vegetable oils are renewable, completely bio-degradable and non-toxic. On top of that, vegetable oils are better performers for certain applications. Due to economic concern and health and environmental considerations, organic chemistry and biotechnology has to play an important role in developing more bioproducts with unrivalled performances.

9.5.9   Lubricants

The market for lubricants is dominated by mineral oil-based products, but their non-ecological characteristics are no longer acceptable due to environmental concerns. It is estimated that about 30% of these products end up in the environment, either because they cannot be collected or because they are subjected to accidental losses or voluntary disposals. On the other hand, oleochemists have a new generation of vegetable oils at their disposal whose characteristics are particularly adapted to lubricant applications and that are able to satisfy both technical requirements and environmental/health issues (Willing, 1999).

The Market

The European lubricant market is estimated at 5×10^6 t. The production of biolubricants in Europe amounts to around 100 000 t representing 2% of the market. In Germany 50 000 t of biolubricants are produced and 1000 t in France. Germany, Austria, Sweden are countries which use the largest amount of environmentally friendly nature-based lubricants proportional to their total consumption.

Production

Traditional lubricants result from petroleum refining. The synthetic oils are prepared by the oligomerisation of ethylene and propylene in the presence of aluminium chloride.

Synthetic esters (e.g. adipate, isostearate, phthalate, dicarboxylic esters) also have a petrochemical origin unlike oleochemical esters that come from fatty acids.

For low demanding applications, non-modified fats are used for lubricant formulations. Vegetable oils formulated with antioxidants can be used as lubricants at moderate temperatures (under 80 °C). In modern lubrication technology, lipochemistry provides the materials for constituting stock oils, lubricating greases or lubrication additives (Erhan and Asadauskas, 2000). The triglyceride skeleton is then replaced by a more resistant polyol such as isosorbitol or neopentylpolyol for which the quaternary carbon improves the thermal stability of the structure. Moreover, the alteration of fatty acid hydrocarbon chains by ozonolysis generates short mono- or di-esters after combination with an alcohol. These reactions lead to different categories of esterified compounds called oleochemical esters.

The selection of crop varieties and recent developments in biotechnology have enabled the development of new varieties of oilseed plants providing oils rich in erucic oleic or lauric acids. For the preparation of one tonne of lubricants, one tonne of vegetable oil is needed, representing about 0.7 ha of rapeseed or 1 ha of sunflower crops. This means that the substitution of 20% of the European lubricant consumption by natural-based lubricants would correspond to about 20% of the total area of rapeseed and sunflower grown in Western Europe.

In recent years, most research has been devoted to studying new fatty acid derivatives with specific functions and their properties as lubricant additives, stability and oxidation resistance, post-consumer collection and recycling. New types of additives are also developing: these are organic compounds with multifunctional characteristics (e.g. friction modifier, anti-wear, anti-oxidant, rust inhibitor, extreme pressure agent) and with good ecological profiles.

Chemical and Physical Characteristics

Vegetable esters possess the natural properties of anti-wear and extreme pressure resistance that confer a remarkable performance in lubrication and thus contribute to the efficient functioning of the equipment. An increased thermal stability, an appropriate viscosity index and a favourable pour point are additional characteristic assets. Advantages linked to the use of natural esters also result from an extended lifetime for the lubricant as well as reduced product consumption. The major features of natural lubricants are:

- natural anti-wear properties;
- good solubility of additives in natural bases;
- absence of deposits linked to the coking phenomenon;
- high viscosity index (low variation of viscosity with temperature) which confer "multigrade character";
- protection against corrosion;
- high biodegradability; and
- absence of ecotoxicity and toxicity.

Table 9.8 presents physicochemical characteristics of different types of esters.

The question of hydrolysis stability arises for esters, contrary to hydrocarbon-based lubricants: this point depends on the presence of traces of impurities and hence on care

Table 9.8 *Chemical and physical characteristics of esters used as lubricants*

Parameters		Oleochemical esters			Synthetic esters	
		Polyols esters	Insaturated TMP esters	Saturated TMP esters	Diesters	Phtha lates
Viscosity at 100 °C	(cSt)	3–6	9.5	4.5	2 à 8	4 à 9
Viscosity at 40 °C	(cSt)	14–35	47.0	20.0	6 à 46	29 à 94
Viscosity index		120–130	190	140	90 à 170	40 à 90
Pour point	(°C)	−60–9	−45	−42	−70 à −40	−50 à −30
Flash point	(°C)	250–310	300	245	200 à 260	200 à 270
Thermal stability		excellent	medium	excellent	good	very good
Primary biodegr-adability	(%)	90–100	90	88	75–100	46–88

during ester manufacturing. Resistance to oxidation and thermal stability can be improved, if necessary, by using additives. The polar group of esters constitute a favourable factor for the affinity towards surfaces, which is in relation to the film strength.

Environmental and Sanitary Stakes

In Europe 20% of lubricants and 30% in the US end up in the environment. Lost oils represent more than 10% of the total consumption of lubricants. These are typically chain saw oils, two-stroke oils, demoulding oils, greases, pneumatic tool oils, coating oils, open gear oils, railway oils and anti-dust oils. Hazardous applications mainly caused by leaks or circuit breaking also result in occasional losses.

Natural oils and their esters are rapidly hydrolysed and biodegraded (>90%) in natural sites where water and bacteria are present (Willing, 2001). The use of naturally based formulations with high biodegradability and low ecotoxicity is thus desirable, since 1 l of lubricant is enough to pollute 1×10^9 l of water.

The biodegradability of synthetic lubricants is much lower since the mineral oil contains branched paraffins, naphtenic or aromatic compounds and hydrocarbons with an uneven number of carbons.

The substitution of mineral oils with vegetable formulations is not only advantageous in the reduction of water and ground pollution, but it also has a health impact that can be measured through the rate of occupational diseases listed by the International Labour Office. For example in France, 2% of the total occupational diseases are linked to the use of lubricants.

Applications

Priority has been focussed on applications in which normal equipment usage entails a frequent loss of oil:

- lubricants in agricultural activities: hydraulic fluids, greases, four-stroke motor oils;
- lubricants for forestry work: chain saw oils, hydraulic fluids, two-stroke engine oils;

- lubricants for water activities and management (mobile equipment, locks, inland shipping, hydro-plants);
- lubricants for food processing equipment;
- demoulding lubricants used in construction; and
- lubricants for textile manufacture.

In several industrial sectors, successful examples of natural lubricant uses are known:

- In forestry, Scandinavian countries widely use nature-based lubricants as chain saw oils and hydraulic fluids. Since 1991, Austria possesses a regulation forbidding the use of non-biodegradable chain saw oil as well as the use of some categories of additives.
- For the construction industry, Germany invited bids including an eco-compatibility clause for demoulding oils. Most building material manufacturers in Europe are willing to choose a release agent that provides better working conditions for the employees. Studies have also shown an improved efficiency of the esters leading to a reduced consumption.
- In water activities, a technology transfer project is in progress (Consortium LLINCWA [lost lubrication in inland and coastal activities] supported by the European DG enterprise, 2003). In this field, expertise of five European countries is gathered to lead pilot projects that attempt to overcome the obstacles.
- Railway transport is interested in biodegradable greases for the lubrication of the lines and switches. Biodegradable esters are primarily used for the switches.
- In the aeronautics field, vegetable-based working fluids have already been adopted in combination with micro pulverisation technology. This technology strongly helps to reduce the volume of used fluids, and also to improve the cleanliness of the working posts.
- In the automobile industry, the ester producers are associated with car manufacturers to make their products approved in engine lubrication. The Mercedes group has recently approved a new ester-based lubricant.
- In the case of food contact, lubricants manufactured in Europe can now get the certification (non-food compounds registration) according to standards defined by the NSF (Public health and safety company).

Much remains to be done in the field of the biolubricants to reach the objective of a 20% market share of biolubricants. In Germany and the Scandinavian countries, the existence of environmental quality labels has been a determining factor ("blue angel", "Nordic swan", "milieukeur"). A European "ecolabel" would allow for the evaluation of environmental criteria that are significant for applications in sensitive natural settings.

For as long as regulations are not in force, the savings and advantages resulting from the performances and the eco-compatibility of biolubricants are the most attractive arguments for promoting their use.

9.6   Minor Constituents in Oils

Crude oils contain varying (sometimes significant) quantities of diverse compounds in addition to triglycerides and partial glycerides. The minor constituents of the oils are phosphatides, cerides, chlorophylls, the unsaponifiable matter (e.g. triterpene, tocopherol, sterol, hydrocarbons) and the alteration products.

9.6.1 Type of Minor Constituents

- Phosphatides present in crude fats are basically phosphoglycerides, which are derivatives of phosphoryl-3-glycerol. Phosphatides include phosphatidic acids, phosphatidyl amino alcohols, phosphatidyl polyols and phosphoglyceride fatty acids.
- Cerides is the general name for esters of fatty acids and of monohydric alcohols, whose molecular weight is high enough to make these alcohols insoluble in water. One can distinguish between natural waxes, abundant in marine animals, from sterides and caro- tenoides that are present in vegetables. The latter class are esters of sterols and fatty acids and hydroxycarotenoid esters.
- Chlorophylls are the green pigments of plants.
- In unsaponifiable matter, two types of compounds are of particular interest: **tocopherols** and **sterols**. They are recovered during the deodorisation stage and at the end of the refining stage for the raw oil. The amount of tocopherols and sterols present in deodistillate depends on the type of the oil refined and on the type and the conditions of desodorisation.

  The tocopherol family are 8-methyl chroman-6-ols substituted at position 1 by a methyl group and a saturated polyisoprepenic chain with 16 carbon atoms (tocopherols) or a triunsaturated one (tocotrienols). Like phenols, tocopherols and tocotrienols are effective antioxidants. Tocopherol is the name for vitamin E. However, all these compounds exhibit a vitamin activity and can be added to edible oils to improve their anti-oxidant properties or can be used in cosmetic formulations.

  Sterols are tetracyclic compounds which generally consist of 27, 28 or 29 carbons atoms. They may have some ethylenic bonds. Sterols are a large part of unsaponifiable matter, account for about 30–60%. It is important to remember that animal fats contain a unique sterol namely cholesterol, whereas vegetable lipids contain two to five major sterols.

  In plants, the most abundant sterol by far is sitosterol followed by campesterol, stigmasterol and isofucosterol. They are raw materials for the synthesis of steroids for pharmaceutical applications (e.g. corticosteroids), and they are increasingly used against cholesterol absorption through human diet.

  Evidently, the proportion of unsaponifiable matter in a fat depends on its biological source and the processing. In general, the concentration of unsaponifiable matter in an unrefined lipid varies from 0.2 to 2%. Its chemical components can vary widely in kind and proportion. The unsaponifiable fraction of soyabean oil represents 1.6% of the crude oil and 0.6–0.7% of the refined oil. It mainly consists of sterols and tocopherols. Other oleaginous crops are of interest for their composition in tocopherol, like maize and also rapeseed and sunflower.
- The adulteration products are composed of free fatty acids and oxidation products that are present in industrially manufactured crude oils and fats, obtained from raw materials which may have been stored for a relatively long time. Free fatty acids are formed as a result of a biochemical degradation of triglycerides, and oxidation phenomena are caused by oxygen in the air.

9.6.2   Utilisation

Sterols are obtained from oils and fats used in industry, which contain about 0.35% of sterols. With an annual consumption of 9.5×10^6 t of oils and fats in the chemical industry, the potential supply of raw materials in the form of sterols is therefore about 33 000 t.

Phytosterols are not synthesised by humans or animals: they are exclusively delivered through diet. Dietary sources are mainly vegetable oils and derivatives like margarine. The plant sterol intake has been associated with a 50% reduction in the risk of lung carcinogenesis. β-sitosterol is an effective agent in the treatment of BHP (Benign Prostatic Hyperplasia) and *tall oil*-derived phytosterols reduce atherosclerosis (Hill *et al.*, 1982).

The importance of sterols to human health is that they inhibit cholesterol absorption (Nigon *et al.*, 2001) and lower the serum total LDL cholesterol levels (Wester, 2000). The use of dietary plant sterols for lowering elevated serum cholesterol values has recently gained much interest, especially after the commercial introduction of margarines containing plant sterols esterified with fatty acids.

Tocopherols (vitamin E) are anti-oxidants naturally occurring in vegetable oils. The anti-oxidant activity of the tocopherols and tocotrienols is mainly due to their ability to donate their phenolic hydrogens to lipid free-radicals (Kamal-Eldin and Appelqvist, 1996). Tocopherol can be synthesised from petrochemicals, but natural vitamin E is more biologically active than a synthetic one. The market for natural vitamin E has increased in the past years, and is mainly used as a dietary supplement in the US and in Japan.

In the vegetable oil refining process, the deodistillate value can be significantly increased because of the tocopherol value:

- distillate value can be up to US$1500/t vs $250–500 for traditional market; and
- global impact per tonne of oil refined between $0.5 and 5/t.

This economic value needs to be balanced with the necessity to:

- keep a high level of tocopherol in the oil (natural anti-oxidant); and
- avoid formation of trans-fatty acids.

9.7   Proteins

The use of proteins for non-food applications is currently more limited compared to the utilisation of carbohydrates and fatty acid derivatives, although proteins do have a great potential use in various applications. The limited application is connected to the higher price compared to, for example, starch. Therefore, most applications that are now being developed are in the area of specialty products with a high added value. Industrial proteins are mainly used in technical applications such as labelling adhesives and paper coatings, although much effort is also been done in the field of packaging.

Proteins are mostly modified in order to adjust the performance of the proteins to the requirement of the application. This is necessary since proteins are water sensitive and thermally labile, therefore they are mostly chemically modified. This modification can involve the incorporation of polar reactive groups, hydrophobisation, hydrophilisation or cross-linking.

Except for the fat/oil and sugar-based surfactants, protein-based surfactants are the most valuable mild surfactants since the structure and properties of the amino acids in the surfactant are similar to the tissue of the skin amino acids. This causes a strong affinity and soft feeling on the skin. Acyl derivatives from glutamic acid and serine are very effective in comparison with conventional surfactants. Also cottonseed protein from waste cottonseed cakes is a cheap source of protein-based surfactants. The production involves the hydrolysis of the protein with sulphuric acid followed by acylation with RCOCl ($R = C_{12} - C_{18}$) to form acyl amino acid sodium salts which are converted into acyl esters of amino acids by esterification with fatty alcohols (Xia and Qian, 2002).

When proteins are considered, a distinction can be made between plant proteins and proteins of animal origin. Plant proteins (wheat protein, soya protein, corn proteins, etc.) are valuable (in Europe there is a need for plant proteins), especially for food applications. Surpluses can be used for feed, resulting in reasonable prices for plant proteins.

On the contrary, proteins coming from animals (keratin, gelatine, etc.) are less valuable for food applications. Some protein sources have now become almost a waste product due to the problems related to scrapie and mad cow disease. This results in surpluses of, for example, animal bone meal. New applications for this type of renewable raw materials need to be developed in the framework of the integral valorisation of renewable resources. In this section, only a few examples of protein applications for the non-food industry will be discussed.

9.7.1 Gelatine

Gelatine is an animal protein that is obtained after partial hydrolysis of the collagen of hides and is mainly obtained from animal skin and bones. The composition is depending on the source of collagen and the type of hydrolysis. It mainly contains glycine, proline, hydroxyproline and glutamic acid. There are two main types of gelatine: gelatine A and gelatine B. Gelatine A is obtained by acid hydrolysis. The hides are washed and soaked with diluted cold mineral acids until maximal swelling of the hides. Then the acid is drained and the gelatine is extracted with hot water, followed by water evaporation and drying of the gelatine. Gelatine B is obtained by a cold alkaline treatment. After soaking, the gelatine is also extracted with hot water, followed by water evaporation and drying of the gelatine (see also p. 90).

An important characteristic of gelatine is that it forms a reversible colloidal gel in water. It swells in cold water and dissolves in warm water. The 3D network is formed by hydrogen bridges, disulphide bridges and ionic interactions between amino groups and carboxylic acid groups. The term "gelatine" is used only when the protein has a food grade. An important area for food-grade gelatine is the pharmaceutical industry since gelatine is widely used for the production of capsules for medicinal compounds.

When the quality is not high enough, technical gelatine can be used as glue. It is used in sandpaper to glue sand to the paper or in the production of matches to glue sulphur to the wooden match. The industrial applications of gelatine are mainly in the photographical sector where it is used as a carrier for silver halide emulsions. Photographical products can consist of up to 20 gelatine layers (mostly gelatine B). Nowadays, with the development of digital photography, the use of gelatine for this application is diminishing.

9.2   Keratin

Keratin is a group of the most important structural proteins of the upper skin and of hair, wool, nails, horns and feathers. The structure of keratin from mammals is called α-keratin and has a α-helix structure, whereas bird and reptile keratin is called β-keratin because of the β-helix.

The mechanical characteristics of the keratin fibres can be explained by the structure of the crystalline fibrils embedded in the amorphous matrix interconnected by the presence of cystine cross-links. Different kinds of keratin can be distinguished depending on the amount of cystine (low, high, very high) and the amount of glycine and tyrosine.

The applications of wool and hair are obvious not only in the textile industry, but exist also in the carpet industry and even in the cosmetic industry. Other applications are diverse, ranging from film formation for wound dressings, anti-skid applications in rubber, to foam stabilisation for foam concrete.

9.3   Wheat Gluten

Proteins have always been interesting from the food packaging point of view because of their non-toxic properties. Therefore, much research has been performed to use proteins or protein fractions to develop packaging materials, coatings or sheets.

Wheat protein has therefore been separated in a glutenin and a gliadin fraction. The gliadin fraction could be used for either glue or coatings. Efforts have been made to use the solid gliadin fraction to develop a coating for protecting cars during transport, which is currently being done by a wax layer. This coating has the advantage that it can be simply removed by washing the car with hot water whereas the wax has to be manually removed. Furthermore, the presence of proteins in washing water does not pose problems in the treatment of wastewater (Lötzbeyer, Wiesmann and Lösche, 1999). Protein modification is being performed by oxidation (to make strong films without losing elasticity), by acetylation (for use as a co-binder in paper coatings) and by deamination in order to improve the foam forming features of the protein.

9.4   Soyabean Proteins

Soyabean proteins are available in large quantities because of the production of soyabean oil. After the extraction of the soyabean oil, the soyabean meal is treated with an alkaline solution and filtered to remove the insoluble saccharides. Then the proteins are precipitated at pH 4.2 with the removal of the water-soluble sugars, followed by neutralisation. This results in a soyabean protein isolate which has a price of about US$3.5/kg which is comparable to the price of polyvinyl acetate.

Most of the soyabean proteins are used for feed, then for food. Only 0.5% of the total production is used for non-food applications. It is mostly used in the paper industry as a co-binder to secure the binding of colourants to paper, which results in the paper having improved printing characteristics, and overall there is a better control of its rheological features. Furthermore, it can be used as a relatively firm and fast-drying glue. It was used in large quantities as an adhesive in the triplex industry but it has since been replaced by synthetic glues. Soyabean proteins can also be used to stabilise latex emulsions, in coatings and as binder in water-based paints and inks.

9.8   Conclusions

The use of fats and lipids to develop industrial products has proven to be successful. Nowadays, two applications have become extremely important, namely the large-scale use of biodiesel and the use of green lubricants.

Biodiesel has several technological advantages compared to fossil fuel, but the price is still not competitive with that of fossil fuels. Therefore, governmental intervention will be required (e.g. through eco-boni or eco-taxes) for biodiesel to break through. Biolubricants are especially advantageous for applications where the lubricant ends up in the environment. Apart from these, numerous other applications are known and are being developed.

The use of proteins as industrial products is smaller and is mostly situated in the packaging and coating sector. Gelatine, however, is probably the most important representative of the protein family and is mostly used in the photographical sector. Because of the biodegradability of these products, industry is looking to include renewable resources in their processes.

References

Akoh, C. and Min, D. (1998). *Food Lipids*, Marcel Dekker Inc., New York.
Bockey, D. and Körbitz, W. (2002). *Stand- und Entwicklungspotential für die Produktion von Biodiesel – Eine internationale Bestandsaufnahme* (eds A. Munack and J. Krahl), Proceedings of the 2nd International Conference on Biodiesel, September 1–17, Braunschweig, Germany, 17–24.
Claude, S., Mouloungui, Z., Yoo, W.-J. and Gaset, A. (2000). US Pat 6 025 504.
Erhan, S.Z. and Asadauskas, S. (2000). Lubricants basestocks from vegetable oils. *Industrial Crops and Products*, **11**, 277.
Franke, B. and Reinhardt, G.A. (2002). http://www.biodiesel.org.
Gunstone, F.D. and Padley, F.B. (1997). *Lipid Technologies and Applications*, Marcel Dekker Inc., New York, 834pp.
Gunstone, F. and Hamilton, R.J. (2001). *Oleochemical Manufacture and Applications*, CRC Press, Sheffield, 325pp.
Gutsche, B. (1997). *Fett Lipid*, **99**, 418–427.
Hill, F., Schindler, J., Schmid, R.D., Wagner, R. and Voelter, W. (1982). Microbial conversion of sterols. 1 Selection of mutants for production of 20-carboxy-pregna-1,4-diene-3-one. *Eur. J. Appl. Microbiol. Biotechnol.*, **15**, 25–35.
Kamal-Eldin, A. and Appelqvist, L.-Å. (1996). Effects of dietary phenolic compounds on tocopherol, cholesterol, and fatty acids in rats. *Lipids*, **35**, 427–430.
Kaltschmitt, M., Reinhardt, G.A. and Stelzer, T. (1997). Life cycle analysis of biofuels under different environmental aspects. *Biomass and Bioenergy*, **12**, 121–134.
Karleskind, A. and Wolff, J.P. (ed) (1992). *Oils & Fats: Properties, Production Applications*, Vol. 2, Technique et Documentation, Lavoisier Paris.

Krahl, J., Bünger, J., Jerebien, H.E., Prieger, K., Schütt, C., Munack, A. and Bahadir, M. (1996). Proceedings of the 3rd Liquid Fuel Conference, Nashville, September 15–17; ASAE, St. Joseph, MI, USA, 149–165.

Lötzbeyer, T., Wiesmann, R. and Lösche, K. (1999). Wheat proteins – A Source for Innovative Biodegradable Materials. *Sixth Symposium on Renewable Resources for the Chemical Industry.* Bonn, Germany, 23–25 March 1999, O25.

Mittelbach, M. and Koncar, M. (1994). Process of Preparing Fatty Acid Alkyl Esters. *European Patent* EP 0708813 B1.

Mittelbach, M. and Trathnigg, B. (1990). Kinetics of Alkaline Catalysed Methanolysis of Sunflower Oil. *Fat Sci. Technol.*, **92**, 145.

Mouloungui, Z., Rakotondrazafy, V., Peyrou, G., Gachen, C. and Eychenne, V. (1998a). *Agro-Food Ind. Hi-Tech.*, 10–14.

Mouloungui, Z. and Gauvrit, C. (1998b). *Ind. Crops and Products*, **8**, 1–15.

National Biodiesel Board (2002). http://www.biodiesel.org.

Nigon F., Serfaty-Lacrosnière, C., Beucler, I., Chauvois, D., Neveu, C., Giral, P., Chapman, M.J. and Bruckert, E. (2001). Plant sterol enriched margarine lowers plasma LDL in hyperlipidemic subjects with low cholesterol intake: effect of fibrate treatment. *Clin. Chem. Lab. Med.*, **39**, 634–640.

Oil World (2003). ISTA Mielke GmbH (eds). Hamburg, Germany (www.oilworld.biz).

Reinhardt, G.A. (1999). *Aktuelle Bewertung des Einsatzes von Rapsöl/RME im Vergleich zu Dieselkraftstoff* (ed. Umweltbundesamt), Berlin, Germany.

Renaud, P. (2002). World Overview of Trends in Oleochemicals. World Oleochemical Conference, Barcelona 14–17 April 2002.

Rodinger, W. (1998). *Toxicology and Ecotoxicology of Biodiesel Fuel,* Proceedings of Plant Oils as Fuels, Potsdam, 16–18 February 1997, Springer, Berlin, 161–180.

Scharmer, K. (2001). Biodiesel- Energie- und Umweltbilanz Rapsölmethylester. UFOP (ed.), Bonn, Germany.

Wester, I. (2000). Cholesterol-lowering effect of plant sterol esters. *Eur. J. Lipid Sci. Technol.*, 37–44.

Willing, A. (1999). Oleochemical esters – environmentally compatible raw materials for oils and lubricants from renewable resources. *Lipid*, **101**, 192.

Willing, A. (2001). Lubricants based on renewable resources – an environmentally comparative alternative to mineral oil products. *Chemosphere*, **43**, 89–98.

Xia, J. and Qian, J. (2002). Protein based surfactants from cotton seed proteins. World Oleochemical Conference, Barcelona 14–17 April 2002.

Zhang, X., Peterson, C.L., Reece, D., Möller, G. and Haws, R. (1998). Biodegradability of biodiesel in the aquatic environment. *Trans. ASAE*, **41**, 1423–1430.

10

High Value-added Industries

Jan Demyttenaere, Jozef Poppe

In co-authorship with *Jan Oszmianski*

10.1 Introduction

In this chapter, attention is paid to the role that renewable resources can play in high value-added industries. Since animals, plants and microorganisms have a very complicated system of enzymes, they contain interesting functionalized molecules, mostly as secondary metabolites. These are formed in minor quantities in order to fulfil a specific function in the organism. There are thousands of interesting compounds in plants and living organisms for different domains of industry (pharmacy, cosmetics, etc.). In this chapter, three very different topics are covered as examples of where renewable resources can add value to agricultural production. This chapter is divided in to three parts corresponding to the three different applications:

1. perfumes, flavours and fragrances and terpenes;
2. mycomedicinals via mushroom cultivation; and
3. natural dyes.

These three parts can serve as an example for many opportunities for industrial crops producing interesting natural products.

10.2 Perfumes

10.2.1 Historical Overview

Perfumes have been part of human history for many thousands of years. The name "perfume" comes from the Latin "per fumum", literally "through smoke". The Romans sent prayers

Renewable Bioresources: Scope and Modification for Non-food Applications. Edited by C.V. Stevens and R. Verhé
© 2004 John Wiley & Sons, Ltd ISBNs: 0-470-85446-4 (HB); 0-470-85447-2 (PB)

to heaven on the wings of sweet-smelling smoke, called "per fumum". Since ancient times, perfumes have been used in religious ceremonies and in the treatment of sickness: besides pleasing odours, numerous aroma chemicals also possess useful bacteriostatic and antiseptic properties. The history of perfumery can be associated to the history of natural perfumes. The introduction of synthetic products in this domain is extremely recent and did not occur until the end of the 19th century. For the ancient civilizations, perfumes were synonymous with crude aromatic raw materials. In China and India, perfumes have already been used in this way for more than 1000 years. Later, perfumed water was used in the Buddhist liturgy. In ancient Egypt, the art of perfumery finally progressed from the use of crude natural objects to become a relatively sophisticated element in the art of healing.

In Roman history, towards 330–350 AD, the appearance of numerous gums has been documented: myrrh, olibanum, which arrived principally from Yemen via the Red Sea. From the year 1000 AD onwards, a large number of aromatic materials in Europe and particularly in France appeared. Spices: cinnamon, pepper, nutmeg, cloves; and aromatic raw materials: musk, civet, benzoin, sandalwood, camphor, the balsams and the gums arrived in Europe through Venice.

The history of distillation began in the East. The first basic principles for the preparation of essential oils had been developed in the Orient: in India, Persia and in Egypt. Turpentine seems the only essential oil for which the method of preparation has been clearly described. Until the Middle Ages, the art of distillation was used mainly for the preparation of distilled waters. An oil appearing on the surface of these waters was regarded as an undesirable by-product. The first description of the distillation of a true essential oil is generally attributed to a Catalonian physician: Arnold de Villanova (1235–1311).

Towards 1500, at the beginning of the Renaissance, Venetian manuscripts mention distilled oils (and not waters) of sage and rosemary. Distillation became known as a way of separating the "essential" from the "non-essential". The Swiss physician, Paracelsus (1493–1541) described his theory of the "Quinta essentia" (quintessence). This leads to the presently used term "essential oil".

Some of the most odoriferous flowers provided no essential oils on steam distillation. Nicholas Lemery published his observations for the flowers of jasmine and violet in his "Cours de Chymie", Paris, 1701. The affinity of fats for odours, particularly at elevated temperatures had already been observed since antiquity. From this invention, the technique of "enfleurage" was born in the 18th century. It is based on low temperature extraction of flowers by direct contact with fats. It was so labour-intensive that it was abandoned after the Second World War.

In 1809, Bertrand indicated that the malaxation of pommades with spirits of wine (85°/86° alcohol) provided products which he called "Extraits de Pommades". At this time the first distillery was created in the South of France which was capable of obtaining the azeotrope of alcohol: 95°/96°.

Extraction from fragrant flowers (daffodil, acacia, rose, jasmine, narcissus, etc.) was done primarily using diethyl ether and carbon disulphide. However, these solvents were too dangerous for industrial uses. From 1880, petroleum ether was used for the extraction of floral products. At the end of the 19th century, an installation for hydrocarbon extraction was developed. Pommades were also extracted with alcohol by Louis-Maximin Roure, who called these products "solids" to distinguish them from the concretes. The essences he obtained for the first time are today called "absolutes".

It was not before Otto Wallach carried out his studies (between 1880 and 1914) on the composition of the essential oils that more was known in this field. Wallach, an assistant of Kekule, was particularly interested in these essential oils. Kekule had already observed the presence of hydrocarbons in turpentine. Consequently, these structures with the molecular formula $C_{10}H_{16}$ had been named "terpenes". Other substances, with the molecular formulas $C_{10}H_{16}O$ and $C_{10}H_{18}O$, and obviously related to terpenes, were known under the name "camphor". The prototype of this group, camphor itself, was known since antiquity in the Orient and at least since the 11th century in Europe.

In 1884, Wallach published work that stated that several terpenes described under different names were, in fact, the same substances. In 1891, he established a list of nine different terpenes that could all be easily characterized. Between 1884 and 1914 he wrote 180 treatises, which were assembled into a book titled *Terpene und Camphor*. Not only terpene hydrocarbons, but also oxygenated compounds in essential oils were described. In 1887, he had already realized that these compounds "must be constructed from isoprene units". Thirty years later, Wallach's isoprene rule was perfected by Robinson (Nobel Prize in chemistry in 1947) who indicated that the joining of these isoprene units must take place in a head to tail fashion. The classification of terpenes according to the number of these units (monoterpenes, sesquiterpenes, diterpenes, triterpenes, etc.) is known as the rule of Ruzicka (p. 262). In 1910, Wallach was honoured with the Nobel Prize in chemistry.

About 200 monoterpenes are currently known, divided into 15 families on the basis of their carbon skeletons. Only around 30 different sesquiterpenes were known 20 years ago and these had nearly 15 different carbon skeletons. Ten years later, about 300 were known and these could be divided into about 40 skeletal families. There are now almost 1000 known sesquiterpenoid compounds belonging to about 200 different skeletal families.

The field of synthesizing odorant substances started long ago. In fact, it was as far back as 1856 that Chiazza made synthetic cinnamic aldehyde by condensing benzaldhyde and acetaldehyde in the presence of hydrochloric acid. Another important example is the synthesis of vanilla, which is one of the most popular flavourings.

Today fragrances are obtained from both natural sources and chemical syntheses. In future a harmonious development between natural substances and synthetic products will continue to be observed. The natural fragrance materials will be used in the perfumery industry for as long as their price does not rise that high that the synthesis of their components becomes profitable. The synthetic products have utility in perfumery as long as they possess a sufficiently high vapour pressure to enable their volatilization and as long as they have an odour that is sufficiently interesting for the perfumer or flavourist to use in his compositions.

10.2.2 Composition and Ingredients of Perfumes

Essential Oils

Essential oils are delicately smelling volatile substances isolated from plant materials. They may be found in roots, flowers, leaves, fruit, seeds or even the bark of the plant. Growing and harvesting conditions are optimized for the production of the best fragrances.

For example, the largest single rose-growing area in the world is located at the foot of the Balkan Mountains in Bulgaria. The damask rose, which only blooms 30 days a year, has been grown in this region for centuries. The blossoms used for the production of the rose oil must be hand-picked individually. This has to be done in the morning because blossoms lose their essential oil content as the sun rises. The best collectors can pluck about 50 kg a day, a harvest that will yield only a few drops of the valuable oil. It still takes 5 t of blossoms to produce less than 1 kg of rose oil, which is worth over US$5000 on the world market. Because of this high price, natural rose oil is hardly used nowadays, and has been replaced by synthetic rose scent. The same applies to jasmine flowers that also have to be collected at early morning before the sun rises. Jasmine is the most frequently used blossom "absolue" today. Around eight million single blossoms have to be processed to obtain one single kilogram of jasmine absolue.

The largest proportion of the scents used for perfumery come from plant origin. Only a few scents come from the animal kingdom, such as ambergris, musk, castor and civet. These ingredients are used as fixatives and will be discussed later. As mentioned before, almost all parts of the plant can be processed to obtain essential oils, from the blossoms to the roots. Leaves and stems are also used (e.g. geranium, patchouli), along with fruits (anise, cardamom, coriander, nutmeg), seeds (celery, dill, parsley), peels (orange, lemon, bergamot), roots (angelica, iris, vetiver), wood (sandalwood, cedar), herbs and grasses (tarragon, sage, thyme, citronella), needles and branches (pine, fir, cypress), resins and balsams (galbanum, myrrh) and even bark (cinnamon). Sometimes the different parts of a plant will yield different fragrants, e.g. the orange tree: its blossoms are a source of neroli oil, the leaves and branches yield petitgrain oil and its peel is used to produce orange oil.

Scents are classified as *notes* based on their olfactory character. We can discern top notes, middle notes and base notes. *Top notes* ("note de tête") give the first impression of the perfume and provide a freshness to the blend; they are detected first, but also fade first. They are light scents that are usually citrus (lemon, bergamot scent), lavendula or grasses lasting 5–10 min. They are light, volatile green notes, sometimes composed of aldehydes and fruity esters. They originate from cajaput, cardamom seeds, basil, bergamot, citronella, coriander, eucalyptus, ginger, grapefruit, lemongrass, lemon, orange, lime, marigold, peppermint, petitgrain, sage, spearmint, tangerine and tea tree. *Middle notes* ("note de coeur") last several hours and are the most prominent within the fragrance. It is the "bouquet" or the heart of a perfume. The heart is the scent that results from the interaction between the fragrance and the skin onto which it is applied. It normally takes 10–20 min for the middle notes to fully develop on the skin. These are usually combinations of spicy, floral (flower) or fruit scents. They may be produced from ambrete seed, black pepper, carrot seed, cassia, chamomile, cinnamon, clove, fir, cypress, juniper, marjoram, melissa, neroli, palmrosa, pine, rose, rosemary, thyme and yarrow. And finally *base notes* ("note de fond") give a perfume depth and warmth. They give the final expression of the fragrance after it has dried. Their long-lasting effect (up to 24 h) is caused by animal substances (fixatives) and they can give a perfume its "woody" and "spicy" impression. Typical ingredients are amyris, anise, angelica root, clary sage, coumarin, fennel, geranium, lavender, lavandin, balsam, cedarwood, frankincense, jasmine, myrrh, patchouli, rosewood, sandalwood, vanillin, vetivert and ylang-ylang. Animal scents are also responsible for the base notes (castoreum, civet, musk).

Carriers and Alcohol

Essential oils are much too strong and too expensive to apply without dilution. Therefore "carrier oils" are used to dilute and carry the fragrance for use as a perfume. These carrier oils are obtained from nuts and seeds, usually by pressing them out of crops. They should be stable and possess little or no fragrance of their own.

The best choices of carrier oils are sweet almond, apricot, grape seed, peach kernel, sesame seed, soya bean, sunflower, avocado, wheatgerm and jojoba. Some of these oils can be used as the only carrier while others are only safe for skin use at 10–25% of the total. Wheatgerm oil should be avoided for use by people with a wheat allergy and soya bean oil can cause acne, allergic reactions or hair damage.

Alcohol (ethyl alcohol) is added to perfumes to carry the fragrance by evaporation as well as to dilute the ingredients. *Perfume* is the most concentrated form of fragrance oil (and the most expensive one) and is the strongest and longest lasting of fragrance forms. Perfumes contain 25–50% fragrance and about 50–75% alcohol and other diluents (carrier oils). *Eau de parfum* is an alcoholic purfume solution containing 10–15% of perfume compound. *Eau de toilette* (*cologne*) is a light form of a specific fragrance with a 3–8% concentration of perfume compound in a water base, it is composed of more than 90% diluent (ethanol) and hence it is less concentrated than *eau de parfum*. *Aftershave* or splashes contain about 99% diluents. A *moisturizing perfume mist* is a fragrance that does not contain alcohol. It is a spray that contains oil and leaves skin with a silky smooth veil of perfumed moisturizers. Since it does not contain alcohol it is less drying to the skin, but on the contrary hydrating. The higher the amount of perfume concentrate in a mixture, the longer lasting the effect of the fragrance and the less frequent that it needs to be reapplied.

Fixatives

Fixatives bind together the ingredients of the composition and lead them from the top to the base notes by regulating the evaporation rate of each individual fragrant component. Perfumes should not break up into three stages, but ought to be well balanced. If perfumes are well formulated and fixatives are properly applied, changes during the evaporation process will be imperceptible and the perfume will be harmonious and rounded-off. Fixatives mainly originate from the animal kingdom. They often provide a base note as well and give a character to the mixture. The four classes of compounds are ambergris, musk, castor and civet.

Ambergris

Ambergris is found in oily grey lumps in tropical seas, primarily in the Indian Ocean. There has been much speculation as to the origin of this material. It is a secretion of the sperm whale, probably resulting from disease. It floats in the sea and it is usually washed up on beaches in clumps or it is found in the ocean in one to seventy pound lumps. The lumps have a strong odour that is very unpleasant in its raw state, therefore it must be dissolved in alcohol. After it is processed, the fragrance is very persistent. Ambergris has

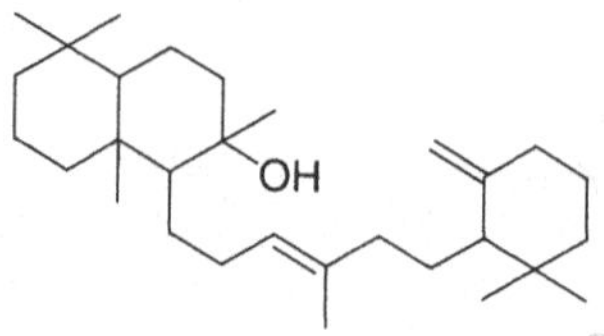

Figure 10.1 *Ambrein, precursor of ambergris odours*

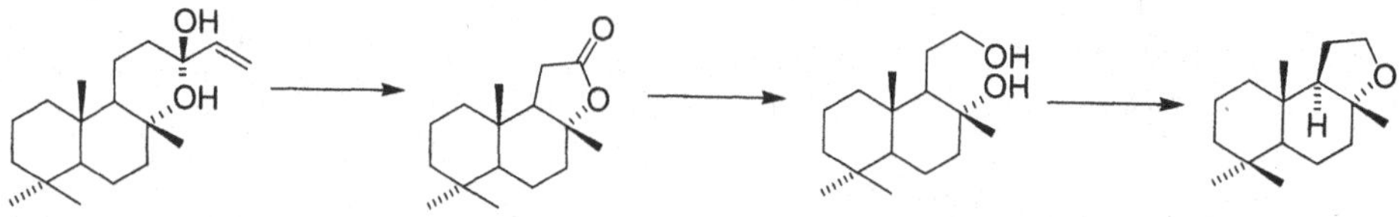

Sclareol

Ambrox

Figure 10.2 *Conversion of sclareol to Ambrox®, creating a typical ambergris note*

been in use since the 6th century. It was used in scented gloves because the odour would last for several years. Today synthetic ambergris is primarily used in the place of genuine ambergris. The major components of ambergris are epicoprosterol (up to 85%) and the odourless triterpene alcohol, ambrein (Figure 10.1). Ambrein is the likely precursor of a number of strongly odoriferous mono-, bi-, and tricyclic compounds that are formed by autoxidation or photoxidation.

One example of a crystalline autoxidation product of ambrein with a typical ambergris odour is Ambrox®. It is prepared from sclareol, a diterpene alcohol obtained from plant waste in the production of clary sage oil. This undergoes oxidative degradation to a lactone, hydrogenation of the latter to the corresponding diol and dehydration yield ambrox (Figure 10.2).

The product is used in perfumery for creating ambergris notes, and it is known under the commercial trade names Ambrox (Henkel), Ambroxan (Firmenich) and Ambroxid.

Musk

Musk is perhaps the most powerful of all perfume fragrances, and the most expensive. It is a substance secreted by the rutting male musk deer that lives in the Himalayas and in the mountainous regions of India, Burma, China and Mongolia – predominantly in areas above 2500 m. The raw material for the scent is found in a walnut-sized sac in the deer's belly (musk glands). Its typically erogenous, dry, woody animal scent is extremely strong and powerful. Musk has been a key constituent in many perfumes since its discovery. It is currently found in 35% of all men's perfumes and fragrances. It is a very good fixative, and is exceptionally long lasting. There are many synthetic musks, and musk is one of the most important ingredients in perfumes. Musk in its natural or synthetic form can be found in 90% of all fine fragrances. Other natural sources include muskrat and ambrette seeds.

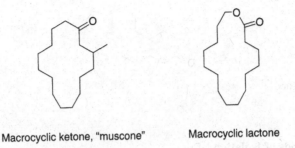

Macrocyclic ketone, "muscone"   Macrocyclic lactone

Figure 10.3 *Substances of the musk group*

Chemically speaking, musk is a group of macrocyclic compounds including ketones and lactones as the most important structures (Figure 10.3). A ketone with the molecular formula $C_{16}H_{30}O$ was isolated by Walbaum in 1906 from natural musk. The substance, which he called "muscone", makes up 0.5–2% of natural musk. Its structure was elucidated in 1926 by Ruzicka as a macrocyclic ketone, 3-methylcyclopentadecanone.

Castor

Castor, or castoreum is a strong-smelling secretion from the preputial follicles of both male and female castor beavers. It has a strong, disagreeable odour until it is considerably diluted. It then becomes highly fragrant, with a warm, sensual scent and a pleasant smell. It is an excellent fixative and gives perfumes a spicy or oriental note. It is very commonly used in men's fragrances due to the sultry, leathery, smoky note. It is also used in oriental women's fragrances. Synthetic castoreums are now available, and can be as good as the natural fragrant.

Civet

Civet is one of the most important animal materials used in perfume. It is taken from a pouch under the tails of male and female Abyssinian civet cats. It is a semisolid, yellowish to brown, unpleasant smelling substance. The isolated compound civettone produces a pleasant musky odour when used in extreme dilutions. Because of its very strong smell, it is only used in minute quantities. The scent is similar to musk, but has a smokier, sweaty aroma. It is an excellent fixative, and used in many top-quality perfumes today. Civet is available in artificial substitutes.

The structure of civettone was also elucidated by Ruzicka in 1926 as a macrocyclic unsaturated ketone, 9-cycloheptadecenone (Figure 10.4). Today beavers and civet cats are raised on special breeding farms whereas real ambergris and musk are hardly used in modern perfume, since there are excellent substitutes of synthetic origin.

Not a single perfume today is composed exclusively of natural scents. All of the well-known names are a fragrant and harmonious blend of natural and synthetic constituents.

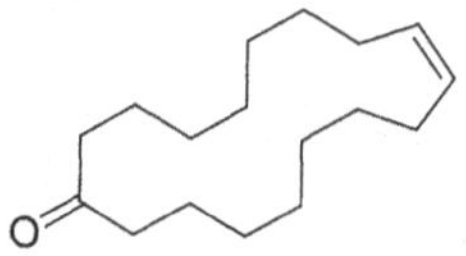

Figure 10.4   *Civettone*

10.2.3   Methods of Isolation

The three most important ways to isolate essential oils from various plant parts are expression, distillation (including steam distillation) and extraction, either with cold fat ("enfleurage à froid"), yielding a "pomade", or with volatile solvents, yielding an "absolute" or "absolue". Thus *pomades* consist of fats that contain fragrant substances and are produced by the hot or cold enfleurage of flowers. An "absolue" is an extract obtained by extraction with volatile solvents (e.g. alcohol) or by enfleurage. It is a liquid or half-liquid oil soluble in alcohol and is mainly composed of the aromatic substances. It is considered the purest and most concentrated form of perfume material, retaining most of the plant's aromatic constituents. Absolutes are used directly as perfume ingredients. A "concrete" on the other hand is the result of extraction of fresh natural plant materials with non-polar organic solvents (e.g. toluene, hexane, petroleum ether), followed by removal of the solvent by vacuum distillation, yielding a fatty semi-solid wax. Apart from the aromatic substances it also contains pigments and other non-volatile natural products including waxy compounds. Concretes are prepared mainly from flowers (rose, jasmine, tuberose, jonquil, ylang-ylang, mimosa, boronia, etc.), but also from other plant materials (lavender, lavandin, geranium, clary sage, violet leaves, oak moss, etc.). The concrete represents the closest odour duplication of the flower, the bark, leaves and so forth. It is the base material for the preparation of the "absolue". The *resin* or *resinoid* is the non-volatile residue obtained from the extract of plant material by the use of alcohol or halogenated hydrocarbons in a floating filter extractor. The products are usually highly viscous and are sometimes diluted to improve their flow and processing properties. Resinoids are primarily used for their excellent fixative properties.

Expression (Citrus Oils)

Expression is a simple technique where the rinds of citrus fruits are cold pressed to extract their essential oils using rollers or sponges. The method applied to lemon and orange oils is therefore also called *sponge process*. There is no heat involved, and as a result the oil's smell is very close to those of the original plant. Only citrus fruits have peels that are rich enough in natural essential oil to make the expression process worthwhile. After the peel has been removed from the fruit, it is pierced with numerous small holes and pressed mechanically. The resulting liquid is allowed to settle and then filtered through wet paper. This separates the aqueous parts from the essential oils. The cold-press process is particularly well suited to oranges, lemons and other citrus fruits whose bright and fresh odour are sensitive to heat treatment.

Distillation (Geranium, Lavender, Neroli, Rose, etc.)

Distillation is the most important method for extracting essential oils and relies on the evaporation of the more volatile constituents of a blend to separate them from the non-volatile parts. Steam distillation can be accomplished in two ways. In its oldest and easiest method, plant material is placed in boiling water, the essential oil will evaporate and will be carried along with the generated steam. After condensation of the steam and the oil, the oil will separate from the water, and can be collected in a Florentine flask. In a second design, steam is produced in a boiler or other vessel away from the still. In this case, the apparatus is fitted with a steam jacket, or a steam coil (for "dry steam" distillation) and also with a "live steam" inlet. By this method, fragile plant parts do not come into contact with boiling water. Plants are crushed to encourage them to release their oils. The plants are steam distilled, and the essential oils vaporize and rise up with the steam. The vapours are captured, and allowed to condense back into liquids. After a period of decantation, the water separates from the odoriferous elements that are collected and named "essences".

Heat sensitive essential-oil constituents and aromatic compounds with very high boiling points can be isolated by evaporation under vacuum using the vacuum distillation technique. A more "high-tech" chemical process is molecular distillation. It is used in producing fragrances that cannot be distilled by conventional methods.

Extraction

Extraction is a method of isolation that has been used since ancient times. Two thousand years ago, the Greeks and Romans produced perfumes by extracting, literally "pulling out" the scent of fragrant flowers and plants with the help of wine, oils or melted fats. Whereas extraction was first performed using cold fat ("Enfleurage à froid" and "maceration"), today it is mostly carried out with the use of volatile solvents.

Enfleurage and maceration (jasmine, tuberose, violet, rose, orange blossom, etc.)

The "Enfleurage" technique is very costly, and is rarely used today. It reached its peak in 1860 and made the reputation of Grasse. It is a labour-intensive process that yields the highest quality of absolutes because it does not involve heat. Heat always alters the fragrance. It is used on delicate flowers that cannot stand the high heat, and that continue to release essential oils after they have been picked. Examples of these flowers are jasmine, violet, tuberose and rose.

Hot enfleurage is the oldest known procedure for preserving plant fragrance compounds. In this method, flowers (or other parts of a plant) are directly immersed in liquid or molten wax. *Cold enfleurage* or the "enfleurage à froid" method was first used in the 19th century, but is now almost completely abandoned. A sheet of glass in a wooden frame, called "chassis" is coated with cold animal fat (refined lard or beef tallow) and the hand-picked petals were placed on top of it. It was used to extract the oils from fragile flowers such as orange blossoms, jasmine or tuberose. Within a few days, the fat absorbs the fragrance from the blossoms. The petals that are extracted completely and have given all their scents are replaced by fresh ones, and this process is repeated until the fat is completely saturated. The resulting fatty substance, which is called "pomade", is then scraped off and the oil retrieved from the fat by dissolving in alcohol. This is mechanically mixed with alcohol for up to one week, and is chilled to −56 °C. The essential oils dissolve in the alcohol while the fat does not. The mixture is chilled and filtered several times to remove all the fat. The result is "essence abolue de pommade", the natural floral essence. The alcohol is then evaporated to leave the pure absolute. Sometimes enfleurage is now carried out with cloth soaked in olive oil or liquid paraffin, which is laid over the frames instead of fat, the resultant perfumed oil being then known as "huile antique".

Maceration is similar to enfleurage. It is used to extract essential oils from animal ingredients, but also rose, vanilla and iris are treated by this process. Maceration consists in the extraction of the flowers by immersion in liquid fats or oils at a temperature of about 60–70 °C. Maceration takes long periods of time (sometimes years).

Extraction with volatile solvents

Today, volatile solvents have replaced animal fats formerly used to extract the scents from plants. This method is used for delicate flowers whose odours are damaged by the high heat needed to boil water. Typical flowers include reseda, rose, jasmine, jonquille, tuberose, violets, cassie, orange flowers, carnations, mimosa and heliotrope. The oils are extracted using solvents that have lower boiling points than water. Depending on the raw material, a variety of solvents can be used, such as petroleum ether, hexane, toluene, methanol and ethanol. Also gases that liquefy under pressure can be applied, including butane and carbon dioxide. The usual method involves placing the fragrant material on perforated metal plates in a container (the extractor); the solvent is passed over them and led into a still, where it evaporates, leaving a semi-solid mass known as *concrete*, which contains the essential oil together with stearoptene. The oil can then be separated from the stearoptene by extraction (shaking) with alcohol in a shaking machine, called "batteuse", producing the substance called *absolute*, which is the purest and most concentrated form of essential oil known.

Apart from these techniques there are various other operations such as rectification, fractional distillation, terpenes removal, and decolourization which improve and refine the numerous raw materials used for the blending and making of perfumes.

10.3   Flavours and Fragrances

10.3.1   Definitions

Fragrances and flavour substances are comparatively strong smelling organic compounds with characteristic, usually pleasant odours. They are used in perfumes and perfumed products (fragrances) or for the flavouring of foods and beverages (flavours). There is a difference between natural, nature-identical and artificial products. Natural products are directly isolated from plant or animal sources by physical procedures, such as extraction, maceration or distillation. Nature-identical compounds are produced synthetically, but are chemically identical to their natural counterparts. Artificial flavour substances are compounds that have not yet been identified in plant or animal products for human consumption. They are made synthetically, have the same smell and other properties as some natural products, but are chemically totally different molecules.

For example, vanillin can be obtained via five different ways: isolation from the orchid (*Vanilla planifolia*), which is a very expensive method; by tissue culture followed by extraction; by microbial transformation of eugenol, the main compound of clove; from lignin through synthesis and from guaiacol, a natural aroma, with a comparable molecular structure. *Only* the vanillin obtained through the first three methods is natural. The other routes yield a nature-identical vanillin. Alcohols, aldehydes, ketones, esters and lactones are classes of compounds that occur most frequently in natural and artificial fragrances.

10.3.2 Volatility

Fragrances must be volatile to be perceived. Important parameters are: the nature of the functional groups, the molecular structure, and the molecular mass, which is normally about 200. Molecular masses of over 300 are exceptions.

A perfume is a mixture of fragrance compounds that differ in volatility. This means that the odour of a perfume composition will change during evaporation. First the top note is smelled, then the middle note or "body" and finally the end note or "dry out", containing the less volatile compounds. Fixatives can be added to prevent the more volatile components from evaporating too rapidly, and hence to lessen the top note and to refine the body.

10.4   Terpenes

10.4.1   Biosynthesis

Terpenoids (often referred to as isoprenoids) constitute the largest group of natural products and are considered to be the most diverse class of substances in nature. They belong to the most important flavour and fragrance compounds, and are found in the microbial, plant and animal kingdom. According to an approximate evaluation, terrestrial plants synthesize annually up to 1.75×10^8 t of the various terpenic compounds, which are returned into the cycle of organic substances mainly as a result of microbial degradation. Terpenoids offer a very wide variety of pleasant and floral scents as they are the main constituents of essential oils, the basis for perfume preparation. Living organisms synthesize a remarkable diversity of isoprenoids (Haudenschild and Croteau, 1998). More than 23 000 different compounds have been isolated and all contain one or more isoprene units, the building block of these compounds. Thus, the structure of terpenoids is built by combining these isoprene units, usually joined head to tail. Terpenes are classified according to the number of isopentenyl units (rule of *Ruzicka*: Figure 10.5), e.g., 2: monoterpenes ($C_{10}H_{16}$), 3: sesquiterpenes ($C_{15}H_{24}$), 4: diterpenes ($C_{20}H_{32}$), 6: triterpenes ($C_{30}H_{48}$) and higher: polyterpenes.

Until 1993, all terpenes were considered to be derived from the classical acetate/ mevalonate pathway (Figure 10 .6a) involving the condensation of three units of acetyl CoA to 3-hydroxy-3-methylglutaryl CoA, the reduction of this intermediate to mevalonic acid

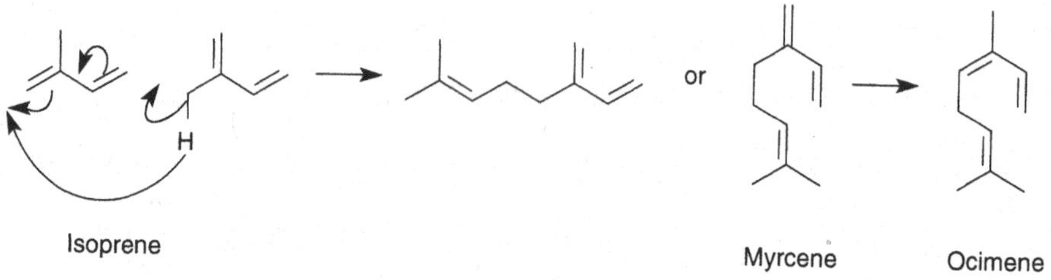

Figure 10.5 *Rule of Ruzicka: head to tail coupling of two isoprene units, giving myrcene*

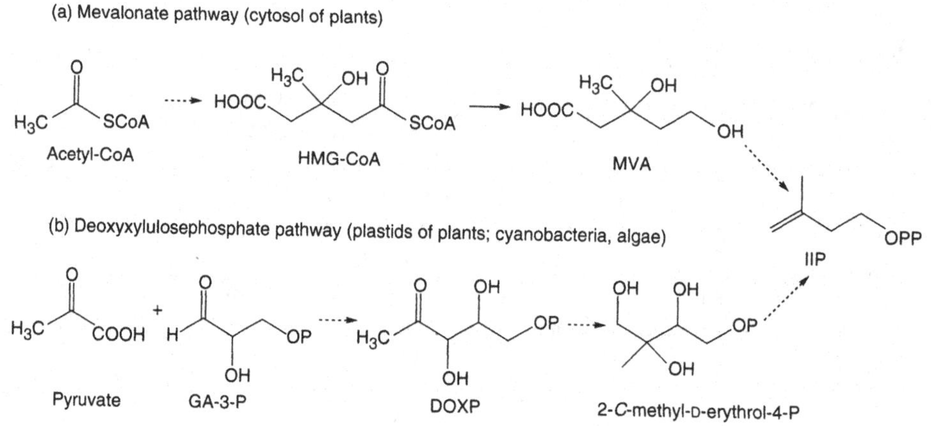

Figure 10.6 *Two Pathways for the biosynthesis of isopentenyldiphosphate (IPP)*

and the conversion of the latter to the essential, biological isoprenoid unit, isopentenyl diphosphate (IPP) (Haudenschild and Croteau, 1998). Recently, a totally different IPP biosynthesis was found to operate in certain eubacteria, green algae and higher plants. In this new pathway glyceradehyde-3-phosphate (GAP) and pyruvate are precursors of isopentenyl diphosphate, but not acetyl-CoA and mevalonate (Lichtenthaler, 1997) (Figure 10.6b). So, an isoprene unit is derived from isopentenyl diphosphate, and can be formed via two alternative pathways, the mevalonate pathway (in eukaryotes) and the deoxyxylulose pathway in prokaryotes and plant plastids.

An extensive overview of the biochemistry and molecular biology of the isoprenoid biosynthetic pathway in plants is given by Chappell (Chappell, 1995). The essential oils are abundant sources of terpenoids. They consist of a complex mixture of terpenes or sesquiterpenes, alcohols, aldehydes, ketones, acids and esters.

The monoterpenes are subdivided into three groups: acyclic, monocyclic and bicyclic (there is only one tricyclic terpene: tricyclene). Each group contains hydrocarbon terpenes, terpene alcohols, terpene aldehydes, ketones and oxides, etc.

The isolation of terpenes from plants causes several problems (e.g., very low concentrations). Therefore other sources of these flavour compounds are being searched for, e.g. microorganisms (especially bacteria and fungi) are used for the production of terpenoids.

Since terpenoids are very important flavour and fragrance compounds, the biotransformation of terpenes offers a very interesting source of novel natural flavours and fragrances, the so-called *bioflavours*. A recent review of the biotechnological production of flavours and fragrances is given by Krings and Berger (1998). The bioconversion of terpenoids by microorganisms for the generation of natural fragrances has been extensively reviewed lately (Demyttenaere, 2001a).

There are many different ways to produce natural fragrances: plant tissue and cell cultures can be used, microorganisms, such as bacteria, yeasts, fungi and algae can produce flavours *de novo* and through *bioconversions*, and enzymes can be applied. The following section will deal with some examples of important biotransformations of terpenoids by microorganisms, more specifically fungi and bacteria, bearing interest to the flavour and fragrance industry.

10.4.2 Bioconversion of Terpenes

As terpenoids, which are functionalized derivatives of terpenes (ketones, alcohols, aldehydes) are extensively applied in industry as flavour and fragrance compounds, in contrast to terpene hydrocarbons, which do generally contribute little to the fragrance, ways are sought to convert cheap terpene hydrocarbons into valuable terpenoids. Since chemical synthesis only produces nature-identical or artificial fragrance compounds, bioconversion of terpenes offers an interesting way of producing natural flavours.

Terpene hydrocarbons, such as (+)-limonene and the pinenes, are the main constituents of some essential oils (e.g. contributing to more than 90% of the composition of orange oil) and are thus ideal starting substrates for the chemical or microbial conversion to terpenoids.

Limonene

Limonene is the most widely distributed terpene in nature after α-pinene. The (+)-isomer is present in *Citrus* peel oils at a concentration of over 90%; a low concentration of the (−)-isomer is found in oils from the *Mentha* species and conifers (Bauer, Garber and Surburg, 1990). The first data on the microbial transformation of limonene date back to the 1960s. A soil bacteria *Pseudomonas* was isolated by enrichment culture technique on limonene as the sole source of carbon.

More recently the biotransformation of limonene by another *Pseudomonas* strain, *P. gladioli* was reported. *P. gladioli* was isolated by an enrichment culture technique from pine bark and sap using a mineral salts broth with limonene as the sole source of carbon. Major conversion products were identified as (+)-α-terpineol and (+)-perillic acid. α-Terpineol is widely distributed in nature and is one of the most commonly used perfume chemicals.

The first data on fungal bioconversion of limonene date back to the late 1960s. A soil microorganism was isolated from a terpene-soaked soil on mineral salts media containing D-limonene as the sole C-source. The strain, identified as *Cladosporium*, was capable of converting D-limonene into an optically active isomer of α-terpineol in yields of approximately 1.0 g/l.

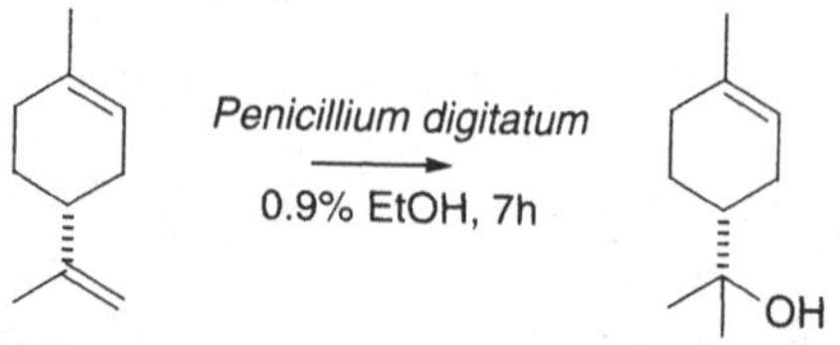

(*R*)-(+)-limonene

(*R*)-(+)-α-terpineol
yield 92%

Figure 10.7 Enantioselective bioconversion of (R)-(+)-limonene to pure (R)-(+)-α-terpineol by P. digitatum

The bioconversion of (*R*)-(+)-limonene to (*R*)-(+)-α-terpineol by immobilized fungal mycelia of *Penicillium digitatum* was described more recently (Tan and Day, 1998). The fungus was immobilized in calcium alginate beads. These beads remained active for at least 14 days when they were stored at 4 °C. α-Terpineol production by the fungus was 12.83 mg/g beads per day, producing a 45.81% bioconversion of substrate.

More recently the biotransformation of (*R*)-(+)- and (*S*)-(−)-limonene by *P. digitatum* was investigated (Demyttenaere, Van Belleghem and De Kimpe, 2001b). One strain of *P. digitatum* was able to convert (*R*)-(+)-limonene to pure (*R*)-(+)-α-terpineol in 8 h with a yield of up to 92% (Figure 10.7). The culture conditions involved, such as the composition of the broth, the type and concentration of co-solvent applied and the sequential addition of substrate, were investigated. It was found that (*R*)-(+)-limonene was converted much better in α-terpineol than (*S*)-(−)-limonene. The highest bioconversion yields were obtained when the substrate was applied as a diluted solution in EtOH.

α-Pinene

The most abundant terpene in nature is α-pinene that is industrially obtained by fractional distillation of turpentine. (+)-α-Pinene occurs, for example, in oil from *Pinus palustris* Mill. at concentrations of up to 65%; oil from *P. pinaster* Soland. and American oil from *Pinus caribaea* contain 70 and 70–80% respectively of the (−)-isomer (Bauer, Garbe and Surburg, 1990).

A commercially important bioconversion of α-pinene by *Pseudomonas fluorescens* NCIMB 11671 was described in the late 1980s (Best *et al.*, 1987). A novel pathway for the microbial breakdown of α-pinene was proposed (Figure 10.8). The attack is initiated

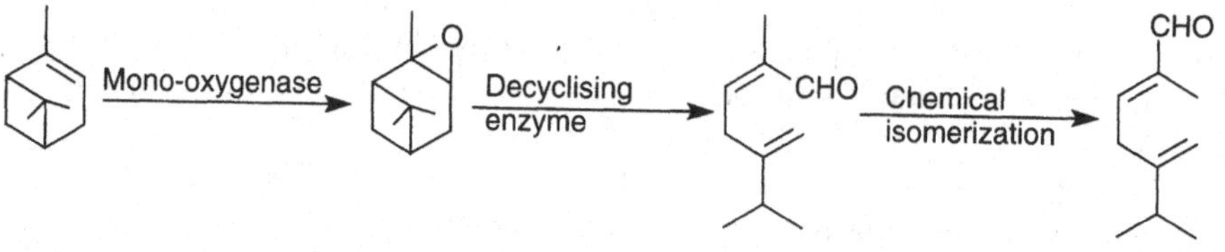

α-Pinene α-Pinene epoxide Isonovalal = *Z*-isomer Novalal = *E*-isomer

Figure 10.8 Bioconversion of α-pinene by Pseudomonas fluorescens NCIMB 11671

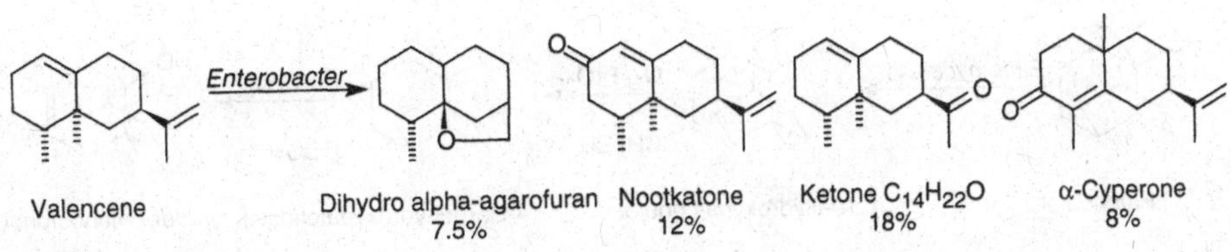

| Valencene | Dihydro alpha-agarofuran 7.5% | Nootkatone 12% | Ketone C$_{14}$H$_{22}$O 18% | α-Cyperone 8% |

Figure 10.9 *Biotransformation of valencene by* Enterobacter *sp.*

by enzymatic oxygenation of the 1,2-double bond to form the epoxide. This epoxide then undergoes rapid rearrangement to produce a novel diunsaturated aldehyde, occurring as two isomeric forms. The primary product of the reaction (Z)-2-methyl-5-isopropylhexa-2,5-dien-1-al (trivial name isonovalal) can undergo chemical isomerization to the E-form (novalal). Isonovalal, the native form of the aldehyde, possesses citrus, woody, spicy notes, whereas novalal has woody, aldehydic, cyclene notes.

Valencene

Valencene, a sesquiterpene hydrocarbon isolated from orange oils is used as starting material for the synthesis of nootkatone, which is used for flavouring beverages (Bauer, Garbe and Surburg, 1990) and which is a much sought-after aromatic substance.

Two bacterial strains, one from soil and the other from infected local beer, which utilized calarene as the sole source of carbon and energy have been isolated by enrichment culture techniques. Both these bacteria were adapted to grow on valencene as the sole carbon source. Fermentations of valencene by these bacteria of the genus *Enterobacter* in a mineral-salts medium yielded several neutral metabolic products: dihydro alpha-agarofuran (7.5%), nootkatone (12%), a ketone (18%) and α-cyperone (8%) (Figure 10.9).

Patchoulol

Patchouli alcohol or *patchoulol* is a major constituent (30–45%) of the patchouli essential oil that is extensively used in perfumery. The essential oil is obtained by steam distillation of the dried leaves of *Pogostemom cablin* (Blanco) Benth. (*Lamiaceae*). Although it is the main component of the patchouli oil, this compound contributes less to the characteristic odour of the oil than nor-pachoulenol, present at a concentration of only 0.3–0.4%.

A process for the production of the latter compound was developed via the microbial 10-hydroxylation of patchoulol and subsequent oxidation of the 10-hydroxypatchoulol (Seitz, 1994). From the 350 microorganisms screened, four strains of *Pithomyces* species carried out regio-selective hydroxylation of patchoulol to 10-hydroxypatchoulol. A method was developed by which 10-hydroxypatchoulol was obtained in 25–45% yields in 1–5-litre fermentation jars at 2–4 g of patchoulol per litre and isolated as pure material in 30% yields. The obtained hydroxylated product can easily be converted chemically to the industrially more important nor-patchoulenol (Figure 10.10).

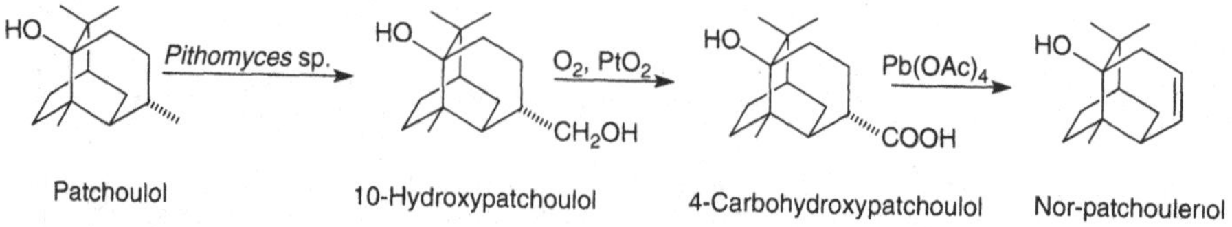

Figure 10.10 *Regioselective 10-hydroxylation of patchoulol by* Pithomyces *sp. and subsequent chemical conversion to nor-patchoulenol*

10.5   From Waste to Mycomedicinals via Mushroom Cultivation

10.5.1   Historical Evolution

The first important step concerning medicines from fungi was realized by Fleming in 1928 while accidentally detecting penicillin exudated by *Penicillium notatum*. This first world-famous antibioticum was very soon prepared on an industrial scale and used as an appreciated medicine against bacterial infections. Later on, other antibiotic pharmaceuticals were developed, e.g. in 1942 terramycin from *Streptomyces rimosus*; in 1943, streptomycin from *Streptomyces griseus*; and in 1947, chloramphenicol from *Streptomyces venezuelae*, etc., all as large-scale produced mycomedicinals prepared from micro-fungi.

Of course when medicinals can be found in micro-fungi, they can surely be found in macro-fungi or mushrooms because they all have a mycelium-thallus on which sporulation occurs directly (micro-fungi) or via a fruit-body (macro- or fleshy fungi or mushroom). It has to be kept in mind that many mushrooms have an asexual microsporulation besides their sexual fruit-body sporulation. So it could be presumed that in the macro-fungi or mushrooms a substantial number of beneficial secondary metabolites would also be detected and developed as commercial pharmaceuticals, moreover since mushrooms have already been used for 2000 years for tea or to make extracts as unpurified medicinal compounds.

The first edition about this exciting subject dates from 1954 when the German Dr Birkfeld published his "Pilze in der Heilkunde", which already included a treasure of curing-methods from micro- and macro-fungi. Indeed, "Pilze" in the German language includes all filamentous fungi with or without fleshy fruit-bodies. Since the edition of Birkfeld, a fast evolution occurred concerning the large-scale preparation of commercial products for the prevention of or recovery from human diseases. This will be detailed in a table that is based on a worldwide review by Professor Dr Wasser dated from 1999 (Birkfeld, 1954; Wasser and Weis, 1999) (section 10.5.3).

10.5.2   The Use of Agro-Forestry Waste Materials for Mushroom Cultivation

Demographic Explosions; A Cry for Food, Health and Medicinal Compounds

The population of our overcrowded world exceeds six billion. Almost 50% of them are poor or hungry. They fight for a parcel of bean soil, coffee field or rice terrace, while in the same village, we smell the burning straw and forest fire or the rotting heaps of organic waste or other agricultural by-products.

Estimating a yearly world harvest of about 6×10^9 kg of mushrooms for six billion humans, the average yearly consumption of 1 kg mushrooms per head is an insufficient weapon against hunger (Chang, 2001a). By a harmonious combination of limiting the world population by family planning and a mushroom-minded use of field- and forestry wastes, starvation, poverty and some diseases could be faced in the near future.

Even in the event that the world population would not further increase, there is an enormous amount of waste from fields, from the agro-industry and the wood industry. By using only 25% of the cereal straw burned annually in the world, this could result in a mushroom yield of 317×10^6 t (317×10^9 kg) of fresh mushrooms per year (Chang and Miles, 1989). At this time, the annual world mushroom production is only 6×10^9 kg for six billion persons or 3 g per day per person (Courvoisier, 1999) (Table 10.1).

In fact, counting the yearly available world waste in agriculture (500×10^9 kg) and forestry (100×10^9 kg), it would be possible to easily grow 360×10^9 kg of fresh mushrooms on the total of 600×10^9 kg of dry waste! This would bring us an annual mushroom food of 60 kg per head per year, containing 4% protein in fresh mushrooms. Furthermore, it is known that 30% of the world population is protein deficient and is lacking essential amino acids vitamins and medicinals. On the other hand, we all know the high risk of a further increase in world population resulting in a greater need for food (field crops) and wood (forest) leading to increasing mountains of wastes.

Numerous Reasons for Growing more Industrial Mushrooms on Waste

There are numerous reasons for growing more industrial mushrooms on waste such as :

- Decreasing the enormous amounts of wastes.
- Converting the waste into mushroom proteins and vitamins, while mushrooms represent one of the world's greatest untapped sources of healthy food in the future, not forgetting the medicinal value of several industrially grown mushroom species.

Table 10.1 World mushroom production in 1986 and 1997 (Sedeyn, Lannoy and Leenkegt, 2002. Reproduced by permission of Proefcentrum Voor De Champignonteelt.)

Species	Weight $\times 10^3$ t				% growth
	1986		1997		
Agaricus bisporus	1227	56.2%	1956	31.8%	59.4
Lentinula edodes	314	14.4%	1564	25.4%	398.1
Pleurotus spp.	169	7.7%	876	14.2%	418.3
Auricularia spp.	119	5.5%	485	7.9%	307.6
Volvariella volvacea	178	8.2%	181	3.0%	1.7
Flammulina velutipes	100	4.6%	285	4.6%	130.0
Tremella spp.	40	1.8%	130	2.1%	225.0
Hypsizygus spp.	–	–	74	1.2%	–
Pholiota spp.	25	1.1%	56	0.9%	124.0
Grifola frondosa	–	–	33	0.5%	–
Others	10	0.5%	518	8.4%	5080.0
Total	2182	100.0%	6158	100.0%	182.2

- No fight for agricultural land is necessary since mushroom cultivation needs only minor space and can even grow in vertical layers or in forests where they do not need extra space.
- Most mushrooms are very productive and can be grown through the year; in temperate zones with heating in winter and in tropical zones with protection against dry and hot seasons.
- The domesticated mushroom possess a rich gamma of enzymes such as cellulases, ligninases, proteases etc. making it possible to biodegrade and assimilate a multitude of organic waste substrates.
- Finally, the development of mushrooms on waste materials is a duty for all of us for the restoration of the damaged ecosystem and it is a way to close the gap between the monocultures and the original balanced growth-planning of the ecosystem on planet Earth.
- More than 1500 macro-fungi are already widely used for human food although only a few are grown on large scale. Because all these macromycetes are active in the biotransformation of lignocellulosic waste, the domestication of at least a hundred desired wild edible species should be stimulated. Also more than 30 different edible mushroom species are already used in commercial medicinal products; based on Chang and Miles (1989) and Poppe (1995).

Worldwide Survey on Agro-Forestry Wastes for Mushroom Cultivation

In this part, the most important wastes that have been used in the past or more recently for large-scale mushroom cultivation are listed, and the best mushroom species for every type of waste are also discussed. Most wastes have a C:N ratio between 32 and 600 and a pH between 5.0 and 7.5 (based on the work of Poppe (2000) in which more than 200 authors are cited).

- Bark: see treebark
- Barley straw: *Hordeum vulgare*, has a biological efficiency of 96% for *Pleurotus*, also suitable for *Agaricus, Pleurotus, Volvariella* and *Stropharia*.
- Bean pods: a substrate component or bulk for *Pleurotus* (Poppe, 1995).
- Bean straw: different genera, for *Agaricus* and as a substrate component; it can also be used as a basic substrate for *Pleurotus* (Poppe, 1995).
- Carton, shredded: suitable for *Stropharia* with or without pasteurization (Poppe, 1995).
- Cereal straw, Graminae: basic substrate for fermented *Agaricus* compost. The most productive combination was cereal straw with bean or pea straw.
- Chicken manure or poultry manure: if the droppings come on the bed of straw or peat then the concentration is lower and we can classify it as a substrate; if the droppings are pure we can consider it as an additive to straw or horse manure. As a substrate component to be co-fermented with the horse manure or in the synthetic straw compost, it gives only variable results. Subsequently, after pasteurization of the chicken manure, it can be used as a very rich N-component for *Agaricus* substrate.
- Coffee parchment, parche de café; suitable with or without pasteurization for *Pleurotus* (Poppe, 1995).

- Corn cobs: hammermilled or crushed, tested first in Hungary in 1956 and gave variable results for *Agaricus*. Generally used for *Pleurotus* and shiitake. Contains 40% of cellulose, 15% of lignin, 0.4% total N, 0.1% P_2O_5, 0.25% K_2O, 0.5% SiO_2, pH 7, C:N 129.
- Corn stalks (*Zea mays*) chopped as a component of *Agaricus* substrate. Also useful for *Pleurotus*. It contains: 48% of cellulose, 16% of lignin, 0.8% total N, 0.35% P_2O_5, 0.4% K_2O, 1.8% SiO_2, pH 7.2 and C:N 63.
- Cotton waste compost: ready for *Volvariella* spawning – pH 7.5, 0.6% total N, 53% total C, C:N 85, 3.4% ash, 8% of hemicellulose, 73% of cellulose, 6% of lignin and 92% of organic matter (Chang and Miles, 1989).
- Grass chaff: for *Stropharia*, single or in a mixture with straw, high production on the grass chaff from Lolium if pasteurized for 2 min in water of 65 °C.
- Flax straw: *Linum usitatissimum*, single or in combination with flax tow, for *Volvariella*, *Pleurotus* and *Auricularia* (Chang and Hayes, 1978).
- Horse manure: 1.1% N, fermented since more than 200 years ago for *Agaricus* (Delmas, 1989). Horse manure is the oldest and best substrate for the heap fermentation for *Agaricus*: 36% of cellulose, 18% of lignin, 1.1% Total N, 0.85 P_2O_5, 0.7% K_2O, pH 7.5 and C:N 42.
- Lignocellulosic by-products of the sugar industry: bagasse, alone or blended with cotton waste and rice straw give a satisfactory *Volvariella* yield compared to sugar cane leaves substrate.
- Newspapers: shredded, when combined with rice bran or with sawdust for *Pleurotus*. Also useful for *Stropharia*. Oak sawdust, supplemented with 10% millet was the best pasteurized or sterilized substrate for shiitake in Canada.
- Paddy straw, see also rice straw. *Tricholoma crassum*, an excellent edible mushroom, could be very satisfactorily grown on paddy straw alone or in combination with horse manure, chicken manure or brewers grain.
- Pennisetum, elephant grass: component of *Volvariella* substrate.
- Pruning wood, chopped, preferentially from broad-leaved wood, for *Stropharia*.
- Rice straw, *Oryza sativa*, immense masses are burned every year or left rotten in moistened fields, intensively used for *Volvariella* and *Pleurotus*, but also used as a chief component of synthetic *Agaricus* compost. It contains 41% of cellulose, 13% of lignin, 0.8% total N, 0.25% P_2O_5, 0.3% K_2O, 6% SiO_2, pH 6.9, C:N ratio 58. Rice straw crushed, is also used for *Pleurotus* in Asia.
- Sawdust can be used in *Agaricus* compost as a component (more than 5%) or as an additive (less than 5%) with straw, but it can also be used as single substrate for *Pleurotus*, *Auricularia*, *Flammulina*, *Tremella*, *Pholiota* and *Hericium*, mostly sterilized.
- Straw – cereal straw: 0.5% total N, 38% of cellulose, 15% of lignin, C:N 90, basic substrate for nearly all cultivated mushrooms, can be enriched with at least 30 different supplement wastes. Straw is especially useful for *Agaricus*. Compost, chopped for *Pleurotus* and *Stropharia*.
- Sugarcane bagasse: *Saccharum officinarum*, sugar cane rubbish, cane trash, 0.7% N, as bulk ingredient in mushroom compost along with horse manure resulted in normal *Agaricus* yield. Good production was also obtained with *Pleurotus*. For *Pleurotus*, the biological efficiency of the pure bagasse is 15%. This is relatively low compared to many other substrates.

- Sunflower husks or Sunflower peels: the sunflower seeds are peeled before the internal seed part can be pressed for oil. Up to now, all the precious waste was burned, which resulted in millions of kg/yr. Very useful for *Pleurotus* without pasteurization and also moderate production for *Stropharia* in open field (Poppe, 1995). Cultivating *Pleurotus ostreatus* in Yugoslavia using the sunflower husks as a supplement on straw or maize stalk resulted in a 8% higher yield.
- Treebark, chopped: can be used alone or in combination with wheat straw, corn cobs and feather meal for *Pleurotus* or as a fermented bulk substrate for *Agaricus*. The origin of most tree bark is from the cellulose paper manufacturers where trees are debarked before chopping and pulp preparation. It was also used as a substrate for *Pholiota*, *Flammulina* and *Schizophyllum* (Delmas, 1989).
- Wheat straw: *Triticum aestivum*, main basic component of fermented *Agaricus* compost, in different percentages, up to 90%; wheat straw contains 1% of protein, 13% of lignin, 39% of hemicellulose and 40% of cellulose. It was burned in voluminous amounts until 1963 in France. Since then, it can be used for *Pleurotus* with a price of US$0.1–0.2 per kg. Straw for *Pleurotus* is only pasteurized and rarely fermented (Delmas, 1989). Wheat straw can also be used for *Volvariella*.

10.5.3   Most Important Waste-grown Mushrooms and their Medicinal Properties

In Table 10.2, a summary of the most commercially grown mushrooms on waste is given. A few micro-fungi such as *Penicillium* and *Streptomyces* are grown on more purified selective organic media. The macro-fungi or mushrooms can be divided into compost-grown species like *Agaricus* and *Volvariella* and wood-inhabiting species like *Pleurotus*, *Lentinus*, *Fomes*, *Ganoderma*, etc. It is impossible to describe all the discovered medicinal molecules up to now for each mushroom. Hobbs needed 252 pages for his book on medicinal mushrooms (Hobbs, 1996) while Lelley used 236 pages for the description of bioactive compounds in mushrooms (Lelley, 1997). We refer here to the famous bilateral table of Wasser and Weis (1999) who compiled a vertical list of 34 mushroom species and a horizontal list of 15 important human diseases. The commercialized mycomedicinals (indicated by x) are clearly separated from the preparates, still being in an experimental stage (indicated by +). Briefly summarized, it may be underlined that a considerable stream of new medicinals can be expected in Europe especially for cancer, immunity and soft antibiotica (Table 10.2). This indeed proves that the production of mycomedicinal compounds from waste products is a high value-added agro-industrial business that will continue its development in the future. Also the development of pharmaceuticals can be associated with the production of protein-rich food.

10.5.4   Some Mycotherapeutic Molecules and their Activities

Antibiotica

Without antibacterial molecules, the mycelium of macro- and micro-fungi is unable to survive the intensive competition with numerous microorganisms always present in all non-sterile organic substrates. Most of the so-called "antibiotica" are active against

Table 10.2 Cross index of the most important mushrooms and their medicinal activity (based on Wasser and Weis 1999, *Medicinal properties of substances occurring in higher Basidiomycetes mushrooms, current perspectives.* Int. J. Med. Mush., **1**, 31–62, updated by Poppe 2003; published by Begell House)

Species	Antifungus	Anti-inflammatory	Antitumour	Antivirus	Antibacteria	Bloodpressure	Cardiovascular	Anticholesterol	Antidiabetes	Immunomodulating	Kidney tonic	Hepatoprotective	Nerve tonic	Sexual potention	Anti-bronchitis
Agaricus bisporus Pilat, button mushroom	–	–	+	–	–	–	×	–	+	×	×	–	–	–	–
Agaricus blazei Murr., sun agaric	+	–	×	–	–	–	–	+	+	×	+	+	–	–	–
Agrocybe aegerita Sing., poplar mushroom	+	–	+	–	–	–	–	+	–	–	–	–	×	–	–
Armillaria mellea Karst., honey mushroom	–	–	–	–	–	×	×	–	–	–	–	×	–	–	×
Auricularia auricula-judae Bull, judahs ear	–	–	+	×	×	+	×	×	–	–	–	×	×	–	×
Cordyceps sinensis Sacc., dear fungus	–	–	×	+	×	×	×	×	–	×	×	–	–	–	–
Coriolus versicolor Quél, turkey tail	+	–	×	–	–	–	×	–	–	×	–	–	–	–	–
Flammulina velutipes Karst, velvet stem	–	×	×	–	+	–	–	–	×	×	–	+	–	–	–
Fomes fomentarius Fr., amadou	–	+	+	–	+	–	–	–	–	+	–	–	–	–	–
Fomitopsis pinicola Karst., quinine conk	–	–	+	–	+	–	–	–	–	–	–	–	–	–	–
Ganoderma applanatum Pat., artist's conk	–	–	+	+	×	–	–	×	–	+	–	×	–	–	×
Ganoderma lucidum Karst., reishi	+	×	×	×	–	×	×	–	×	×	×	+	×	–	+
Grifola frondosa Gray, maitake	–	–	×	×	–	×	–	–	–	×	×	×	–	–	×
Hericium erinaceus Pers., hedgehog	–	–	+	–	–	–	–	–	–	×	–	–	×	–	–
Hypsizygus marmoreus Bigel., shimeji	–	–	×	–	–	–	–	–	–	–	–	–	–	–	–
Inonotus obliquus Pilat, birch inonotus	–	×	×	–	–	–	–	–	–	×	×	×	–	–	–
Laetiporus sulphureus Murr., sulphur fungus	+	–	+	–	–	–	–	–	–	–	–	–	–	–	–

Table 10.2 Continued

	Antifungus	Anti-inflammatory	Antitumour	Antivirus	Antibacteria	Bloodpressure	Cardiovascular	Anticholesterol	Antidiabetes	Immunomodulating	Kidney tonic	Hepatoprotective	Nerve tonic	Sexual potention	Anti-bronchitis
Lentinus edodes Sing., shiitake	–	×	×	×	×	×	–	×	×	×	×	×	–	×	–
Lenzites betulina Fr., gilled polypore	–	–	+	–	–	–	+	–	–	–	–	–	–	–	–
Marasmius androsaceus Fr., hairy marasm.	–	×	–	–	–	–	–	–	–	–	–	–	×	–	–
Oudemansiella mucida Höhn., beech roter	×	–	–	–	–	–	–	–	–	–	–	–	–	–	–
Penicillium griseofulvum Dierickx	×	–	–	–	–	–	–	–	–	–	–	–	–	–	–
Penicillium notatum Link	–	–	–	–	×	–	–	–	–	–	–	–	–	–	–
Piptoporus betulinus Karst., birch conk	+	–	+	–	+	–	–	–	–	–	–	–	–	–	–
Pleurotus ostreatus Kumm., oyster mush.	–	–	+	+	+	–	–	×	–	–	–	–	–	–	–
Pleurotus pulmonarius Quél., oyster mush.	+	–	+	–	–	–	–	+	–	–	–	–	–	–	–
Polyporus umbellatus Pers., umbel polypor.	–	×	×	+	×	–	–	–	–	×	×	×	×	–	×
Schizophyllum commune Fr., split gill	–	–	×	–	×	–	–	–	–	×	–	×	–	–	–
Streptomyces griseus Waksman	–	–	–	–	×	–	–	–	–	–	–	–	–	–	–
Streptomyces venezuelae Hen.	–	–	–	–	–	–	–	–	–	–	–	–	–	–	–
Tremella fuciformis Berk., white fungus	–	+	+	–	–	+	–	+	+	+	–	+	–	–	×
Tremella mesenterica Retz., white butter	–	–	–	–	–	–	–	–	–	–	–	–	–	–	+
Volvariella volvacea Sing., rice straw mush.	–	–	+	+	+	–	–	+	–	–	–	–	–	–	–

Legend   × = commercial; + = not yet commercial; – = no data.

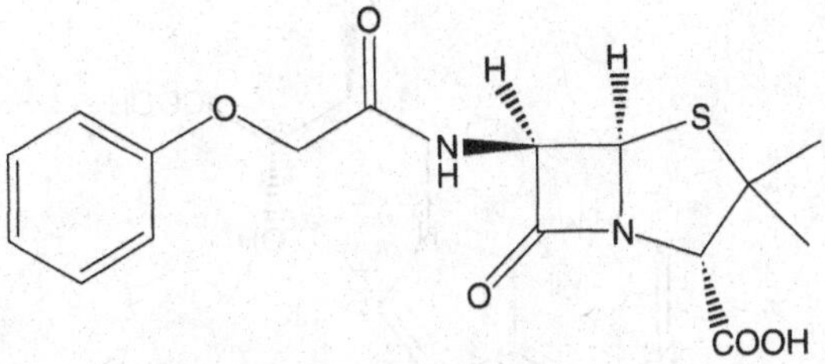

Penicillin V

Figure 10.11   *Structural formula of penicillin V*

Gram-positive as well as Gram-negative bacteria and even against some fungi. The best known antibiotics are penicillin from *Penicillium* (Figure 10.11), streptomycin from *Streptomyces*, cinnabarin, ganoderic acid and triterpenes from *Ganoderma*, and pleurotin and pleuromutilin from *Pleurotus*, etc. Antibiotics from macro-fungi can be used against microbes that are becoming resistant against antibiotics from micro-fungi (Odura *et al.*, 1976; Wasser and Weis, 1999).

Antidiabetica

Extracts from button mushroom *Agaricus* and the dear fungus *Cordyceps* showed active sugar-lowering effects against diabetes mellitus, probably by a hypoglycemic poly-saccharide in such mushrooms. Exopolymers from shiitake *Lentinus edodes* showed also a hypoglycemic effect and are already commercialized. From reishi *Ganoderma lucidum* several exopolymers called ganoderan A, B and C were isolated showing an active hypoglycemic effect by reducing the glucose level and enhancing the plasma-insuline level.

β-D-Glucan

The best immunomodulating polysaccharide is composed of 1,3- and 1,6-betaglucans. In 1969 the Japanese doctor Chihara discovered this antitumour molecule in the shiitake *L. edodes* and called it "lentinan". The same glucans were later also detected in maitake *Grifola frondosa* and in the sponge mushroom *Sparrassis crispa* and later in some other mushrooms. The molecular weight of the β-D-glucan is 1200 kDa. Lentinan is binding to specific proteins in order to strengthen the T-cells, macrophages and natural killer cells. They also enhance the production of cytokines, interleukines and interferons, which stimulate the antibody-production. It was pointed out that already a great part of the immuno-modulating and tumour inhibiting effect happens in the human intestine and abdomen.

Chitin

The most important fibre component in shiitake *L. edodes* and other mushrooms. In the small intestine this large molecule reduces the re-uptake of the secreted bile-cholesterol by binding it and exhausting it via the stools (Flynn, 1991). The total chitin-mass in

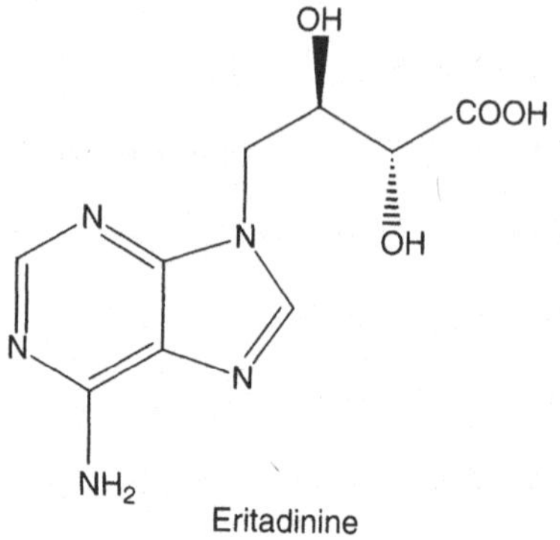

Figure 10.12   *Structured formula of eritadinine*

fungi is at least a hundred times higher compared to Crustacea, because it is a building-stone of almost all fungi. It is a strong absorber for intestinal toxins and heavy metals. It is applied as a carrier of medicaments, an artificial skin or wound cover, for chronic kidney insufficiency, stomach pains and eye problems (Burdiukova and Gorovoj, 2001).

Eritadinine

A hypolipidemicum with strong cholesterol-lowering (up to 25%) activity; especially by binding cholesterol followed by excretion. Eritadinine behaves as a phytosterol. Recently, it was proved that eritadinine also has an antithrombose-effect (Lelley, 1997; Wasser and Weis, 1999) (Figure 10.12).

Ganoderic Acids

These could be isolated from reishi *G. lucidum*; these are lanosta-derivates, or triterpenes useful as antihistaminicum and antihypertensicum, and due to their cytotoxic activity against strange cells, they also act as antitumour (Lelley, 1997; Wasser, 1999; Chang and Mshigeni, 2001b) (Figure 10.13).

Krestin

A high molecular protein-bound polysaccharide containing 10% peptide and 90% β-glucan. Krestin is isolated from turkey tail *Coriolus versicolor* and is prescribed in Japan against breast cancer, intestine cancer and lung cancer. The oral medicament already has a turn-over of more than €550 million in 1990 (only in Japan). Other activities of krestin are immunomodulation, pro-interferon, antiviral, and antimetastase. The authors claim that krestin performs comparably as an antitumour to the classical western chemotherapy. In the Far East, breast and prostate cancer levels are ten times lower than in the rest of the world, which is also due to the population's regular diet of soja beans being so rich in isoflavones (Lelley, 1997; Miles and Chang, 1997).

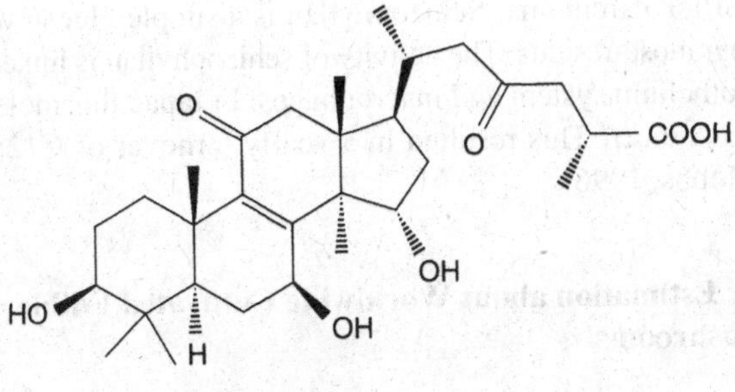

Ganoderic acid

Figure 10.13 *Structural formula of ganoderic acid*

Lentinan

This historical supermolecule extracted from the shiitake *L. edodes* is one of the most described and best acting polysaccharides of the mushroom world. The Japanese Prof. Dr. Goro Chihara was the first who detected, and published as early as 1969, the antitumour and immunomodulating activities of this shiitake mushroom (Chihara *et al.*, 1969). Chihara described lentinan as a β-1,3-glucan showing strong activation of thymus-lymphocytes (T-cells), with an activity against sarcoma 180, fibrosarcoma and against neoplasms (Chihara, 1978). According to Flynn (1991): "Lentinan stimulates T-cells, T-helper cells, cytotoxic macrophages; interleukine 1, 2 and 3; interferon and antibody formation are enhanced; this means a wide immunising activity." Lentinan also acts as an antiviral molecule by interferon-enhancement in the body cells so that virus replication is inhibited; the anti-tumour activity of lentinan is fortified by the immune messenger interleukine. Lentinan is officially recognized as an injection preparate against intestinal cancer in Japan, with a yearly turnover of €85 million.

Lovastatine

The best-known producer of lovastatine is the genus *Pleurotus*. It was proved that lovastatin was most concentrated in the gills of the *Pleurotus* fruit-body. It is the fibre pulp complex of the oyster mushroom which limits the resorption of cholesterol in the gastro-intestinal tract so that up to 75% of the cholesterol can be removed. Besides this, some authors also report a 30% anti-atherogenic effect due to the hypolipidemic activity of *Pleurotus* as a food supplementation (Wasser and Weis, 1999; Gunde-Cimerman and Plementas, 2001).

Schizophyllan

β-1,6-Glucan, a very viscous polysaccharide isolated from the splitgil *Schizophyllum commune*; antitumour activity against subcutaneous mice-tumours namely sarcoma 180, sarcoma 37 and Ehrlich carcinoma. Schizophyllan is a simple glucan with repeated units of several β-glucopyranose-residus. The activity of schizophyllan is linked to its stimulation of the reticulo-endothelium system and macrophages. In Japan this molecule is prescribed for injection against cancer. This resulted in a yearly turnover of €128 million by 1990 (Komatsu, 1974; Hobbs, 1996).

10.5.5   Economic Estimation about Worldwide Cultivated Edible and Medicinal Mushrooms

Yearly Costs for Wastes

As stated earlier, about 6×10^9 kg waste is used for 6×10^9 kg mushrooms. Except for cereal straw, the only cost is the transport of the waste from the waste location to the mushroom production farm (€±0.08). The straw used to grow mushrooms costs an average of €0.1/kg and is applied for 50% of the total world production of mushrooms. The other 3×10^9 kg of waste is free so that we have to count the transport of €0.24 billion. Subtotal for wastes

   €0.30 billion for straw
   €0.24 billion for straw transport
   €0.24 billion for other wastes transport
   €0.74 billion as total cost for 6×10^9 kg wastes

Costs and Benefits in Mushroom Production

It is impossible to know the exact production cost in all temperate and tropical climate zones of the world, but we can estimate a worldwide average production cost of €1/kg. The mushrooms can be sold at €±2/kg, depending on the species. So, the subtotal for production costs equals €6 billion and the market value is equal to €12 billion.

Special Case of Mushrooms Used for Pharmaceutical Preparations

We may estimate at least 2000 commercial producers of mushroom medicinals in Asia, 200 in the USA and 50 in Europe. The part of the mushrooms for pharmaceutical use can be estimated at 1 pro mille or 6×10^6 kg. From this basic material, an expensive protocol is used to produce well-dosed pills or vials for injection. The cost of the production of the medicine is difficult to estimate but some turnover figures are published (Miles and Chang, 1997). From these data, a considerable benefit is achieved, starting from waste via mushrooms and the preparation of medicinals. Indeed, considering only the anti-cancer mycoproducts extracted from turkey tail *Coriolus versicolor* (krestin), from shiitake *L. edodes* (lentinan) and from split gill *S. commune* (schizophyllan), a total turnover of €769 million in 1990 was generated (Table 10.3).

Concerning the medicinal mushroom products in China, in 2002 already 700 commercial health products were on the market. These include about 25% of all Chinese medicaments. A large part of the products are derived from *Ganoderma*, *Cordyceps* and shiitake.

Table 10.3 *Pharmaceuticals developed from mushrooms in Japan (Miles and Chang, 1997. Reproduced by permission of World Scientific Publishing Company)*

Name	Krestin	Lentinan	Schizophyllan
Abbreviation	PSK/PSP		
Mushroom species	*Coriolus versicolor* (Fr.) Quel.(mycelium)	*Lentinula edodes* (Berk.) Pegler (fruiting bodies)	*Schizophyllum commune* Fr. (mycelium)
Polysaccharide	Glucan with β-1,6, β-1,3 branching; β-1,4 main chain	Glucan with β-1,6 branching; β-1,3 main chain	Glucan with β-1,6 branching; β-1,3 main chain
Molecular weight	ca. 100 000	ca. 500 000	ca. 450 000
Specific rotation $[\alpha]_d$	–	+14 at 22 °C in NaOH	+18 at 24 °C in H_2O
Administration	Oral	By injection	By injection
Indication	Cancer of digestive system, breast cancer, pulmonary cancer	Gastric cancer	Cervical cancer
1990 sale value	US$556 million	US$85 million	US$128 million

Almost half of the products are officially registered and approved by the Chinese Ministry. The total Chinese yearly turnover of mycomedicinals is €2.8 billion from which 564 million was for export.

10.5.6 Conclusions

A Non-green Revolution for Human Welfare ("Non-green" Because Mushrooms Lack Chlorophyll)

Chang published in 2001 a broad article entitled "A 40 year journey through bioconversion of lignocellulosic wastes to mushrooms dietary supplements" in the unique international *Journal on Medicinal Mushrooms*. The non-stop mushroom cultivation on field- and forestry wastes results in a non-green revolution including efficient mycomedicinals. Indeed, from a mushroom production of 170×10^6 kg in 1960, we climbed up to 1257×10^6 kg in 1981, and to 6158×10^6 kg in 1997. The use of wastes for this recent mushroom production is more than 6×10^9 kg of which 80% comes from straw from wheat, rice and corn. Counting about 20 000 kinds of natural mushrooms, only 2000 have been grown as mycelium or fruit-bodies. About 300 species proved to have medicinal properties while approximately 1700 others are considered to contain potential medicinal molecules. Of course some of the medicinal mushrooms were and are still directly picked from nature.

The most important mycomedicinal molecules are found in the classes of polysaccharides, triterpenes, antibiotics, anticholesterolics, antidiabetics and melanins. All mycomedicinals together are estimated by Chang to be increasing in value year-on-year (1991: €1.2 billion, 1994: €3.6 billion and 1999: €5.9 billion). The author underlines that more than 25% of the total yearly turnover is realised by *Ganoderma* alone! Finally while up to now, almost 80% of the mycomedicinals originated in the Far East, we may expect very soon also in Europe a real explosion of new mycomedicinals especially developed for the better prevention of and recovery from cancer, viral diseases, and heart problems.

10.6   Natural Dyes

10.6.1   Natural Colourants: Their Sources and Isolation Methods

Natural colourants can be prepared by extracting vegetable raw materials and can be used in food, cosmetics and pharmaceuticals. The use of colourants in food in the European Community (EC) is controlled by EC Directive 2645/62 (as amended) and in the USA by the Colour Additives Amendments of 1960 (Public Law 86–618) to the food Drug and Cosmetic Act 1938. These statutes make no provision for the terms "natural", "artificial" and "nature identical" although these descriptions are finding increasing use in the labelling and advertising of foods (Francis, 1986).

Manufacturers of natural colourants have concentrated their efforts on maximizing the yields from conventional sources. Extraction methods usually involve disruption of the plant material followed by acidified aqueous extraction for anthocyanins, beetroot red and cochineal or by solvent extraction for chlorophyll, carotenoids, annatto and spice extracts. Stabilizers such as citric acid, ascorbic acid and antioxidants help to reduce pigment losses during the extraction.

Selection of a suitable cultivar of the plant is a cost-effective way of improving the pigment yield. Plant tissue cultures will, in the future, assist suppliers in selecting highly pigmented cultivars. Enzymes offer a means of enhancing natural colourant extraction but their use has been limited, presumably by the low value of pigments. Pectinase and cellulases have, however, been used to produce a high-carotene content in carrot juice. Bromelain and β-glucosidase, too, have been used in extracting pigments from gardenia fruits, while protease treatment of *Dactylopius coccus* produces a better quality of cochineal.

Natural colourant extracts, in particular those produced by aqueous extraction from or direct pressing of plant materials, contain a high proportion of other dissolved solids including salts, organic acids, phenolic compounds and sugars. Drying these high-solid extracts produces hygroscopic powders, which have certain drawbacks. In the case of beetroot, the powder has a characteristic odour (geosmin), a high nitrate and nitrite content and therefore a limited application in foods. In anthocyanin extracts, coextracted sugars may degrade to furfural and 5-hydroxymethylfurfural during dehydration and contribute to accelerated pigment degradation. The problem can be minimized by treating the extracts with microorganisms.

Plant cell culture in fermentors has potential for the production of secondary metabolites such as pharmaceuticals, perfumes, flavours and colourings. Commercial success has been achieved with shikonin, a red naphthoquinone dye produced in Japan and used to promote healing. The potential of plant cell cultures has resulted in the development of techniques ranging from relatively low-yielding heterogeneous cell suspension, through immobilization techniques and the use of "hairy root" cultures.

Anthocyanins

Anthocyanins (E163) are a chemical class of red to blue pigments that are commonly found in mature fruit (e.g. strawberries, blueberries, cherries, grapes), vegetables (e.g. onions, cabbages), seeds (e.g. purple sunflower) and flowers.

A red anthocyanin colourant can be made more readily from grape pomace by-products, primarily the skin remaining from the manufacture of wine. The procedures involve grinding the pomace or marc and then extracting with water, usually in the presence of sulphite or acids. Enzymes such as pectinase and others may be added to make the pigment extraction more efficient. The mixture is filtered, chilled to remove tartrates and acidified, followed by distillation to lower the SO_2 content. The extract is concentrated and used directly, or after purification with ion-exchange resins. The extract is placed on an affinity ion-exchange column to which the pigment adheres, whereas impurities are removed with the effluent. The column is then eluted with an ethanol/water mixture to obtain a colourant solution with an intense purplish-red colour. Alcohol may be used as well instead of water for the initial extraction, but it requires another processing step to remove the alcohol. An anthocyanin concentrate can be prepared from grape fermentation marc by ultrafiltration and reverse osmosis.

Membrane processing has found wide application in concentrating and separating food ingredients of a thermally labile nature. Three main overlapping operations can be defined according to the size of the molecules or particles that can be retained and to the operating pressures used:

1. Reverse osmosis 0.0001–0.001 μm (20–60 bar);
2. Ultrafiltration 0.001–0.02 μm (1–10 bar); and
3. Microfiltration 0.02–10 μm (1–10 bar).

For natural colouring production, membrane processing offers a number of benefits including concentration of pigments from dilute juices, removal of microorganisms from fermented or infected juices and removal of cell debris from plant extracts. Ultrafiltration of anthocyanin containing extracts using cellulose acetate membranes with molecular weight cut-off of 1000 and 500 Da respectively has been shown to remove 75–90% of water and 50–60% of sugars in a single pass. This represents a twofold increase in pigment content and thus a viable extraction method. The aqueous liquid of colourant concentrate is dried to produce a dry red colourant powder by spray-drying methods.

Anthocyanin-type colourants can also be extracted from other different natural products, such as purple corn, elderberries, black mulberries, chokeberry, black currants, hollyhock flowers, morning glory flowers, sorghum grain, sunflower seeds, beans, cabbage and so forth. A red anthocyanin pigment can be produced by aerobic culture of callus tissue of grapes or other plants. The cultured cells are extracted and purified by different methods.

Betalains

Betanin (E162) is the predominant pigment in red beets (beetroots). Betalaines are the red-purple pigments of beetroot, *Beta vulgaris*. They are water soluble and exist as internal salts (zwitterions) in vacuoles of plant cells. Betalaines are made up of red beta-cyanins and yellow betaxanthins. Betacyanins comprise about 90% of beetroot betalains. The extraction methods of betalains from red beets involve grinding, pressing to express the juice, acidification, ultrafiltration, concentration and drying. A fermentation step to lower the sugar content may be included.

Approximately 80% of beetroot juice solids are carbohydrate and nitrogenous compounds. Fermentation of this juice using *Candida utilis* under partial anaerobic conditions reduces the solids substantially, giving five- to sevenfold increase in pigment content on a dry-weight basis. In addition, the dried product does not contain the characteristic beetroot odour and has a reduced nitrate content. Yeast cells, produced during the fermentation, which takes about 7 h at 30 °C, may be harvested by centrifugation and used as a source of biomass for animal feed supplement.

Ripe fruits of the red-ink weed pokeberries (*Phytolacca*) also contain betalains, which are also known to contain the toxic saponin, phytolaccanin; a hexane solvent step is useful to lower the saponin content. Pigments are extracted by heating with water. The liquid is removed by pressing and then filtered and extracted with hexane or ether. The aqueous liquid is concentrated under vacuum to a dry solid content of 40–50%. Betalains are very unstable pigments; the stabilization procedures involve the addition of a variety of additives such as starch, maltodextrin, pectic acid, carmel, gum Arabic, isoascorbic acid, citric acid, sodium chloride, EDTA and phosphate salts.

Carotenoids

Many raw materials are known as sources for carotenoid pigments production. The carotenoid pigments can be prepared from citrus, tomatoes and carrots. The main carotenoids in citrus are β-citraurin and cryptoxanthin, lycopene is the major pigment in red tomatoes. The major carotenoid in carrots is β-carotene (80%) and α-carotene (20%). Raw material from all three sources can be treated with a solvent such as hexane or alcohol to yield a suspension of carotenoid pigments in oil. This may be used directly as a colourant. Red peppers (paprika) can also be used for carotenoid pigment preparation by alcohol or oil extraction.

Annatto is a natural pigment derived from the seed of the tropical bush *Bixa orellana*. The major colour present is *cis*-bixin, the monomethyl ester of the diapocarotenoic acid norbixin, which is found as a resinous coating surrounding the seed itself. Also present, as minor constituents, are *trans*-bixin and *cis*-norbixin. The annatto bush is native to Central and South America where its seeds are used as a spice in traditional cooking.

Gardenia fruits contain the carotenoids crocetin and crocin. Crocetin is a carotenoid with two carboxylic acid residues. Crocin is the digentiobioside ester of crocetin. The same pigments also occur in saffron from the stigmas of *Crocus sativus*. The colourant annatto is produced by abrasion of the seeds with continuous aqueous counter-current extraction, followed by concentration and drying of the liquid. The colourant from gardenia is produced by aqueous extraction, solvent or resin purification and drying.

Natural renewable sources of carotenoid pigments also include krill, yeast, algae, such as *Chlorella pyrenoidosa* and corn. The colourants from these raw materials can be prepared by different methods of extraction and purification. One way of reducing the content of co-extractants during pigment purification is the use of supercritical carbon dioxide as an extracting solvent. Some success has been obtained in extracting carotene and bixin from natural sources.

To make carotenoids commercially useable in foods, it was necessary to formulate them to provide water dispersibility and stability as well as convenience and stability for fat-soluble forms. Natural preparation containing carotenoids is mixed with a protective film-forming colloid and dried by spray drying, fluid bed, roller, thin layer evaporator or freeze drying. The colloid can be gelatine, gum arabic, dextrin or pectin.

Chlorophylls

The chlorophylls of green leaves are composed of four pyrrole rings with magnesium in the centre of the molecule and a phytoene side chain. Removal of the magnesium yields pheophytin, whereas removal of the phytoene side chain yields chlorophyllin. Removal of both magnesium and phytoene yields pheophorbide. Chlorophyll and chlorophyllin (E140 and E141): chlorophyll is the natural green pigment participating in the photosynthetic process (1×10^9 t are naturally degraded each year); chlorophyllin is a water-soluble derivative. The green leaves of barley and wheat, grass leaves, green vegetable and algae can be used as raw material for green pigment production.

Chlorophyll colourant can be prepared by extracting raw vegetable material with an organic solvent saponified with an alkali hydroxide, acidified, stabilized, converted to a water-soluble form and dried. Extraction is often conducted with an organic solvent such as alcohol, acetone and trichloroethylene. Green iron chlorophyllin compounds are produced by treatment of phytochlorine with iron in the presence of a reductant. The bright green water-soluble alkali metal salts of iron chlorophyll are prepared by reacting iron chlorophyll with alkali metal salts in alcoholic media.

Flavonoids and Polyphenols

Flavonoids are water-soluble yellow colourants produced from different plants such as *Coreopsis tinctoria*, *Rubiaceae*, *Gardenia*, *Caesalpina crista*, *Plantago asiatica*, *Geranium nepalense* shea nuts and pollen. Polyphenol dark colourant is produced from cocoa bean shells, husks and stalks. They are also extracted from other sources such as tea, persimmons and tamarinds. The flavonoid and polyphenol colourants can be obtained from a raw plant material by grinding, multiple extraction with an organic solvent such as alcohol or acetone and hot water after concentration in vacuum. The extracts can be purified, concentrated and used directly as colourants.

Other Novel Sources

The red microalgae constitute a division of the eukaryotic algae that are characterized by their chlorophyll (a pigment), the number of carotenoids, as well as the phycocyanins and phycoerythrins (photosynthetic accessory pigments, collectively known as phycobiliproteins which are red or blue). These algal pigments have potential as natural colourants for use in food, cosmetics and pharmaceuticals, particularly as substitutes for synthetic dyes.

Phycobiliproteins

These are red or blue pigments that are characteristic of three types of algae: the *Rhodophyta*, the *Cyanophyta* and the *Cryptophyta*. They are built up of bilins, which are open-chain tetrapyrroles, covalently linked via one or two thioether links to the cysteine residues in the apoproteins. Phycobiliproteins absorb light in the visible region of 450–650 nm. The pigments are soluble in water; – a convenient way of supplying the water-soluble dyes for use in the food and cosmetics industries would be as concentrated solutions. The concentration of the dye required to colour food or cosmetic products is 0.1%.

Solutions of phycobiliproteins were found to be stable at pH values in the range of 5–9. At lower pH values the phycobiliproteins precipitate. The phycobiliproteins are, generally, sensitive to high temperatures, although *c*-phycoerythrin and *c*-phycocyanin from thermophilic blue-green algae are reasonably heat-stable. Phycoerythrin in nori, a dried red macroalga of the genus Porphyra, is more stable during storage than phycocyanin, chlorophyll and carotenoids. Even at the higher water activity of 0.6, phycoerythrin is stable during storage for six months; phycocyanin is less stable, and chlorophyll and carotenoids can be stored only in a dry environment ($a_w = 0$).

Vitisins

Anthocyanins are a widespread source of naturally occurring colourants of foods. Their use as an added colour to foods and drinks has been limited due to a number of drawbacks, such as sensitivity to bleaching by sulphur dioxide and limited colouring capability at pH values above 3.5. Vitisin A, vitisin B, vitisidin A and vitisidin B are four new anthocyanin pigments found in both red table wines and fortified red wines in trace amounts and are thought to be formed in wines during maturation. All four anthocyanins have their four positions substituted, and it is anticipated that this substitution affects the properties of these anthocyanins in solutions. Substitution at the C-2 position gives greater resistance to colour loss with sulphur dioxide, greater colour expression at higher pH values and increased stability.

Carminic Acid

Cochineal and cochineal carmine, the red colouring substance from coccid insects, contains carminic acid (7-D-glucopyranosyl-3,5,6,8-tetrahydroxy-1-methyl-9,10-dioxoanthracene-2-carboxylic acid) which has a pale yellow colour but complexes readily with tin, aluminium and other metals into brilliant red pigments, as does the structurally related dye alizarin; the complex is relatively stable under the conditions of food processing but it is easily bleached by sulphur dioxide.

Curcumin

This is the principal colour present in the rhizome of the turmeric plant (*Curcuma longa*). Turmeric has been used as a spice for many thousands of years and is today still one of the principal ingredients of curry powder. Turmeric is cultivated in many tropical countries including India, China and Pakistan and is usually marketed as the dried rhizome, which is subsequently milled to a fine powder. This imparts both flavour and colour to a food product.

10.6.2   Properties of Natural Colourants and their Stabilization

Anthocyanins

Anthocyanins are widely distributed in plants, and are responsible for the pink, red, purple and blue hues seen in many flowers, fruits and vegetables. They are water-soluble flavonoid derivatives, which can be glycosylated and acylated.

The aglycone is referred to as an anthocyanidin. There are six commonly occurring anthocyanidin structures. However, anthocyanidins are rarely found in plants – rather they are almost always found as the more stable glycosylated derivatives, referred to as anthocyanins. Sugars are present most commonly at the C-3 position, while a second site for glycosylation is the C-5 position and, more rarely, the C-7 position. The sugars that are present include glucose, galactose, rhamnose and arabinose. The sugars provide additional sites for modification as they may be acylated with acids such as *p*-coumaric, caffeic, ferulic, sinapic, acetic, malonic or *p*-hydroxybenzoic acid. Because of the diversity of glycosylation and acylation, there are at least 300 naturally occurring anthocyanins.

Recently, there has been interest in anthocyanins, not only for their colour properties, but also for their activity as antioxidants. The colour and stability of an anthocyanin in solution is highly dependent on the pH. They are most stable and most highly coloured at low pH values and gradually lose colour as the pH is increased. At around pH 4–5, the anthocyanin is almost colourless. This colour loss is reversible, and the red hue will return upon acidification. This characteristic limits the application of anthocyanins as a food colourant to products that have a low pH.

The loss of colour, as pH is increased, can be monitored by measuring the absorption spectrum of the pigment using a spectrophotometer. There is a decrease in the peak at 515 nm as pH is increased, indicating that the red hue is being lost. The loss of red colour with an increase in pH implies that there is an equilibrium between the two forms of anthocyanin. These are the red flavylium cation and the colourless carbinol base. The flavylium cation, as the name implies, has a positive charge associated with it, while the carbinol base is the hydrated form of anthocyanin (Figure 10.14).

The vast majority of the anthocyanin in solution is accounted for by these two species. But there are actually two additional species or forms of anthocyanin in solution – the blue quinoidal base, which is also in a pH-dependent equilibrium with the flavylium cation and the colourless chalcone. There are very small amounts of the quinoidal base or the chalcone present at any pH. As well, at high pH values, there may be irreversible changes brought to the structure of the anthocyanin causing permanent loss of the red hue, even at acidic pH values.

Anything which interrupts the conjugated double-bond system of the anthocyanin causes a loss of colour. The presence of the positive oxonium ion next to the C-2 position makes anthocyanins particularly susceptible to nucleophilic attack by compounds such as sulphur dioxide or hydrogen peroxide. Anthocyanins form a colourless addition complex with sulphur dioxide by forming a flaven-4-sulphonic acid. This addition can be reversed under low pH conditions (pH 1) to yield the coloured anthocyanin. While this loss of colour is not desirable in many foods, sulphur dioxide is used to decolourize maraschino cherries during their processing.

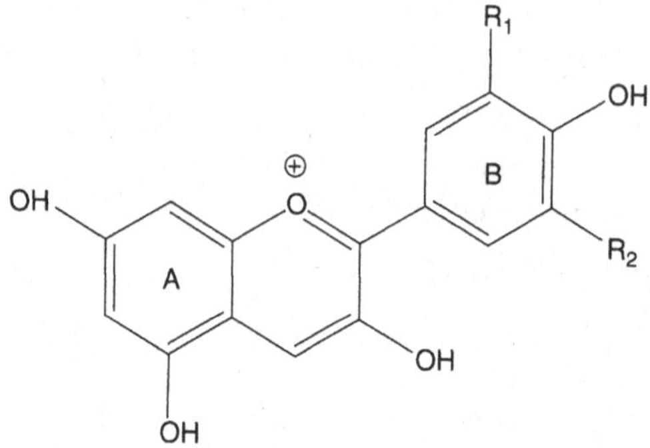

Anthocyanes

$R_1=R_2=H$	Pelargonidin
$R_1=OH; R_2=H$	Cyanidin
$R_1=R_2=OH$	Delphinidin
$R_1=OCH_3; R_2=H$	Peonidin
$R_1=OCH_3; R_2=OH$	Petunidin
$R_1=R_2=OCH_3$	Malvidin

Figure 10.14 *Structural formula of the anthocyanins*

Sulphur dioxide is commonly used as a food preservative. It is very effective in inhibiting the growth of some yeasts and in preventing enzymatic browning reaction catalyzed by polyphenol oxidase. Although still widely used during food processing, the use of sulphur dioxide is restricted, as it may cause serious reactions in sensitive individuals such as those suffering from asthma. It is illegal to use sulphur dioxide on fresh fruits and vegetables. Oxidizing agents such as hydrogen peroxide can effectively decolourize anthocyanins, causing ring cleavage at the C-2 and C-3 positions to form an *o*-benzoyloxyphenyl acetic acid ester under acidic conditions. One possible source of hydrogen peroxide is the oxidation of ascorbic acid.

Some metals such as Fe^{3+} and Al^{3+} form deeply coloured coordination complexes with anthocyanins that have *ortho*-dihydroxy groups on the B-ring. Such metalo-anthocyanin complexes have been found to produce discolouration in some canned fruit products, including pears and peaches. Anthocyanins are sensitive to thermal processes, yielding a loss in the desirable hue and an increase in a brown hue as the pigment degrades and polymerizes. The extent of colour loss at any temperature is dependent on the pH of the environment surrounding the anthocyanin.

Summary of anthocyanins

- Responsible for the red to blue colours in most plants (>300 identified).
- Water soluble, but better solubility in alcohols.
- Always present as glycosides.
- Reversible structural transformation with change in pH.

Degradation during processing and storage:

1. Hydrolysis of the glycosidic substitute – the aglycon is unstable.
2. Enzymatic hydrolysis by glycosidases.
3. Alkaline saponifies acyl sugar ester.
4. Metal ion complexes and colour change.
5. Condensation with other phenolics to form polymeric tannin pigments – wine ageing and insolubilization.
6. Oxidation by polyphenoloxidase (PPO).

Betalains

The most important of the betalains is betanin. This makes up 75–95% of the total colouring matter found in the beet. They are all either glycosides or free aglycones of betanidine or its C-15 isomer. Betaxanthines, in beetroot represented by the vulgaxanthines, are characterized by the lack of an aromatic ring system attached to N-1, and the absence of sugar residues. The other coloured compounds are the isobetanin, betanidin, isobetanidin and isobetanin. In addition to these, the sulphate monoesters of betanin and isobetanin are prebetanin and isoprebetanin respectively.

Betalains are fairly stable under food processing conditions, although heating in the presence of air at neutral pH values causes breakdown to brown compounds. The colour of betalanin solutions is stable at low temperature. Half-life is 413.6 min at 25 °C versus 83.5 min at 60 °C.

Ascorbic acid has been shown to protect the red colour even when it is exposed to drastic treatments such as sterilization. Half-life is doubled in 0.1% ascorbic acid. Metals decrease the stability of betalain pigment: the effect of iron was bigger than that of chromium. In short storage tests, minimal colour changes occurred. The betalain pigment has the potential for use in food processing at low temperature, such as in the dairy industry for the production of ice cream and dairy drinks (Figure 10.15).

The betalain pigments are hardly affected by pH changes in the range normally encountered in foodstuffs. They are stable between pH 4.0 and 7.0. Below pH 3.5 the absorption maximum of betanine solutions is 535 nm, between pH 3.5 and 7.0 it is 538 nm and at pH 9.0 it rises to 544 nm. The thermostability in model systems is dependent on the pH and is greatest in the range of pH 4.0 and 5.0. In beet juice or puree the thermostability is greater than that in the model systems. This makes it appear that a protective system is present in the beets. Both air and light have a degrading effect on betanin, which, in addition, is cumulative. The maximum colour intensity of beet red is reached at pH 5.0, and the pigment cannot be used in acidic medium because of its heat sensitivity.

Summary of betalains

- Not as widely distributed.
- Water soluble, charged phenolic glycosides.
- No structural transformation with pH change – red coloured at wide range.

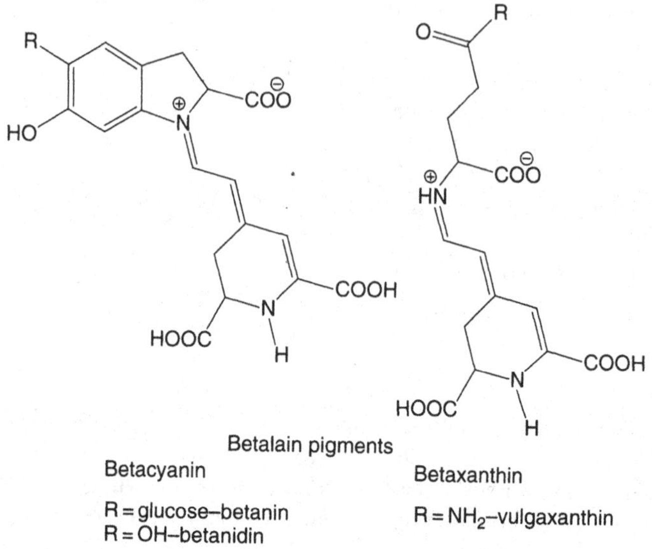

Betalain pigments

Betacyanin

R = glucose–betanin
R = OH–betanidin

Betaxanthin

R = NH₂–vulgaxanthin

Figure 10.15   *Structural formula of betalain pigments*

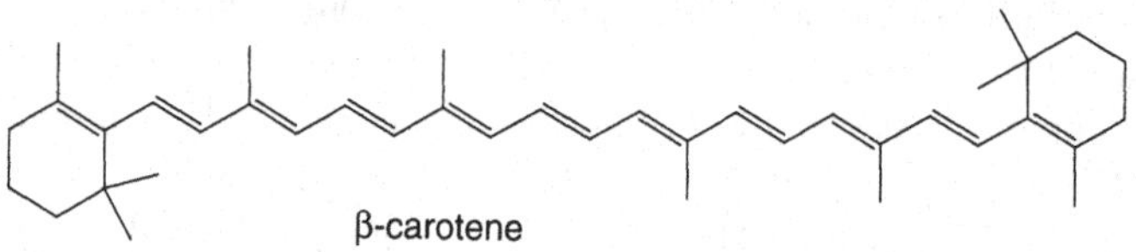

β-carotene

Figure 10.16   *Structural formula of β-carotene*

Carotenoids

Carotenoids are a group of aliphatic fat-soluble compounds widely distributed in plants and are also deposited in the tissue of some animals through feed sources. The word "carotenoid" is derived from the name "carotene" which was given to the yellow pigment of the carrot from which it was first isolated in 1831. The colour range of carotenoids can vary from yellow through orange and red. Just think of the colours found in bananas, carrots, tomatoes, pink grapefruit and watermelon. In concentrated crystalline forms, some may even appear violet or black. In addition, the carotenoid astaxanthin may be associated with a protein, resulting in a blue or green hue, which is found in live lobsters or crabs (Figure 10.16).

Chemically they are terpenoid compounds composed of isoprene units. There are two main groups of carotenoids:

1. The carotenes, which contain only hydrogen and carbon, may be cyclic or linear. Of the carotenes, β-carotene is well known as a vitamin A precursor. Lycopene, one of the red pigments in tomatoes, is a carotene.

2. Oxycarotenoids, contain oxygen in the form of hydroxy, epoxy or oxy groups, as well as hydrogen and carbon. These compounds are also referred to as xanthophylls. Lutein, found in egg yolk, is a modified β-carotene that is di-hydroxylated.

Carotenoids are not soluble in water and have limited solubility in lipids. However, even with limited lipid solubility, they can yield sufficient colour in foods due to their high tinctorial strength. The presence of multiple, unsaturated carbon–carbon double bonds present in carotenoids makes them especially susceptible to oxidation, which can be accelerated by light, metals or hydroperoxides. Oxidation of the double bonds can lead to bleaching of colour, loss of vitamin activity (for β-carotene) and formation of off-odours. Oxidation is desirable in some food processing, since bleaching of carotenoids in wheat flour yields a whiter bread which consumers find attractive. However, oxidation of carotenoids in flour used for pasta is not beneficial, as the golden colour of the pasta is desirable.

The enzyme lipoxygenase (LOX, E.C.1.1.11.13) has been used to bleach flours. Although LOX does not act on carotenoids directly, it produces hydroperoxy fatty acids during its catalytic reaction, which in turn may co-oxidize and decolourize the carotenoids. Carotenoids exhibit good stability to thermal treatments, provided that oxygen and light are not present. However, with heating, the naturally occurring double-bond configuration can rearrange to form the other *conformation*. This causes a slight shift in the absorbance maximum. Carotenoids are not stable under acid conditions, but are more stable under alkaline conditions.

Summary of carotenoids

- Carotenes – hydrocarbons (β-carotene, lycopene); Xanthophylls – oxygenated carotenoids (lutein), acyclic (lycopene) or cyclic (β-carotene) precursor of vitamin A and antioxidant
- Naturally all *trans*, heat and acid induce *cis* (paler colour).
- Oxidative degradation has the greatest impact on carotenoid quality.

Chlorophylls

Chlorophyll occurs in plants in two forms, designated "a" and "b" which have blue-green and yellow-green hues respectively. In plant tissue, chlorophyll is found in association with proteins, which provides an additional complexation site for the magnesium, above or below the plane of the ring. Because of the phytol chain, chlorophyll is lipid soluble, but loss of the chain confers greater water solubility (Figure 10.17).

Chlorophyll is not that important as a food colourant, but is very important as an indicator of food freshness and ripeness. For example, even without any other sensory cues, it is obvious from the colour whether beans or peas have been canned or frozen.

Chlorophylls are relatively unstable pigments, and are affected by both alkaline and acid conditions. Under acidic conditions, the magnesium is replaced by a proton, yielding the pigment pheophytin. Pheophytin is olive green in colour. This reaction occurs readily

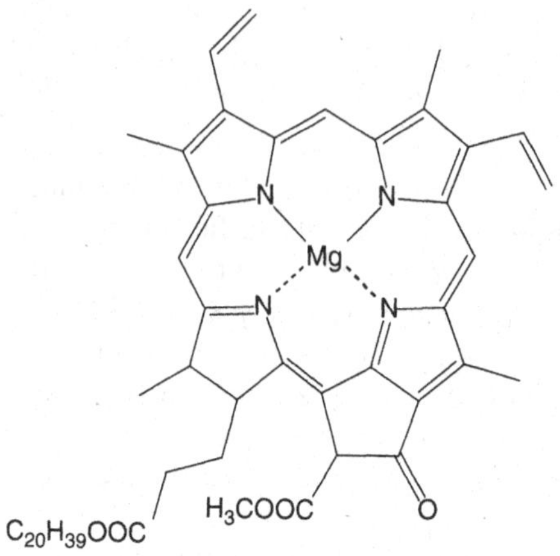

R = CH₃–Chlorophyll A
R = COH–Chlorophyll B

Figure 10.17   *Structural formula of chlorophylls*

when green vegetables are given a thermal treatment. When heated, the cell membranes begin to deteriorate, liberating acids and decreasing the pH in the cell. It is this decrease in pH which results in the transformation of the chlorophyll into pheophytin. If the decrease in pH which normally occurs during heating is prevented, the green colour can be retained. However, under alkaline conditions, the phytol chain can be cleaved, yielding chlorophyllide. Native chlorophyll is soluble in the chloroform layer, chlorophyllide is soluble in the upper water phase. Chlorophyllide, although bright green in colour, is hydrophilic in nature. Therefore, the pigment can be easily lost by blanching or soaking in water. Both pheophytin and chlorophyllide can be converted into pheophorbide, an olive green pigment, which has lost both the magnesium ion and the phytol chain.

While alkaline and acid conditions can cause alterations to the structure of chlorophyll, there are two enzymatic activities, chlorophyllase and magnesium de-chelatase, which act to cleave the phytol chain and remove the magnesium respectively. Replacement of the magnesium with copper yields a bright green pigment that can be used as a food colourant.

Summary of chlorophyll

- Four pyrrole rings (dihydroporphyrin) and a cyclopentanone ring, which are the chromophore responsible for light absorption (10 double bonds), and Mg in the centre.
- Degradation during postharvest storage, with formation of pheophytin (olive green).
- Mild alkaline prevents pheophytin formation – Blair process.
- Cu and Zn replace Mg to form bright green colour – Veri–Green process.
- Chlorophyllase catalyze the removal of phytol chain.

10.6.3   Advantages and Disadvantages of Natural Colourants and Synthetic Dyes

The current consumer preference for naturally derived colourants is associated with their image of being healthy and of good quality. Natural colourants have become increasingly popular with consumers because synthetic colourants tend to be perceived as undesirable and harmful; some are considered to be responsible for allergenic and intolerance reactions (Askar, 1999).

Compared to natural colours, the artificial colours have many advantages including high tinctorial strength, cost efficiency, good thermal, light and chemical stability, consistent quality and ample supply. Therefore, the utilization of artificial colours in the food processing industry is not a problem. The use of natural colourants in the food industry needs careful studies concerning their chemical and physical properties; besides, their stability during processing and storage is essential.

Factors which limit the use of natural colours in the food industry are:

- high cost;
- low tinctorial strength; and
- poor stability during food processing and storage.

In order to have an equivalent colour value, natural colour additives are 30–100 times more expensive than the artificial colours. To provide natural colourants in a convenient, stable form, it is necessary to disperse them in a suitable solvent or dry them with a carrier. Both operations result in a further dilution of the colourant.

The main problem with natural colours is their instability during food processing and storage. Natural colours are affected, more or less, by:

- pH
- oxygen
- temperature
- metal contamination
- other additives.

There has been much interest in the development of new natural colourants for use in the food industry due to strong consumer demand for more natural products (Downham and Collins, 2000). Because of the various drawbacks of the existing commercial natural food colourants (instability against light, heat or adverse pH), the demand for natural colourants with better characteristics is repeatedly raised by the food industry. The diversity of tropical and subtropical vegetation offers a promising range of unknown plant compounds that might prove applicable for the needs of humans. However, current toxicological studies and the huge cost involved in testing limit the introduction of new colourants to our food system.

Artificial colourants, i.e. those commonly referred to as "coal tar dyes" or "food dyes", have been ideal to replace natural colourants which are destroyed during food processing, to reduce colour variations in product batches and to provide consumers' appeal. However, in recent years the number of dyes suitable for food use has been drastically reduced as a result of toxicological studies and, in Europe, by harmonization of European Community legislation. In addition, there has been consumer pressure via major supermarket chains to use more natural food products incorporating natural ingredients.

Natural colourants as a rule are more expensive, less stable and possess lower tinctorial power. In addition, they are frequently present as mixtures in the host materials and vary with season, region, variety, etc. Disadvantages associated with natural colourants have been identified at low yields: colour instability resulting from effects of pH, light, heat and freezing; and possible action with other properties that may be undesirable. However, they do have the advantage of being perceived as safer than synthetic colours.

Since colourants are generally added to food at a very early stage in the processing, they must be stable during heating and cooling, in an acidic medium or in an oxygenated atmosphere. In particular, they must remain stable during storage of the food when they are often exposed to light. Many natural colourants lack this stability so that, despite other advantages, their use is limited. Sulphur dioxide, which is used in a number of foods, can destroy many colours.

Pigments that are to be used as colourants must be extracted from the host material and prepared in some form, such as dried powdered beadlets, which increase its ease of use in a food product. The breakthrough in the use of natural pigments occurred with the synthesis of the carotenoids, β-carotene, β-apo-8′-carotenal and canthaxanthin. A synthetically pure pigment can also be modified or prepared into a suitable market form to enhance its stability and to meet the wide variety of applications required by food manufacturers. One obvious success in this area has been the production of β-carotene forms that are pH stable, resistant to chemical reduction and water dispersible.

Commercial exploitation of these procedures is unlikely to develop further because the yields are usually low, resulting in high-cost additives; in addition, naturally identical colourants have to be confirmed as toxicologically safe before they are adopted by regulatory authorities, and then this occurs only if there is a perceived need.

The natural food colourant industry is experiencing annual growth rates of 5–10% (compared with 3–5% for its synthetic counterpart). The demand for natural colourants is expected to continue. Producers of confectionery, soft drinks, alcoholic beverages, salad dressings and diary products are the most significant users of natural colourants (Wissgott and Bortlik, 1996). The preference for natural colourants worldwide is greater than that for the synthetic ones, as indicated by the patent literature.

10.7   Conclusions

Next to the use of renewable resources for bulk applications in industry, renewable resources offer a great potential to high value applications. In this chapter, three examples for high value applications have been discussed, e.g. the use of perfumes and fragrances, the use of mycomedicinals and the use of natural colourants. In these three applications, there are both advantages and disadvantages to using renewable resources. In most of the applications, the cost price of the interesting components is a major issue because the interesting compounds are formed in the plants or animals together with hundreds of other compounds. The separation of the interesting compound mostly leads to the high price of the compounds. On the other hand, the demand for natural compounds coming from the consumer population will certainly lead to the further development and optimization of several processes so that the actual use of natural components of renewable resources will increase in the future. The genetic modification of microorganisms and plants could certainly play an important role in this development.

References

Askar, A. (1999). Application of natural colorants in the food industry. *Fruit Proces.*, **2**, 42–44.

Bauer, K., Garbe, D. and Surburg, H. (eds) (1990). *Common Fragrance and Flavor Materials: Preparation, Properties and Uses*. Second, revised edition. VCH Publishers, New York, NY (USA), 218pp.

Best, D.J., Floyd, N.C., Magalhaes, A., Burfield, A. and Rhodes, P.M. (1987). Initial enzymatic steps in the degradation of alpha-pinene by *Pseudomonas fluorescens* NCIMB 11671. *Biocatalysis*, **1**, 147–159.

Birkfeld, A. (1954). *Pilze in der Heilkunde*. Ziemsen, Wittenberg, 52pp.

Burdiukova, L. and Gorovoj, L. (2001). Fungal chitin in medicine, prospects for its application. *Int. J. Med. Muh.*, **3**, 126–127.

Chang, S.T. (2001a). A 40 year journey through bioconversion of lignocellulosic wastes to mushrooms and dietary supplements. *Int. J. Med. Mush.*, **3**, 299–310.

Chang, S.T. and Hayes, W.A. (1978). *The Biology and Cultivation of Edible Mushrooms*. Academic Press, New York, London. 819pp.

Chang, S.T. and Miles, P.G. (1989). *Edible Mushrooms and their Cultivation*. CRC Press Florida, 355pp.

Chang, S.T. and Mshigeni, K.E. (2001b). *Mushrooms and Human Health*. University of Namibia, 79pp.

Chappell, J. (1995). Biochemistry and molecular biology of the isoprenoid biosynthetic pathway in plants. *Annu. Rev. Plant Physiol. Plant Pol. Biol.*, **46**, 521–547.

Chihara, G. (1978). Antitumor and immunological properties of polysaccharides from fungal origin. *Mushroom Science*, **10**, 797–784.

Chihara, G. *et al.* (1969). Inhibition of mouse sarcoma 180 by polysaccharides from Lentinus edodes. *Nature*, **222**, 637–648.

Courvoisier, M. (1999). Les champignons comestibles dans le monde. *Bul. Fed. Nat. Syn. Champ*, **82**, 829–837.

Delmas, J. (1989). Les champignons et leur culture. La maison rustique, Flammarion Paris, 970pp.

Demyttenaere, J.C.R. (2001a). Biotransformation of Terpenoids by Microorganisms. In: *Studies in Natural Products Chemistry. Volume 25: Bioactive Natural Products (Part F)* (ed. Atta-ur-Rahman), Elsevier Science Publishers, UK, 125–178.

Demyttenaere, J.C.R., Van Belleghem, K. and De Kimpe, N. (2001b). Biotransformation of (*R*)-(+)- and (*S*)-(−)-limonene by fungi and the use of Solid Phase Microextraction for screening. *Phytochemistry*, **57**, 199–208.

Downham, A. and Collins, P. (2000). Colouring our foods in the last and next millennium. *Int. J. Food Sci. Tech.*, **35**, 5–22.

Flynn, V.T. (1991). Is the shiitake mushroom an aphrodisiac and a cause of longevity? *Science and Cultivation of Edible Fungi* (ed. Maher), Rotterdam: Balkema, 345–361.

Francis, F.J. (1986). *Handbook of Food Colorant Patents*, Food & Nutrition Press, Inc., Westport, Connecticut, pp. 1–150.

Gunde-Cimerman, N. and Plementas, A. (2001). Hypocholesterolic activity of the genus Pleurotus. *Int. J. Med. Mush.*, **3**, 395–397.

Hobbs, C. (1996). *Medicinal mushrooms, an exploration of tradition, healing and culture*. Interweave Press, Loveland, 3rd edition, 252pp.

Haudenschild, C.D. and Croteau, R.B. (1998). Molecular engineering of monoterpene production. *Genet. Eng.*, **20**, 267–280.

Komatsu, N. (1974). Biological activation of schizophyllan. *Mushr. Science*, **9**, 867–869.

Krings, U. and Berger, R.G. (1998). Biotechnological production of flavours and fragrances. *Appl. Microbiol. Biotechnol.*, **49**, 1–8.

Lelley, J. (1997). Die Heilkraft der Pilze, Gesund durch Mycotherapie. *Ekon & List*, 236pp.

Lichtenthaler, H.K., Rohmer, M. and Schwender, J. (1997). Two independent biochemical pathways for isopentenyl diphosphate and isoprenoid biosynthesis in higher plants. *Physiol. Plant.*, **101**, 643–652.

Miles, P.G. and Chang, S.T. (1997). *Mushroom Biology.* World Scientifics, Singapore, London, Hongkong, 194pp.

Odura, K., Munnecke, D., Sims, J. and Keen, N. (1976). Isolation of antibiotics produced in culture of *Armillaria mellea. Trans. Brit. Myc.*, **66**, 195–199.

Poppe, J. (1995). Twenty wastes for twenty cultivated mushrooms. In: *Mushr. Sci.* (ed. Elliot), **14**, 171–180.

Poppe, J. (2000). Use of agricultural waste materials in the cultivation of mushrooms. In: *Mushr. Sci.* (ed. Van Griensven), **15**, 4–23.

Poppe, J. (2003) In: *Proceedings van het vijfole Internationaal Congres over Orthomoleculaire Geneeskunde* (ed. Faché, W.), Gent, 11 Oktober 2003, 53pp.

Sedeyn, P., Lannoy, P. and Leenkegt, L. (2002). Oesterzwamteelt, Provincie West-Vlaanderen, Roeselaere-Rumbeke, 194pp.

Seitz, E.W. (1994). In: *Bioprocess Production of Flavor, Fragrance, and Color Ingredients* (ed. Gabelman, A.), John Wiley & Sons, Inc., 95.

Tan, Q. and Day, D.F. (1998). Bioconversion of limonene to α-terpineol by immobilized *Penicillium digitatum. Appl. Microbiol. Biotechnol.*, **49**, 96.

Wasser, S.P. and Weis, A.L. (1999). Medicinal properties of substances occurring in higher Basidiomycetes mushrooms, current perspectives. *Int. J. Med. Mush.*, **1**, 31–62.

Wissgott, U. and Bortlik, K. (1996). Prospects for new natural food colorants. *Trends Food Sci. Tech.*, **7**, 298–302.

11

Renewable Resources: Back to the Future

Christian V. Stevens

11.1 Introduction

The final chapter of this book will try to summarise the concept of renewable resources and will try to look ahead and comment on the future of renewable resources. Of course, the development of the use of renewable resources itself and the speed of the development are dependent on many factors and trying to predict the future is much too ambitious. Some major aspects will be highlighted in an attempt to point out the trends that will be of importance in the future.

11.2 Sustainability and Renewables

One of the major changes of the last two decades in Western societies in relation to production of goods has been the growing realisation and concern that resources are limited, and that we have to take care of the environment if we want our children to live in a well-functioning society. This broad idea is covered by the term "sustainable development". Although the term "sustainability" has been used and misused many times, it has become an important aspect in laying out industrial goals, in setting up new processes, in taking care of personnel in large companies and in everyday industrial activity.

Although the concept of sustainability has also been used to condemn all sorts of industrial activities, it should not be seen as a "back to nature" as some activist groups would like to use it. On the contrary, sustainable development concerns the development of a complex interaction between the best available technology, consumer awareness, welfare and social behaviour. It is a challenge to companies to maintain and improve

Renewable Bioresources: Scope and Modification for Non-food Applications. Edited by C.V. Stevens and R. Verhé
© 2004 John Wiley & Sons, Ltd   ISBNs: 0-470-85446-4 (HB); 0-470-85447-2 (PB)

quality of life in harmony with environmental considerations and the use of resources. This development is possible only by the interplay of technology, chemistry, biology, chemical engineering, genetic engineering and so forth. The negative attitude of policy makers and activist groups towards new directions in science (e.g. genetic engineering) is certainly not the way to develop a sustainable society. A careful and ethical exploration of these new domains will lead to an improved and more sustainable functioning of our industry.

Renewable raw materials can help in the development of a more sustainable industry. It will certainly not be the ultimate solution to all problems but it certainly deserves research efforts and attention in order to further explore the wide variety of compounds that are produced by nature and by agriculture. In the 20th century, the enormous possibilities of the petrochemistry and related chemical industries turned away interest in chemistry and in the availability of renewable raw materials. Because of pressure on the environment, interest is slowly shifting back to the important role that renewable resources can play in sustainable development. This will probably lead to a synergistic functioning of an industry based on fossil fuels and renewable resources, certainly for the coming decades. The transition between the two types of resources will also depend on the speed of depletion of fossil fuels. Although there are several scenarios, nobody really knows when the reserve of fossil fuels will be exhausted since it is a complex equilibrium between consumption, the discovery of new fields and the price of crude oil. It is price that makes it possible to use new and more expensive techniques to gain or to exploit oil from areas that are not currently competitive.

Further, the interaction between renewable and fossil resources will be influenced in an important way by other types of evolving technologies, certainly in the area of energy production (hydrogen technology, solar energy systems, nuclear fusion, wind energy, etc.). The most serious challenge will be to replace fossil fuels in the area of energy production and related to this aspect, renewable resources will be able to produce a certain percentage of the energy supply. In most of the energy scenarios, biomass will cover approximately 20–25% of future energy supply. For energy production, a multitude of technologies will have to cover this supply, certainly since nuclear power will not be accepted in the long run.

11.3 Policy and Renewables

The interaction between policy and renewable resources is complex and only a couple of aspects are being commented on in this section.

First of all, the definition of renewable resources is not general and has lead to numerous discussions. The definition of renewable resources in the European policy context differs to that of other countries. For example, bonemeal can be considered as a renewable source of material in certain European countries, whereas it is defined as a waste product in other legislations. Recently, manure was transported from Western Europe to Russia to fertilise, but it was not allowed to enter the country since it was considered a waste product. Therefore, if global problems need to be addressed, a common vocabulary is absolutely necessary.

The discussion on what materials can be considered as renewable resources or biomass has been quite stringent, since it is very important in order to develop national regulations that are in agreement with each other.

The impact of policy on the development of the use of renewable resources is tremendous, certainly in the developing stage of the technology. The start of a technology is always accompanied with high costs resulting in a period in which the technology is not cost-effective. Therefore, industry is not always eager to invest money in very new technologies. In cases where the new technology has advantages for society (e.g. a smaller environmental footprint of the technology), governmental support can be crucial to overcome the first difficult period. Stimulating research in the area of renewable resources can therefore help to develop new products and technologies and bring the new technology to a level where it becomes interesting for companies to invest in the new technology. University research can play a crucial role in this technology-transfer process.

In cases where the social benefit or the burden to society is great, the development of new laws or taxes can stimulate the transition to new products or technologies at a time when it would not otherwise be profitable. A good example is the obligation to use renewable-based lubricants in applications where the lubricants immediately enter the environment (e.g. in forestry applications where chain saws are used). This is independent of the price of the renewables-based lubricant compared to a lubricant derived from fossil resources. The damage to the environment is considered so important that the government decided that lubricants with poor biodegradability should no longer be used. However, it has to be added that products derived from renewable resources do not *per se* have a better biodegradability than ones derived from fossil resources.

A similar type of system developed by governments is the creation of eco-taxes or eco-boni, in order to discourage or to stimulate consumers to use certain products. However, this is rather an artificial system and would probably lose its effect after being in place for a long period.

The driving force for governments to introduce such measures is mainly of an environmental nature. In these cases, the governments try to protect their own environment. However, one issue is that environmental problems do not end at national borders and that governments do not take the same decisions. Examples are abundant: the Kyoto protocol is not signed by all countries; clean air can be purchased in industrially less developed countries; economic principles and the protection of unique economies are also very important and so forth. International institutions need to play an important role in order to try to coordinate solutions to these problems. However, it is difficult to be optimistic when we are faced with our inability to coordinate efforts in an international manner even when dealing with deathly diseases such as HIV.

It is noteworthy that Western countries are now trying to introduce more renewable resources into their industries, whereas developing countries, which used to have a more renewable-based tradition, are still trying to copy an industry completely based on fossil fuels.

11.4 Integral Valorisation of Resources

For a long time, production processes have been focussed on products. The aim was to produce the product in the cheapest way with a sufficient quality. When side products were formed or if the wanted constituent was extracted from the resources, the residues were dumped or used for low-value applications.

With an increasing awareness of the limitation of resources, more and more attention is paid to the integral use of resources and an evaluation of the production chain rather than the production process alone. This idea also gave rise to the LCA methods that truly consider the product from cradle to grave in order to obtain a measure of the impact and energy demand of a product. However, LCA focusses only on one product.

Under the concept of renewable resources, attention is paid to evaluate all parts of the plant or animal. Some part of the plant may be used for high value applications, whereas other parts might be valuable for lower value applications. The valorisation of some side products in particular needs further development. The residues of renewable resources can ultimately be used as a source of bio-energy.

As was discussed for the concept of bio-energy, the integral use of renewable resources in small-scale companies can also create opportunities to create jobs in more remote areas and it helps to make small societies become more self-supporting. Looking at the world including developing countries, this way of organising small communities can help to reduce the migration of people from the countryside to large cities where they face unemployment. For those countries in particular, small projects rather than large projects can help to improve the situation.

The development of renewable resources will also help to support local agricultural communities in which people immediately see the result of their work.

Also in industrialised countries, more and more people are convinced that a further increase in the scale of production can be connected to some problems (diseases in monoculture, the rapid development of infectious diseases in cattle, etc.). Of course, it is unrealistic to believe that modern agriculture in Western countries will shift back to small-scale companies. A well-balanced mix between the two systems in relation to the potential of the area under consideration is probably the best choice. In more remote areas, for example in Northern Europe, more small-scale projects would be advisable compared to some agricultural areas in France, which have an excellent soil quality.

11.5 Production of Primary Raw Materials

Natural raw materials have a lot of potential for developing new products, but of course the first requirement is that plants need to be cultivated in the region under consideration. Climate, soil quality and water availability – these need to be in agreement with the needs of the plant. Also the variety of crop needs to be evaluated in detail. The evaluation of each crop is therefore a complex interdisciplinary study to integrate variety selection, breeding or growing conditions and harvesting technology.

On top of all the technological conditions needed for a profitable process, consumer attitude is becoming more and more important. Next to the fact that the consumer is in favour of the use of natural products, the attitude towards the use of genetically modified crops is very different in different regions of the world. The use of modified crops is accepted in the US; on the contrary, Europe is completely banning the import of genetically modified crops, leading to serious debates between the two sides. This discussion is now far beyond the scientific and technological aspects and has become in some cases an emotional and ethical discussion. Opponents are launching terms such as "Frankenstein food" in order to sway public opinion. In this debate, scientists have missed the chance to put forward a clear, unbiased and scientific argument to the general public.

Concerning the use of renewable resources for non-food applications, production requires the use of productive arable land that cannot be used for food production. This aspect is also a point of debate since some experts maintain that this arable land will be necessary for producing food for the growing number of people in the world. This discussion is certainly beyond the scope of this book. Many aspects will influence this issue because the production capacity of arable land can rise considerably in many regions of the world. Also, bearing this issue in mind, it makes sense to use non-food application crops (which can also be used as food crops) as much as possible. This not only gives a great deal of flexibility to switch from food use to non-food use according to demand, but also saves on equipment and the existing technology.

The use of natural compounds for industrial applications is still a domain where a lot of progress can be expected. Also minor chemical modification of natural products leading to functional, but still biodegradable, compounds will be further established in the future. These two trends will certainly be stimulated by consumer attitude.

11.6 Bio-Energy

The energy from renewable resources certainly grabbed most attention under the broad concept of renewables. The awareness of the limited amount of fossil fuels has raised the question of where society will obtain energy from in the coming centuries in a world where nuclear energy is no longer much appreciated.

Currently, no technology is ready to take over energy production from fossil fuels. Some technologies, such as solar energy, nuclear fusion, hydrogen production, photovoltaic cells and biomass, are very promising.

Renewable resources will not be able to close the gap if fossil fuels run out, but in the majority of potential schemes, renewables will be an important source of energy. The use of side products or waste products to produce energy by burning or biogas by anaerobic digestion is certainly of great importance. As illustrated, it also leads to the creation of new jobs at the locations where renewables are produced.

Growing energy crops can also help to reduce carbon dioxide emissions, which need better control in the near future according to the Kyoto protocol. In countries that are now very dependent on the import of fossil fuels, more attention should be paid to the growth of some energy corps. In developing countries, in tropical or subtropical areas, the production of sugar cane is very attractive since the energy needed for sugar production can be supplied completely by the burning of dried bagasse. There is even a considerable surplus of energy produced during this process.

11.7 Carbohydrates, Lipids and Proteins

The three major classes of constituents of renewable resources are carbohydrates, lipids and proteins. Up to now most of the attention on these classes has been focussed on their food applications. However, these three classes could also be of major importance for non-food applications. The classes of carbohydrates (starch, cellulose, etc.) and lipids (biodiesel, lubricants, etc.) have been exploited to some degree, but both classes still have a lot of potential.

Chemical modification of saccharides can lead to new classes of biodegradable functional compounds that can be of use as a bulk chemical as well as a fine-chemical or a pharmaceutical agent. Since the natural availability of the saccharides is high and the prices are quite low compared to other functionalised molecules, there is still a great deal of potential for the application of saccharides. In this field, the use of waste materials can be made more efficient and more research should be undertaken to develop the technology to upgrade some waste streams into useful intermediates.

In the field of lipids, there is a widespread use of lipids as natural components but there is also a lot of potential for modified lipids as industrial products. Most of the modification is being performed on the carboxylic acid part, and the modification in the aliphatic chain needs to be developed considerably. Further modification could result in very useful building blocks for several industries.

For proteins, only some applications for non-food applications are known. Therefore, the necessary chemical modification of proteins for non-food applications and the study of their valuable characteristics remains a relatively unexplored area of research.

In summary, during the last century the development of industry was intensively connected to products derived from fossil fuels. With the changing needs and a changing attitude, it should be an invitation to many research institutes to get more involved in the development of applications in this area.

11.8   High Value-added Industries

For a couple of million years, nature has developed a very efficient way of producing highly complicated organic structures with very specific functions. Therefore, the use of these highly efficient catalysts in nanoplants is attractive. This has lead to the phrase "the plant is the plant". There are several well-known examples where interesting molecules can be produced, e.g. taxol for the treatment of several types of cancer, quinines from the bark of trees, and polyphenols as anti-oxidants.

Furthermore, complicated structures can also be produced using cheap sugar as a carbon source for microorganisms to develop pharmaceutically attractive molecules. Genetic modification can lead to a multiplication of the range of attractive secondary metabolites. Therefore it is very necessary to explore further the opportunities of genetic engineering, although it must be done under stringent conditions to prevent abuse or to prevent research for unethical purposes.

The high value-added products could also be an important stimulus for the agricultural community all over the world, because the agricultural population is diminishing almost everywhere. One negative example that demonstrates that agriculture can take advantage of the production of high value-added products in developing countries is the production of narcotic drugs and opium. If agricultural communities have the opportunity to breed plants for the production of valuable metabolites, it will not only help agriculture, but create a local industry on the basis of their own work, which could help to close the gap between North and South. Therefore, in these areas, renewables can help to build a future based on tried and tested concepts. So, let us consider going "Back to the future" on the basis of interdisciplinary research and developments in the fields of chemistry, biochemistry, fermentation, genetic engineering and agriculture, in order to build a more sustainable world with an improved level of welfare.

Index

Note: Page numbers given in italics refer to figures and numbers in bold refer to tables.

Renewable Bioresources: Scope and Modification for Non-food Applications. Edited by C.V. Stevens and R. Verhé
© 2004 John Wiley & Sons, Ltd   ISBNs: 0-470-85446-4 (HB); 0-470-85447-2 (PB)